Mark Grote
mngrote@ucdavis.edu

Random Integral Equations with Applications to Life Sciences and Engineering

To

A. T. Bharucha-Reid

...γηράσκω 'αεί διδασκόμενος... Socrates

This is Volume 108 in
MATHEMATICS IN SCIENCE AND ENGINEERING
A series of monographs and textbooks
Edited by RICHARD BELLMAN, *University of Southern California*

The complete listing of books in this series is available from the Publisher upon request.

RANDOM INTEGRAL EQUATIONS WITH APPLICATIONS TO LIFE SCIENCES AND ENGINEERING

Chris P. Tsokos
DEPARTMENT OF MATHEMATICS
UNIVERSITY OF SOUTH FLORIDA
TAMPA, FLORIDA

W. J. Padgett
DEPARTMENT OF MATHEMATICS AND
COMPUTER SCIENCE
UNIVERSITY OF SOUTH CAROLINA
COLUMBIA, SOUTH CAROLINA

ACADEMIC PRESS   New York and London   1974
A Subsidiary of Harcourt Brace Jovanovich, Publishers

COPYRIGHT © 1974, BY ACADEMIC PRESS, INC.
ALL RIGHTS RESERVED.
NO PART OF THIS PUBLICATION MAY BE REPRODUCED OR
TRANSMITTED IN ANY FORM OR BY ANY MEANS, ELECTRONIC
OR MECHANICAL, INCLUDING PHOTOCOPY, RECORDING, OR ANY
INFORMATION STORAGE AND RETRIEVAL SYSTEM, WITHOUT
PERMISSION IN WRITING FROM THE PUBLISHER.

ACADEMIC PRESS, INC.
111 Fifth Avenue, New York, New York 10003

United Kingdom Edition published by
ACADEMIC PRESS, INC. (LONDON) LTD.
24/28 Oval Road, London NW1

Library of Congress Cataloging in Publication Data

Tsokos, Chris P
    Random integral equations with applications to life sciences and engineering.

    (Mathematics in science and engineering, v. 108)
    Bibliography: p.
    1.   Stochastic integral equations.   I.   Padgett,
W. J., joint author.   II.   Title.   III.   Series.
QA274.27.T76   515'.45   73-2079
ISBN 0–12–702150–7

AMS (MOS) 1970 Subject Classifications: 60H20, 45G99, 93E15

PRINTED IN THE UNITED STATES OF AMERICA

Contents

PREFACE ix

General Introduction 1

Chapter I. Preliminaries and Formulation of the Stochastic Equations

1.0 Introduction 6
1.1 Basic Definitions and Theorems from Functional Analysis 7
1.2 Probabilistic Definitions 12
1.3 The Stochastic Integral Equations and Stochastic Differential Systems 18
    Appendix 1.A 22

Chapter II. Some Random Integral Equations of the Volterra Type with Applications

2.0 Introduction 29
2.1 The Random Integral Equation 30
    2.1.1 Existence and Uniqueness of a Random Solution 30
    2.1.2 Some Special Cases 33
    2.1.3 Asymptotic Stability of the Random Solution 38
2.2 Some Applications of the Equation 39
    2.2.1 Generalization of Poincaré–Lyapunov Stability Theorem 40
    2.2.2 A Problem in Telephone Traffic Theory 42
    2.2.3 A Stochastic Integral Equation in Hereditary Mechanics 46

2.3	The Random Integral Equation	49
2.4	Applications of the Integral Equation	55
	2.4.1 A Stochastic Integral Equation in Turbulence Theory	55
	2.4.2 Stochastic Models for Chemotherapy	57

Chapter III. Approximate Solution of the Random Volterra Integral Equation and an Application to Population Growth Modeling

3.0	Introduction	65
3.1	The Method of Successive Approximations	66
	3.1.1 Convergence of the Successive Approximations	68
	3.1.2 Rate of Convergence and Error of Approximation	71
	3.1.3 Combined Error of Approximation and Numerical Integration	74
3.2	A New Stochastic Formulation of a Population Growth Problem	78
	3.2.1 The Deterministic Model	79
	3.2.2 The Stochastic Model	81
	3.2.3 Existence and Uniqueness of a Random Solution	84
3.3	Method of Stochastic Approximation	87
	3.3.1 A Stochastic Approximation Procedure	87
	3.3.2 Solution of Eq. (3.0.1) by Stochastic Approximation	89
	3.3.3 Numerical Solution for a Hypothetical Population	94

Chapter IV. A Stochastic Integral Equation of the Fredholm Type and Some Applications

4.0	Introduction	97
4.1	Existence and Uniqueness of a Random Solution	98
4.2	Some Special Cases	110
4.3	Stochastic Asymptotic Stability of the Random Solution	113
4.4	An Application in Stochastic Control Systems	115
4.5	A Random Perturbed Fredholm Integral Equation	120

Chapter V. Random Discrete Fredholm and Volterra Systems

5.0	Introduction	132
5.1	Existence and Uniqueness of a Random Solution of System (5.0.1)	133
5.2	Special Cases of Theorem 5.1.2	136
5.3	Stochastic Stability of the Random Solution	139
5.4	An Approximation to System (5.0.1)	141
5.5	Application to Stochastic Control Systems	148
	5.5.1 A Discrete Stochastic System	148
	5.5.2 Another Discrete Stochastic System	152

Chapter VI. Nonlinear Perturbed Random Integral Equations and Application to Biological Systems

6.0	Introduction	156
6.1	The Random Integral Equation	157
	6.1.1 Existence and Uniqueness of a Random Solution	157
	6.1.2 Some Special Cases	159
6.2	Applications to Biological Systems	165
	6.2.1 A Random Integral Equation in a Metabolizing System	165
	6.2.2 A Stochastic Physiological Model	170
	6.2.3 A Stochastic Model for Communicable Disease	176

Chapter VII. On a Nonlinear Random Integral Equation with Application to Stochastic Chemical Kinetics

7.0	Introduction	180
7.1	Mathematical Preliminaries	181
7.2	An Existence and Uniqueness Theorem	194
7.3	A Stochastic Chemical Kinetics Model	197
	7.3.1 The Concept of Chemical Kinetics	198
	7.3.2 Stochastic Interpretation of the Rate of Reaction	201
	7.3.3 Rate Functions of General Reaction Systems	201
	7.3.4 A Stochastic Integral Equation Arising in Chemical Kinetics	204

Chapter VIII. Stochastic Integral Equations of the Ito Type

8.0	Introduction	207
8.1	Preliminary Remarks	208
8.2	On an Ito Stochastic Integral Equation	212
8.3	On Ito–Doob-Type Stochastic Integral Equations	214
	8.3.1 An Existence Theorem	215

Chapter IX. Stochastic Nonlinear Differential Systems

9.0	Introduction	217
9.1	Reduction of the Stochastic Differential Systems	219
	9.1.1 Stochastic System (9.0.1)–(9.0.2)	219
	9.1.2 The Random Differential System (9.0.3)–(9.0.4)	220
	9.1.3 The Stochastic System (9.0.5)–(9.0.6)	221
	9.1.4 The Random Differential System (9.0.7)–(9.0.8)	222
9.2	Stochastic Absolute Stability of the Differential Systems	225
	Appendix 9.A	239
	9.A.1 Stochastic Differential System (9.0.1)–(9.0.2)	239
	9.A.2 Stochastic Differential System (9.0.3)–(9.0.4)	240
	9.A.3 The Reduced Stochastic Integral Form of Systems (9.0.1)–(9.0.2) and (9.0.3)	240

Chapter X. Stochastic Integrodifferential Systems

10.0	Introduction	241
10.1	The Stochastic Integrodifferential Equation	243
	10.1.1 Asymptotic Behavior of the Random Solution	247
	10.1.2 Application to a Stochastic Differential System	250
10.2	Reduction of the Stochastic Nonlinear Integrodifferential Systems with Time Lag	251
	10.2.1 The Integrodifferential System (10.0.2)–(10.0.3)	251
	10.2.2 The Random Integrodifferential System (10.0.4)–(10.0.5)	253
10.3	Stochastic Absolute Stability of the Systems	255

Bibliography 260

Index 275

Preface

Random or stochastic integral equations arise in virtually every field of scientific endeavor. Recently, attempts have been made by many scientists and mathematicians to develop and unify the theory of stochastic or random equations using the concepts and methods of probability theory and functional analysis.

We have two main objectives in this book. First, we wish to give a complete presentation of various aspects of some of the most general forms of non-linear stochastic integral equations of the Volterra and Fredholm types which have been studied, including the problems of existence, uniqueness, stochastic stability, and approximation of random solutions of the equations. In addition, we investigate stochastic integral equations of the Ito–Doob type. The second objective is to apply the theory developed to some very important problems in the life sciences and engineering.

With respect to the applications, stochastic models for various phenomena in the biological, engineering, and physical sciences are obtained. For example, applications of the theory to the following areas are given: telephone traffic theory, hereditary mechanics, turbulence theory, chemotherapy, population growth, stochastic control systems, metabolizing systems, physiological models, communicable diseases, chemical kinetics, and stochastic integrodifferential systems.

The book will be of value to mathematicians, probabilists, statisticians, and engineers who are working in the theoretical and applied aspects of random integral equations. The book can be used for a beginning graduate course on random integral equations with emphasis being placed on probabilistic modeling of various problems in life sciences and engineering.

It should be pointed out that this book differs in purpose considerably from the book of A. T. Bharucha-Reid [7]. He is concerned primarily with the overall theory of random integral equations, whereas we emphasize the stochastic modeling aspects and applications of certain types of random or stochastic integral equations. Thus, the two books are complementary.

This book was written with the direct and indirect help of many people. We are grateful to Dr. J. Susan Milton for her helpful and stimulating discussions during the preparation of the manuscript. We would also like to acknowledge Dr. A. N. V. Rao for his assistance and valuable comments and Ms. Debbi Beach for her excellent typing of the manuscript. In addition we would like to express our appreciation to Professor Richard Bellman for his encouragement and interest in the subject matter.

Finally, we would like to express our sincere thanks to our families for their understanding and patience during the writing of the book.

General Introduction[†]

Due to the nondeterministic nature of phenomena in the general areas of the biological, engineering, oceanographic, and physical sciences, the mathematical descriptions of such phenomena frequently result in random or stochastic equations. It is the aim of this book to present theoretical results concerning certain classes of stochastic or random equations and then to apply those results to problems that arise in the general areas just mentioned. In order to understand better the importance of developing such a theory and its application, it is of interest to consider first the various ways in which these equations may arise.

Usually the mathematical models or equations used to describe physical phenomena contain certain parameters or coefficients which have specific physical interpretations but whose values are unknown. As examples, we have the diffusion coefficient in the theory of diffusion, the volume-scattering coefficient in underwater acoustics, the coefficient of viscosity in fluid mechanics, the propagation coefficient in the theory of wave propagation, and the modulus of elasticity in the theory of elasticity, among others. The mathematical equations are solved using as the value of the parameter or coefficient the *mean value* of a set of observations experimentally obtained. However, if the

[†] Adapted from Padgett and Tsokos [12] with permission of Taylor and Francis, Ltd.

experiment is performed repeatedly, then the mean values found will vary, and if the variation is large, the mean value actually used may be quite unsatisfactory. Thus in practice the physical constant is not really a constant, but a random variable whose behavior is governed by some probability distribution. It is thus advantageous to view these equations as being random rather than deterministic, to search for a random solution, and to study its statistical properties.

There are many other ways in which random or stochastic equations arise. Stochastic differential equations appear quite naturally in the study of diffusion processes and Brownian motion (Gikhmann and Skorokhod [1]). The classical Ito stochastic integral equation (Ito [1]) may be found in many texts, for example, Doob [1]. Integral equations with random kernels arise in random eigenvalue problems (Bharucha-Reid [7]). Stochastic integral equations describe wave propagation in random media (Bharucha-Reid [7]) and the total number of conversations held at a given time in telephone traffic theory (Fortet [1] and Padgett and Tsokos [4]). In the theory of statistical turbulence, stochastic integral equations arise in describing the motion of a point in a continuous fluid in turbulent motion (Bharucha-Reid [7], Lumley [1], Padgett and Tsokos [3]). Integral equations were used in a deterministic sense by Bellman, Jacquez, and Kalaba [1-3] in the development of mathematical models of chemotherapy. However, due to the nondeterministic nature of diffusion processes from the blood plasma into the body tissue, the stochastic versions of these equations are more realistic and should be used (Padgett and Tsokos [1, 2, 10]). Stochastic integral equations also arise in problems in chemical kinetics and metabolizing systems (Milton and Tsokos [1, 4]). Random equations are also frequently encountered in a natural way in systems theory (for example, Morozan [1-5] and Tsokos [1-5]).

These examples point out the importance of random equations in diverse areas. However, in many instances the scientist tends to use a deterministic model to represent a process under investigation with the philosophy that there is a deterministic but unknown function $x(t)$ which describes the phenomenon he observes. He then attempts by experimental methods to determine as accurately as possible the form of this function. A standard procedure is to obtain, at several specified values of t, observations on the value of $x(t)$ and then to use as the "true" value of $x(t)$ some estimate based on these observations, the usual estimate being the mean. In this way a *single* trajectory can be constructed which is then taken as the true form of $x(t)$ and is subsequently used in working with the model. This general technique characterizes the deterministic approach to a physical situation. However, if this procedure were repeated many times, *even under the most carefully controlled conditions*, the trajectories so obtained will differ, and in most cases this variation could be *quite significant*. If this is indeed the case, then there is evidence that there is more than mere measurement error entering into the picture, and that, in fact, the function which governs the process is not a fixed unknown entity, but a random one. Thus it is more realistic in this situation to construct a stochastic model for the system rather than a deterministic model. This entails the basic assumption that at each point t, $x(t)$ is not a fixed unknown value which should be estimated, but rather a random variable which we denote by $x(t;\omega)$, where $\omega \in \Omega$, the supporting set of a complete probability measure space $(\Omega, \mathscr{A}, \mathscr{P})$. An important point to be made is that if a stochastic model is assumed when a deterministic model could be justified, nothing is lost; but if a deterministic model is assumed when in fact the process is random, then the results obtained could be quite unsatisfactory.

Recently attempts have been made by many scientists and mathematicians to develop and unify the theory of stochastic or random equations: Adomian [1–4], Ahmed [1], Anderson [1, 2], Bharucha-Reid [1–7], Hans [1], Tsokos [4], Padgett and Tsokos [1, 3, 5–9, 11, 12, 15], Rao [1], Milton *et al.* [1]. It was Antonin Spacek from Prague, Czechoslovakia, who began this work, utilizing the concepts and methods of probability theory and functional analysis. In fact, Bharucha-Reid [5] refers to *probabilistic functional analysis* as being concerned with the applications and extensions of the methods of functional analysis to the study of the various concepts, processes, and structures which arise in the theory of probability and its applications.

Random or stochastic equations have been categorized into four main classes as follows:

(1) Random or stochastic algebraic equations.
(2) Random differential equations.
(3) Random difference equations.
(4) Random or stochastic integral equations.

In this book we will be concerned with certain classes of random or stochastic integral equations of the Volterra and Fredholm types and a class of random integrodifferential equations of the Volterra type. For example, in Chapters II and III we will study various aspects of the stochastic integral equation of the Volterra type in the form

$$x(t;\omega) = h(t;\omega) + \int_0^t k(t,\tau;\omega) f(\tau, x(\tau;\omega))\, d\tau \qquad (0.1)$$

for $t \geq 0$, where the integral is interpreted as a mean-square integral. We will consider the *existence*, *uniqueness*, *asymptotic behavior*, and *approximation* of random solutions of each type of stochastic integral equation studied in this book. In addition, the second aim of the book is to present numerous applications of the theory of such equations to the problems in chemotherapy, chemical kinetics, physiological systems, population growth, telephone engineering, turbulence, and systems theory as previously mentioned. Furthermore, these equations are more general than any random Volterra or Fredholm equations of these forms that have been studied to date. The generality consists primarily in the choice of the stochastic kernel and the nonlinearity of the equations. This book includes the recent work of the authors, Padgett and Tsokos [1–15], and generalizes the work of Hans [1], Bharucha-Reid [1, 3–5], and Anderson [1, 2].

In the area of systems theory we will apply the general theory which is presented concerning random integral equations to certain problems in random differential systems and random integrodifferential systems (Tsokos [1–3, 5], Tsokos and Hamdan [1]). The nonlinear stochastic differential and integrodifferential systems will be reduced in a unified way to stochastic nonlinear integral equations. Then in each case the existence and uniqueness of a random solution of the stochastic system will be investigated. In addition, we will consider the concept of *stochastic absolute stability* of the systems. This type of stability has been studied in the deterministic case by many scientists, but to the knowledge of the authors it has not been utilized for random systems.

The concept of absolute stability arose in the context of differential control systems and the general theory of stability of motion. The primary mathematical technique which was universally used to study absolute stability was Lyapunov's direct method. However, in the late 1950's when Lyapunov's method appeared to be exhausted V. M. Popov developed a new approach, obtaining very elegant and powerful results. Popov's method is known as the frequency response method. In Chapter IX we successfully utilize the frequency response method with a random parameter to investigate the stochastic absolute stability of several stochastic differential systems. These results generalize the recent results of Morozan [1] in that he chose a specific form of the stochastic kernel, namely an exponential form.

In Chapter I we present preliminary notations, definitions, lemmas, and theorems which are essential to the aims of this book. Further, we define and formulate the types of stochastic integral equations and the stochastic differential systems which will be investigated in later chapters. Finally, in Appendix 1.A of Chapter I the proofs of some of the fixed-point theorems that are utilized throughout the book are given. As already mentioned, in Chapters II and III we will investigate the existence, uniqueness, asymptotic properties, and approximation of random solutions of certain stochastic integral equations of the Volterra type. In addition, several applications and examples of such equations will be presented in the areas of chemotherapy, telephone traffic theory, turbulence theory, hereditary mechanics, and stochastic systems theory. Chapter IV will be devoted to stochastic integral equations of the Fredholm type. Certain random or stochastic discrete Volterra and Fredholm equations will be considered in Chapter V along with their approximate solutions and applications. Chapter VI will deal with perturbed versions of the random equations studied in Chapters II and IV and their application to biological systems. In Chapter VII an application of a nonlinear random integral equation to a problem in stochastic chemical kinetics is presented. A connection between Ito's equation and certain classes of stochastic integral equations studied in earlier chapters is given in Chapter VIII; that is, Ito's equation is studied by applying some aspects of the "theory of admissibility" (Corduneanu [1]). Several stochastic nonlinear differential systems and their stochastic absolute stability are studied in Chapter IX. Chapter X is devoted to the investigation of a class of random or stochastic integrodifferential equations and its application to nonlinear stochastic systems.

CHAPTER I

Preliminaries and Formulation of the Stochastic Equations

1.0 Introduction

In an attempt to make this book essentially self-contained, one purpose of the present chapter is to give some of the basic definitions and theory of functional analysis which will be used throughout the text. Therefore Section 1.1 will consist of the statements of several definitions concerning linear topological spaces and some important theorems which are needed in later discussions. In Appendix 1.A we will give the proofs of some of the classical fixed-point theorems of functional analysis for the interested reader, but otherwise most of the theorems will be stated without proof for the sake of brevity.

The second purpose of this chapter is to present the probabilistic definitions, basic assumptions, and notations that are essential to the development of the theory in later chapters. These will be given in Sections 1.2 and 1.3. Section 1.2 will be devoted to certain probabilistic definitions and notations, while the specific types of stochastic or random integral equations which will be investigated are given in Section 1.3. Some definitions and notations concerning the stochastic differential systems to be studied also will be presented in Section 1.3.

1.1 Basic Definitions and Theorems from Functional Analysis

The following basic definitions and theorems are stated for the convenience of the reader.

Definition 1.1.1 A real-valued measurable function $f(x)$ defined on a closed interval $[a, b]$ is said to be a *square-summable* function if

$$\int_a^b |f(x)|^2 \, dx < +\infty.$$

We shall designate the class of all such functions by the symbol L_2.

Definition 1.1.2 A real number associated with $f \in L_2$, denoted by

$$\|f\| = \left\{ \int_a^b |f(x)|^2 \, dx \right\}^{\frac{1}{2}},$$

is called the *norm* of f.

Definition 1.1.3 The element f of the space L_2 is called a *limit of the sequence* $f_1, f_2, f_3, \ldots$ of elements of the same space if for every $\varepsilon > 0$ there exists a nonnegative integer N such that

$$\|f_n - f\| < \varepsilon$$

for all $n > N$.

Definition 1.1.4 A nonempty set H is called a *metric space* if to an arbitrary pair x, y of elements in H there corresponds a real number $\rho(x, y)$ possessing the following properties:

(i) $\rho(x, y) \geq 0$, where $\rho(x, y) = 0$ if and only if $x = y$.
(ii) $\rho(x, y) = \rho(y, x)$.
(iii) $\rho(x, z) \leq \rho(x, y) + \rho(y, z)$ for any $x, y, z \in H$ (triangle inequality).

The number $\rho(x, y)$ is the distance between the elements x and y.

Definition 1.1.5 A sequence $\{x_n\}$ of elements in a metric space is said to be *convergent in itself* or a *Cauchy sequence* if for every $\varepsilon > 0$ there exists an

N such that for $n > N$ and $m > N$ we have

$$\rho(x_n, x_m) < \varepsilon.$$

Definition 1.1.6 A metric space H is said to be *complete* if every sequence of its elements which is convergent in itself has a limit in H.

Definition 1.1.7 A set H of elements $x, y, z, \ldots$ is said to be a *linear space* if:

(i) To every pair of elements x and y of H there corresponds a third element of H, $z = x + y$, called the *sum* of x and y.

(ii) To every element $x \in H$ and every scalar a there corresponds an element, $ax \in H$, which is called the product of a and x.

(iii) The operations introduced possess the following properties for every $x, y, z \in H$ and scalars a and b:

(1) $x + y = y + x$, i.e., addition is commutative.
(2) $(x + y) + z = x + (y + z)$, i.e., addition is associative.
(3) $x + y = x + z$ implies $y = z$.
(4) $1x = x$.
(5) $a(bx) = (ab)x$.
(6) $(a + b)x = ax + bx$.
(7) $a(x + y) = ax + ay$.

Definition 1.1.8 A linear space H is said to be *normed* if to each $x \in H$ there corresponds a real number $\|x\|$, called the *norm* of this element, possessing the following properties for every $y \in H$ and every scalar a:

(i) $\|x\| \geq 0$, where $\|x\| = 0$ if and only if $x = 0$.
(ii) $\|ax\| = |a| \cdot \|x\|$, and in particular $\|-x\| = \|x\|$.
(iii) $\|x + y\| \leq \|x\| + \|y\|$.

Definition 1.1.9 A complete normed linear space is called a *Banach space*. A *Fréchet space* is a complete linear metric space.

Definition 1.1.10 Let H be a given set, and let $\mathscr{F}$ be a set of subsets of H having the following properties:

(i) $H \in \mathscr{F}$.
(ii) $\phi \in \mathscr{F}$.
(iii) The union of any nonempty family of sets from $\mathscr{F}$ belongs to $\mathscr{F}$.
(iv) The intersection of any two sets of $\mathscr{F}$ belongs to $\mathscr{F}$. The ordered pair $(H, \mathscr{F})$ is then called a *topological space* and $\mathscr{F}$ is the *topology of the space*. The sets belonging to $\mathscr{F}$ are called open sets.

Definition 1.1.11 If the set H in Definition 1.1.10 is a linear space and the two basic operations (addition and multiplication by a constant) are continuous, then H is a *linear topological space* or *vector space*.

1.1 BASIC DEFINITIONS AND THEOREMS

Definition 1.1.12 Let H and H_1 be metric spaces and let T be a rule which associates some point $y \in H_1$ with every point $x \in H$. Such a rule is called an *operator* which is defined on the space H and maps H into the space H_1. If $y \in H_1$ is the point which the operator T assigns to the point $x \in H$, we write $y = T(x)$ and call y the *value* of the operator T at the point x.

Definition 1.1.13 Let the operator T map the metric space H into itself. If there exists a real number $q, 0 \leq q < 1$, such that for arbitrary points x and x' of the space H we have

$$\rho(T(x), T(x')) \leq q \cdot \rho(x, x'),$$

then we call T a *(strict) contraction operator*.

Definition 1.1.14 Let $f, g \in L_1(-\infty, \infty)$. The function

$$h(x) = \int_{-\infty}^{\infty} f(x-y) g(y) \, dy = \int_{-\infty}^{\infty} f(y) g(x-y) \, dy$$

is defined almost everywhere on the real axis and is called the *convolution* of the functions f and g.

Definition 1.1.15 Let $(H, \mathscr{F})$ be a topological space. We shall say that a collection of subsets $\mathscr{F}' \subset \mathscr{F}$ is a *base* for this topology if for every $0 \in \mathscr{F}$ there exists a subset $0' \in \mathscr{F}$ such that $0' \subset 0$.

Definition 1.1.16 A linear topological space is said to be *locally convex* if it possesses a base for its topology consisting of convex sets.

We now state several theorems which will be needed in later chapters. The proofs of the classical fixed-point theorems are presented in Appendix 1.A for the convenience of the interested reader.

Theorem 1.1.1 (*Minkowski's inequality*) (Natanson [1]) If $f(x) \in L_2$ and $g(x) \in L_2$, then

$$\left\{ \int_a^b |f(x) + g(x)|^2 \, dx \right\}^{\frac{1}{2}} \leq \left\{ \int_a^b |f(x)|^2 \, dx \right\}^{\frac{1}{2}} + \left\{ \int_a^b |g(x)|^2 \, dx \right\}^{\frac{1}{2}}.$$

Theorem 1.1.2 (*S. Banach's fixed-point principle*) (Natanson [2]) If T is a contraction operator on a complete metric space H, then there exists a unique point $x^* \in H$ for which

$$T(x^*) = x^*.$$

Theorem 1.1.3 (*Closed-graph theorem*) (Goldberg [1]) A closed linear operator mapping a Banach space into a Banach space is continuous.

Theorem 1.1.4 (Halanay [1]) If $f(x), g(x) \in L_1(-\infty, \infty)$, then the convolution $h(x)$ is defined for almost every x, $h(x) \in L_1(-\infty, \infty)$, and we have

$$\int_{-\infty}^{\infty} |h(x)|\, dx \leqslant \int_{-\infty}^{\infty} |f(x)|\, dx \int_{-\infty}^{\infty} |g(x)|\, dx.$$

Theorem 1.1.5 (Halanay [1]) The Fourier transform of the convolution $h(x)$ is the product of the Fourier transforms of the functions $f(x)$ and $g(x)$.

Theorem 1.1.6 (*Parseval equality*) (Bochner [1]) Let

$$f(t) \in L_1(-\infty, \infty) \cap L_2(-\infty, \infty) \quad \text{and}$$

$$\tilde{f}(i\lambda) = \int_{-\infty}^{\infty} e^{-i\lambda t} f(t)\, dt, \quad \text{for } \lambda \text{ real.}$$

Then

$$\int_{-\infty}^{\infty} |\tilde{f}(i\lambda)|^2\, d\lambda = 2\pi \int_{-\infty}^{\infty} |f(t)|^2\, dt.$$

Lemma 1.1.7 (Barbalat [1]) If

  (i) $f(t)$ is a continuous function, and its derivatives $f'(t)$ are bounded for $t \geqslant 0$;
  (ii) $G(x)$ is a continuous function, $G(x) > 0$ for any $x \neq 0$, $G(0) = 0$; and
  (iii) $\int_0^{\infty} G[f(t)]\, dt < \infty$;

then

$$\lim_{t \to \infty} f(t) = 0.$$

Definition 1.1.17 A continuous operator T from a Banach space H into a Banach space H_1 such that the image of closed bounded sets in H is compact is called a *completely continuous* operator.

Two other useful fixed-point theorems are due to Schauder and Krasnosel'skii.

Theorem 1.1.8 (*Schauder's fixed-point principle*) (Krasnosel'skii [1]) Let W be a closed, bounded convex set in a Banach space, and let T be a completely continuous operator on W such that $T(W) \subset W$. Then T has at least one fixed point in W. That is, there is at least one $x^* \in W$ such that

$$T(x^*) = x^*.$$

Theorem 1.1.9 (*Krasnosel'skii's fixed-point theorem*) (Krasnosel'skii [1]) Let S be a closed, bounded convex subset of a Banach space, and let U and V be operators on S satisfying:

1.1 BASIC DEFINITIONS AND THEOREMS

(i) $U(x) + V(y) \in S$ whenever $x, y \in S$.
(ii) U is a contraction operator on S.
(iii) V is completely continuous.

Then there is at least one point $x^* \in S$ such that

$$U(x^*) + V(x^*) = x^*.$$

That is, there is at least one point in S which is a fixed point of the operator $U + V$.

Note that Schauder's fixed-point theorem is a special case of Theorem 1.1.9.

Definition 1.1.18 Let H be a linear space. A mapping (x, y) taking points x and y in H into the real (or complex) numbers is called an *inner product* if for each $x, y, z \in H$ and scalar α we have

(i) $(x + y, z) = (x, z) + (y, z)$.
(ii) $(\alpha x, y) = \alpha(x, y)$.
(iii) $\overline{(x, y)} = (y, x)$, the bar denoting complex conjugate.
(iv) $(x, x) > 0$ if x is not the zero element of H.

In this case H is called an *inner product space*. The norm of an element $x \in H$ may be defined in terms of the inner product by

$$\|x\| = (x, x)^{\frac{1}{2}}.$$

Definition 1.1.19 A Banach space H whose norm is defined in terms of the inner product as just given is called a *Hilbert space*.

The following theorems will also be needed in the sequel.

Theorem 1.1.10 (Dunford and Schwartz [1]) If $\{T_n\}$ is a sequence of continuous linear operators from a Fréchet space H into a Fréchet space H_1 such that for each $x \in H$, $\lim_{n \to \infty} T_n(x) = T(x)$ exists, then $\lim_{x \to 0} T_n(x) = 0$ uniformly for $n = 1, 2, \ldots$, and T is a continuous linear operator from H into H_1.

Theorem 1.1.11 (Horváth [1, p. 114]) A locally convex Hausdorff space X whose topology δ is defined by an increasing sequence of semi-norms $\rho_n(x)$, $n = 1, 2, 3, \ldots$, is metrizable with metric

$$\rho(x, y) = \sum_{n=1}^{\infty} \frac{1}{2^n} \frac{\rho_n(x - y)}{1 + \rho_n(x - y)}.$$

Theorem 1.1.12 (Yosida [1, p. 76]) A linear space X can be topologized by a family of semi-norms satisfying the axiom of separation in such a way that the space is locally convex.

Theorem 1.1.13 (Horváth [1, p. 96]) The locally convex topology defined on a linear space X by a family of semi-norms is Hausdorff if and only if the family satisfies the axiom of separation.

1.2 Probabilistic Definitions

We shall denote by $(\Omega, \mathscr{A}, \mathscr{P})$ a probability measure space; that is, Ω is a nonempty abstract set, $\mathscr{A}$ is a σ-algebra of subsets of Ω, and $\mathscr{P}$ is a complete probability measure on $\mathscr{A}$.

The following spaces of functions are basic to this investigation.

Definition 1.2.1 $C = C(R_+, L_2(\Omega, \mathscr{A}, \mathscr{P}))$ will denote the space of all *continuous and bounded functions* on $R_+ = [0, \infty)$ with values in $L_2(\Omega, \mathscr{A}, \mathscr{P})$.

Definition 1.2.2 We shall denote by $C_g = C_g(R_+, L_2(\Omega, \mathscr{A}, \mathscr{P}))$ the space of all continuous functions from R_+ into $L_2(\Omega, \mathscr{A}, \mathscr{P})$ such that

$$\left\{ \int_\Omega |x(t;\omega)|^2 \, d\mathscr{P}(\omega) \right\}^{\frac{1}{2}} \leq Zg(t), \qquad t \in R_+,$$

where Z is a positive number and $g(t)$ is a positive continuous function defined on R_+.

Definition 1.2.3 We shall further define the following space: $C_c = C_c(R_+, L_2(\Omega, \mathscr{A}, \mathscr{P}))$ is the space of all continuous functions from R_+ into $L_2(\Omega, \mathscr{A}, \mathscr{P})$ with the topology of uniform convergence on the intervals $[0, T]$ for $T > 0$. This space, C_c, is a locally convex space (Yosida [1, pp. 24–26]) whose topology will be defined by means of the following family of *semi-norms*:

$$\|x(t;\omega)\|_n = \sup_{0 \leq t \leq n} \left\{ \int_\Omega |x(t;\omega)|^2 \, d\mathscr{P}(\omega) \right\}^{\frac{1}{2}}, \qquad n = 1, 2, 3, \ldots. \qquad (1.2.1)$$

These semi-norms satisfy the following conditions:

(i) $\|x(t;\omega)\|_n \geq 0$, for $n = 1, 2, 3, \ldots$; if $\|x(t;\omega)\|_n = 0$ for all n, then $x(t;\omega) = 0$ a.e., i.e., $x(t;\omega)$ is the zero element of C_c.
(ii) $\|\alpha x(t;\omega)\|_n = |\alpha| \cdot \|x(t;\omega)\|_n$.
(iii) $\|x(t;\omega) + y(t;\omega)\|_n \leq \|x(t;\omega)\|_n + \|y(t;\omega)\|_n$.

We now proceed to verify that the manner in which we have defined the semi-norms (1.2.1) in the space C_c satisfies Conditions (i)–(iii). Condition (i) is obviously satisfied from the definition of semi-norm. Condition (ii) can be

1.2 PROBABILISTIC DEFINITIONS

shown as follows:

$$\|\alpha x(t;\omega)\|_n = \sup_{0 \leq t \leq n} \left\{ \int_\Omega |\alpha x(t;\omega)|^2 \, d\mathcal{P}(\omega) \right\}^{\frac{1}{2}}$$

$$= \sup_{0 \leq t \leq n} |\alpha|^2 \left\{ \int_\Omega |x(t;\omega)|^2 \, d\mathcal{P}(\omega) \right\}^{\frac{1}{2}}$$

$$= |\alpha| \sup_{0 \leq t \leq n} \left\{ \int_\Omega |x(t;\omega)|^2 \, d\mathcal{P}(\omega) \right\}^{\frac{1}{2}}$$

$$= |\alpha| \cdot \|x(t;\omega)\|_n.$$

Next we must show that the triangular inequality is satisfied, that is,

$$\|x(t;\omega) + y(t;\omega)\|_n \leq \|x(t;\omega)\|_n + \|y(t;\omega)\|_n. \tag{1.2.2}$$

Applying Minkowski's inequality and the fact that

$$\sup_{0 \leq t \leq n} [f(t) + g(t)] \leq \sup_{0 \leq t \leq n} f(t) + \sup_{0 \leq t \leq n} g(t),$$

it follows from the definition of the semi-norm that

$$\|x(t;\omega) + y(t;\omega)\|_n = \sup_{0 \leq t \leq n} \left\{ \int_\Omega |x(t;\omega) + y(t;\omega)|^2 \, d\mathcal{P}(\omega) \right\}^{\frac{1}{2}}$$

$$\leq \sup_{0 \leq t \leq n} \left\{ \int_\Omega |x(t;\omega)|^2 \, d\mathcal{P}(\omega) \right\}^{\frac{1}{2}}$$

$$+ \sup_{0 \leq t \leq n} \left\{ \int_\Omega |y(t;\omega)|^2 \, d\mathcal{P}(\omega) \right\}^{\frac{1}{2}},$$

which means that

$$\|x(t;\omega) + y(t;\omega)\|_n \leq \|x(t;\omega)\|_n + \|y(t;\omega)\|_n.$$

Therefore we have shown that it is valid to define the semi-norms by (1.2.1).

One can define the topology on this space by the following distance function:

$$\rho(x, y) = \sum_{n=1}^\infty \frac{1}{2^n} \frac{\|x - y\|_n}{1 + \|x - y\|_n}.$$

With respect to this distance function $\rho(x, y)$, the space $C_c(R_+, L_2(\Omega, \mathcal{A}, \mathcal{P}))$ is a complete metric space. That is, every Cauchy sequence is convergent. In this space a sequence of functions is convergent if and only if it is convergent on every compact interval $[0, T]$, $0 < T$.

Note that the following inclusions hold: $C \subset C_g \subset C_c$. Note also that the space $C = C(R_+, L_2(\Omega, \mathscr{A}, \mathscr{P}))$ is the space of all *second-order* stochastic processes (Prabhu [1]) defined on R_+ which are bounded and continuous in mean-square.

Let B and D be a pair of Banach spaces such that B, $D \subset C_c$ and let T be a linear operator from C_c into itself. Then with respect to B, D, and T we state the following definitions.

Definition 1.2.4 The pair of spaces (B, D) will be called *admissible* with respect to the operator $T: C_c(R_+, L_2(\Omega, \mathscr{A}, \mathscr{P})) \to C_c(R_+, L_2(\Omega, \mathscr{A}, \mathscr{P}))$ if and only if $T(B) \subset D$.

Definition 1.2.5 An operator T is said to be closed if from

$$x_n(t;\omega) \xrightarrow{B} x(t;\omega) \quad \text{and} \quad (Tx_n)(t;\omega) \xrightarrow{D} y(t;\omega)$$

it follows that

$$(Tx)(t;\omega) = y(t;\omega).$$

Definition 1.2.6 The operator T is said to be *continuous* on $C_c(R_+, L_2(\Omega, \mathscr{A}, \mathscr{P}))$ if and only if

$$(Tx_n)(t;\omega) \to (Tx)(t;\omega)$$

in $C_c(R_+, L_2(\Omega, \mathscr{A}, \mathscr{P}))$ for every sequence $\{x_n(t;\omega)\}$ such that $x_n(t;\omega) \to x(t;\omega)$ in the same space.

Definition 1.2.7 By the space $L_\infty(\Omega, \mathscr{A}, \mathscr{P})$ we mean the space of all measurable and $\mathscr{P}$-essentially bounded functions, i.e., a function is in $L_\infty(\Omega, \mathscr{A}, \mathscr{P})$ if it is bounded in the ordinary sense on a set $\Omega - \Omega_0$, where $\mathscr{P}(\Omega_0) = 0$.

Definition 1.2.8 By stating that the Banach space B is *stronger* than the space $C_c(R_+, L_2(\Omega, \mathscr{A}, \mathscr{P}))$ we mean that every convergent sequence in B, with respect to its norm, will also converge in C_c (but the converse is not true in general).

Definitions 1.2.9–1.2.11 and Theorem 1.2.1 are due to Bharucha-Reid, Mukherjea, and Tserpes [1].

Definition 1.2.9 By the space $L_p(\Omega, \mathscr{A}, \mathscr{P})$, $1 \leqslant p < \infty$, we mean the space of all measurable functions defined on R_+ such that for each $t \in R_+$ we have

$$\int_\Omega |x(t;\omega)|^p \, d\mathscr{P}(\omega) < +\infty.$$

1.2 PROBABILISTIC DEFINITIONS

The norm in this space is defined by

$$\|x(t;\omega)\|_p = \left\{\int_\Omega |x(t;\omega)|^p d\mathscr{P}(\omega)\right\}^{1/p} < \infty.$$

If $x(t;\omega)$ is a vector-valued function with m components, then we define, as usual,

$$|x(t;\omega)| = [x_1^2(t;\omega) + x_2^2(t;\omega) + \cdots + x_m^2(t;\omega)]^{\frac{1}{2}}.$$

Definition 1.2.10 Let $q = p/(p-1)$, $1 < p < \infty$. The sequence $\{x_n(\omega)\}$, $\omega \in \Omega$, converges weakly to $x(\omega)$ in $L_p(\Omega, \mathscr{A}, \mathscr{P})$ if

$$\lim_{n\to\infty} \int_\Omega x_n(\omega) h(\omega) d\mathscr{P}(\omega) = \int_\Omega x(\omega) h(\omega) d\mathscr{P}(\omega)$$

for every $h \in L_q(\Omega, \mathscr{A}, \mathscr{P})$. If $p = 1$, then $\{x_n(\omega)\}$ converges weakly to $x(\omega)$ in $L_1(\Omega, \mathscr{A}, \mathscr{P})$ if

$$\lim_{n\to\infty} \int_\Omega x_n(\omega) h(\omega) d\mathscr{P}(\omega) = \int_\Omega x(\omega) h(\omega) d\mathscr{P}(\omega)$$

for every bounded measurable function h on Ω.

Definition 1.2.11 A family of measurable functions $\{x_n(t;\omega)\}_{n=1}^\infty$ is said to be an *equicontinuous* family if for every $\varepsilon > 0$ there is a $\delta > 0$ such that $\|x_n(t_1;\omega) - x_n(t_2;\omega)\| < \varepsilon$ whenever $|t_1 - t_2| < \delta$ for all $n = 1, 2, \ldots$.

The following random version of the Arzela-Ascoli theorem will be used in Chapter II.

Theorem 1.2.1 Let Φ be the class of all functions $x(t;\omega)$ which are product-measurable on $[0, 1] \times \Omega$ and satisfy $|x(t;\omega)| \leq N_0$. Suppose that for every n the map $x_n(t;\omega) \in \Phi$ is continuous from $[0, 1]$ into $L_p(\Omega, \mathscr{A}, \mathscr{P})$. Assume further that the family of maps $\{x^*[x_n(t;\omega)]\}$, $n = 1, 2, \ldots$, from $[0, 1]$ into the reals R is eventually equicontinuous for every x^*, that is, given $\varepsilon > 0$, there is an M and a $\delta > 0$ such that $|t - t_0| < \delta, n \geq M$, implies that

$$|x^*[x_n(t;\omega)] - x^*[x_n(t_0;\omega)]| < \varepsilon.$$

Then there exists a subsequence $\{x_{n_i}(t;\omega)\}$ such that for some map $x(t;\omega) \in \Phi$ this subsequence converges in the weak topology of $L_p(\Omega, \mathscr{A}, \mathscr{P})$ to $x(t;\omega)$ for every $t \in [0, 1]$.

The following definitions will also be needed in the study.

Definition 1.2.12 Let H be the set of all functions $x(t;\omega)$ in $C_c(R_+, L_2(\Omega, \mathscr{A}, \mathscr{P}))$ such that:

(i) $\|x(t;\omega)\|^2_{L_2(\Omega,\mathscr{A},\mathscr{P})}$ is integrable on R_+.

(ii) For any function $y(t;\omega)$ satisfying (i), $y(t;\omega)\in H$ if the inner product $(x(t;\omega), y(t;\omega))_{L_2(\Omega,\mathscr{A},\mathscr{P})}$ is integrable on R_+.

For $M > 0$ let $B_M, D_M \subset H$ be Hilbert spaces with the inner product on B_M defined by

$$(x,y)_{B_M} = \int_0^M (x(t;\omega), y(t;\omega))_{L_2(\Omega,\mathscr{A},\mathscr{P})}\, dt,$$

and that on D_M, $(x,y)_{D_M}$ is defined likewise. These are valid inner products, as can easily be shown. Since $L_2(\Omega, \mathscr{A}, \mathscr{P})$ is an inner product space, we have for any scalar α

$$(\alpha x, y)_{B_M} = \int_0^M (\alpha x(t;\omega), y(t;\omega))_{L_2(\Omega,\mathscr{A},\mathscr{P})}\, dt$$

$$= \alpha \int_0^M (x(t;\omega), y(t;\omega))_{L_2(\Omega,\mathscr{A},\mathscr{P})}\, dt$$

$$= \alpha(x,y)_{B_M}.$$

Also

$$\overline{(x,y)_{B_M}} = \int_0^M \overline{(x(t;\omega), y(t;\omega))}_{L_2(\Omega,\mathscr{A},\mathscr{P})}\, dt$$

$$= \int_0^M (y(t;\omega), x(t;\omega))_{L_2(\Omega,\mathscr{A},\mathscr{P})}\, dt = (y,x)_{B_M},$$

$$(x_1 + x_2, y)_{B_M} = \int_0^M (x(t;\omega), y(t;\omega))_{L_2(\Omega,\mathscr{A},\mathscr{P})}\, dt$$

$$+ \int_0^M (x_2(t;\omega), y(t;\omega))_{L_2(\Omega,\mathscr{A},\mathscr{P})}\, dt$$

$$= (x_1, y)_{B_M} + (x_2, y)_{B_M},$$

and if $x(t;\omega) \neq 0$ for almost all $\omega \in \Omega$ and $t \in R_+$,

$$(x,x)_{B_M} = \int_0^M (x(t;\omega), x(t;\omega))_{L_2(\Omega,\mathscr{A},\mathscr{P})}\, dt > 0.$$

The norm of an element of B_M is then defined by

$$\|x(t;\omega)\|_{B_M} = (x,x)_{B_M}^{\frac{1}{2}} = \left\{\int_0^M \|x(t;\omega)\|_{L_2(\Omega,\mathscr{A},\mathscr{P})}^2\, dt\right\}^{\frac{1}{2}}$$

and that of an element of D_M, $\|x(t;\omega)\|_{D_M}$, is similarly defined. Since we have that $\|x(t;\omega)\|_{L_2(\Omega,\mathscr{A},\mathscr{P})}$ is integrable on R_+, for every $M > 0$ the norms

1.2 PROBABILISTIC DEFINITIONS

defined here exist and are finite. If $M \to \infty$, then the norm of an element of B_∞ is given by

$$\|x(t;\omega)\|_{B_\infty} = \left\{\int_0^\infty \|x(t;\omega)\|_{L_2(\Omega,\mathscr{A},\mathscr{P})}^2 \, dt\right\}^{\frac{1}{2}} < \infty$$

and the norm of an element of D_∞ is defined likewise.

Note that Hilbert spaces such as those just given exist, since we may take C_g, with $g(t) = e^{-\beta t}$, $\beta > 0$, $t \in R_+$, and with the appropriate inner product, as the space B_∞ (or D_∞).

In order to study random discrete equations in Chapter V, we must define the following spaces.

Definition 1.2.13 We denote by X the space of all functions **x** from N, the positive integers, into $L_2(\Omega, \mathscr{A}, \mathscr{P})$. That is, for each $n = 1, 2, \ldots$, the value of **x** at n is $x_n(\omega) \in L_2(\Omega, \mathscr{A}, \mathscr{P})$. The topology of X is the topology of uniform convergence on every set

$$N_m = \{1, 2, \ldots, m\}, \quad m = 1, 2, \ldots,$$

that is, $\mathbf{x}_i \to \mathbf{x}$ as $i \to \infty$ in X if and only if

$$\lim_{i \to \infty} \|x_{i,n}(\omega) - x_n(\omega)\|_{L_2(\Omega,\mathscr{A},\mathscr{P})} = 0$$

uniformly on every set N_m, $m = 1, 2, \ldots$.

Note also that X is a locally convex space (Yosida [1, pp. 24–26]), with the topology defined by the following family of semi-norms:

$$\|x_n(\omega)\|_m = \sup_{0 \leq n \leq m} \|x_n(\omega)\|_{L_2(\Omega,\mathscr{A},\mathscr{P})}, \quad m = 1, 2, \ldots.$$

Definition 1.2.14 We let X_g be the Banach space of sequences in X for which there exist positive numbers $g_n < \infty$ and some constant $Q > 0$ such that

$$\|x_n(\omega)\|_{L_2(\Omega,\mathscr{A},\mathscr{P})} \leq Q g_n, \quad n = 1, 2, \ldots.$$

The norm in X_g is defined by

$$\|\mathbf{x}\|_{X_g} = \|x_n(\omega)\|_{X_g} = \sup_n \left\{\frac{\|x_n(\omega)\|_{L_2(\Omega,\mathscr{A},\mathscr{P})}}{g_n}\right\}.$$

When $g_n = 1$ for $n = 1, 2, \ldots$ we obtain the Banach space X_1 of all *bounded* functions from N into $L_2(\Omega, \mathscr{A}, \mathscr{P})$. The norm in X_1 is defined by

$$\|\mathbf{x}\|_{X_1} = \|x_n(\omega)\|_{X_1} = \sup_n \|x_n(\omega)\|_{L_2(\Omega,\mathscr{A},\mathscr{P})}.$$

Definition 1.2.15 We shall denote by X_{bv} the Banach space of all functions in X of *bounded variation*, that is,

$$\|x\|_{X_{bv}} = \|x_1(\omega)\|_{L_2(\Omega, \mathscr{A}, \mathscr{P})} + \sum_{i=1}^{\infty} \|x_{i+1}(\omega) - x_i(\omega)\|_{L_2(\Omega, \mathscr{A}, \mathscr{P})} < \infty,$$

which defines the norm in X_{bv}.

The definitions of X, X_g, and X_{bv} are stochastic generalizations of some spaces considered by Petrovanu [1] in the nonstochastic case.

1.3 The Stochastic Integral Equations and Stochastic Differential Systems

In this section we will give the main types of stochastic integral equations to be investigated and state some of the assumptions that are made. Also, certain definitions concerning the stochastic differential systems to be studied will be presented.

The main types of equations which will be studied in Chapters II, III, and IV are those of the Volterra type in the form

$$x(t;\omega) = h(t;\omega) + \int_0^t k(t,\tau;\omega) f(\tau, x(\tau;\omega)) \, d\tau \qquad (1.3.1)$$

and those of the Fredholm type in the form

$$x(t;\omega) = h(t;\omega) + \int_0^\infty k_0(t,\tau;\omega) e(\tau, x(\tau;\omega)) \, d\tau, \qquad (1.3.2)$$

where $t \geq 0$ and (i) ω is a point of Ω; (ii) $h(t;\omega)$ is the *stochastic free term* or *free random variable* defined for $0 \leq t$ and $\omega \in \Omega$; (iii) $x(t;\omega)$ is the unknown random variable for each $t \geq 0$; (iv) the *stochastic kernel* $k(t,\tau;\omega)$ is defined for $0 \leq \tau \leq t < \infty$ and $\omega \in \Omega$; (v) the *stochastic kernel* $k_0(t,\tau;\omega)$ is defined for $0 \leq t < \infty$ and $0 \leq \tau < \infty$ and $\omega \in \Omega$; (vi) $f(t,x)$ and $e(t,x)$ are scalar functions defined for $0 \leq t$ and scalars x.

The integrals in Eqs. (1.3.1) and (1.3.2) will be interpreted as mean-square integrals (Prabhu [1]).

We shall assume that $x(t;\omega)$ and $h(t;\omega)$ are functions of the argument $t \in R_+$ with values in the space $L_2(\Omega, \mathscr{A}, \mathscr{P})$. The functions $f(t, x(t;\omega))$ and $e(t, x(t;\omega))$ under convenient conditions will also be functions of $t \in R_+$ with values in $L_2(\Omega, \mathscr{A}, \mathscr{P})$. The stochastic kernels $k(t,\tau;\omega)$ and $k_0(t,\tau;\omega)$ will be essentially bounded functions with respect to $\mathscr{P}$ for every t and τ such that $0 \leq \tau \leq t < \infty$ and $0 \leq t < \infty$, $0 \leq \tau < \infty$, respectively. The values of the

1.3 STOCHASTIC INTEGRAL EQUATIONS AND DIFFERENTIAL SYSTEMS 19

stochastic kernel $k(t, \tau; \omega)$ for fixed t and τ will be in $L_\infty(\Omega, \mathscr{A}, \mathscr{P})$, so that the product of $k(t, \tau; \omega)$ and $f(t, x(t; \omega))$ will always be in $L_2(\Omega, \mathscr{A}, \mathscr{P})$. A similar assumption holds for $k_0(t, \tau; \omega)$ for fixed t and τ.

With respect to the stochastic kernel $k(t, \tau; \omega)$, we shall assume that the mapping

$$(t, \tau) \to k(t, \tau; \omega)$$

from the set $\Delta = \{(t, \tau): 0 \leqslant \tau \leqslant t < \infty\}$ into $L_\infty(\Omega, \mathscr{A}, \mathscr{P})$ is continuous. That is, whenever $(t_n, \tau_n) \to (t, \tau)$ as $n \to \infty$ we have

$$\mathscr{P}\text{-ess}\sup_\omega |k(t_n, \tau_n; \omega) - k(t, \tau; \omega)| \to 0 \quad \text{as} \quad n \to \infty$$

or, equivalently,

$$\inf_{\Omega_0} \left\{ \sup_{\Omega - \Omega_0} |k(t_n, \tau_n; \omega) - k(t, \tau; \omega)| \right\} \to 0 \quad \text{as} \quad n \to \infty$$

where $\mathscr{P}(\Omega_0) = 0$. Likewise, for the stochastic kernel $k_0(t, \tau; \omega)$ we will assume that the mapping

$$(t, \tau) \to k_0(t, \tau; \omega)$$

from the set $\Delta_1 = \{(t, \tau): 0 \leqslant t < \infty, \ 0 \leqslant \tau < \infty\}$ into $L_\infty(\Omega, \mathscr{A}, \mathscr{P})$ is continuous. Further assumptions will be given at appropriate points in the text.

We will also study in Chapter IV a stochastic integral equation of the mixed Volterra–Fredholm type of the form

$$x(t; \omega) = h(t; \omega) + \int_0^t k(t, \tau; \omega) f(\tau, x(\tau; \omega)) \, d\tau$$

$$+ \int_0^\infty k_0(t, \tau; \omega) e(\tau, x(\tau; \omega)) \, d\tau \qquad (1.3.3)$$

for $t \geqslant 0$. Equation (1.3.3) is of interest because Eqs. (1.3.1) and (1.3.2) arise as special cases of it.

The following perturbed random Volterra and Fredholm integral equations will be investigated in addition to the above types:

$$x(t; \omega) = h(t, x(t; \omega)) + \int_0^t k(t, \tau; \omega) f(\tau, x(\tau; \omega)) \, d\tau \qquad (1.3.4)$$

and

$$x(t; \omega) = h(t, x(t; \omega)) + \int_0^\infty k_0(t, \tau; \omega) e(\tau, x(\tau; \omega)) \, d\tau, \qquad (1.3.5)$$

where $t \geq 0$ and $h(t, x)$ is a scalar function of t and x possessing certain continuity properties which will be stated later.

We now give the following definitions.

Definition 1.3.1 By a *random solution* of any one of the stochastic integral equations (1.3.1)–(1.3.5) we will mean a function $x(t;\omega)$ which belongs to $C_c(R_+, L_2(\Omega, \mathscr{A}, \mathscr{P}))$ and satisfies the equation $\mathscr{P}$-a.e.

Definition 1.3.2 A random solution $x(t;\omega)$ is said to be *stochastically asymptotically exponentially stable* if there exist constants $\rho > 0$ and $\beta > 0$ such that

$$\{E|x(t;\omega)|^2\}^{\frac{1}{2}} = \left\{\int_\Omega |x(t;\omega)|^2 \, d\mathscr{P}(\omega)\right\}^{\frac{1}{2}} \leq \rho e^{-\beta t}, \qquad t \in R_+.$$

Definition 1.3.3 A random solution $x(t;\omega)$ is said to be *asymptotically stable in mean square* if

$$\lim_{t \to \infty} E|x(t;\omega)|^2 = 0.$$

Definitions 1.3.2 and 1.3.3 are important in applications in which the behavior of a stochastic system as time becomes large is of interest. That is, conditions are needed for which the system remains stable in some sense.

In many applications random nonlinear Volterra integral equations arise in the form

$$x(t;\omega) = h(t;\omega) + \int_0^t K(u, x(u;\omega); \omega) \, du, \qquad (1.3.6)$$

where $t \geq 0$ and (i) $K(u, x; \omega)$ is the random kernel defined for $0 \leq u \leq t < \infty$ and $\omega \in \Omega$; (ii) the random function $x(t;\omega)$ is unknown, $t \in R_+$; and (iii) the random function $h(t;\omega)$ is known, $t \in R_+$. Equations such as (1.3.6) have been studied by Bharucha-Reid, Mukherjea, and Tserpes [1] utilizing the spaces $L_p(\Omega, \mathscr{A}, \mathscr{P})$ given in Definition 1.2.9. In Chapter II we will present some theory concerning the random equation (1.3.6) and two important applications of such equations.

Definition 1.3.4 A random function $x(t;\omega)$ is said to be a *random solution* of Eq. (1.3.6) if for every $t \in R_+$ it satisfies the equation $\mathscr{P}$-a.e.

In Chapter V we will use the spaces given by Definitions 1.2.13–1.2.15 in order to investigate the existence and asymptotic behavior of random solutions of stochastic discrete equations of the Volterra type in the form

$$x_n(\omega) = h_n(\omega) + \sum_{j=1}^n C_{n,j}^*(\omega) f_j(x_j(\omega)) \qquad (1.3.7)$$

F_ε be defined on $\overline{M}$ as given by Eq. (1.A.2). If $x \in M$, then

$$\|T(x) - F_\varepsilon T(x)\| < \varepsilon.$$

PROOF By definition

$$\|T(x) - F_\varepsilon T(x)\| = \left\|T(x) - \left\{\sum_{i=1}^{p} m_i[T(x)]v_i \bigg/ \sum_{i=1}^{p} m_i[T(x)]\right\}\right\|$$

$$= \left\|\left\{\sum_{i=1}^{p} m_i[T(x)]T(x) - \sum_{i=1}^{p} m_i[T(x)]v_i\right\} \bigg/ \sum_{i=1}^{p} m_i[T(x)]\right\|$$

$$\leq \sum_{i=1}^{p} m_i[T(x)]\|T(x) - v_i\| \bigg/ \sum_{i=1}^{p} m_i[T(x)]$$

$$< \varepsilon \sum_{i=1}^{p} m_i[T(x)] \bigg/ \sum_{i=1}^{p} m_i[T(x)] = \varepsilon$$

by definition of $m_i(x)$, completing the proof.

Theorem 1.A.5 (*Schauder's fixed-point theorem*) Let M be a convex, bounded, closed set in a Banach space and let T be a compact transformation such that $T(M) \subset M$. Then T has a fixed point in M. That is, there exists an $x_0 \in M$ so that $T(x_0) = x_0$.

PROOF Since $T(M) \subset M$, we have $\overline{T(M)} \subset \overline{M}$. The fact that M is closed implies that $M = \overline{M}$, and hence $\overline{T(M)} \subset M$. Let $\{\varepsilon_n\}$ be a monotone decreasing sequence such that $\lim_{n \to \infty} \varepsilon_n = 0$. Let $T_n = F_{\varepsilon_n} T$ be defined on M as described in Theorem 1.A.4. For $x \in M$ we have

$$T_n(x) = F_{\varepsilon_n}(T(x)).$$

But $T(x) = y \in M$; therefore $F_{\varepsilon_n}(T(x)) = F_{\varepsilon_n}(y)$. Suppose $\{v_1, \ldots, v_{p_n}\}$ is an ε_n-net of $\overline{T(M)}$. The function

$$F_{\varepsilon_n}(y) = \sum_{i=1}^{p_n} m_i(y)v_i \bigg/ \sum_{i=1}^{p_n} m_i(y),$$

which means that

$$F_{\varepsilon_n}(y) = \frac{m_1(y)v_1}{m_1(y) + \cdots + m_{p_n}(y)} + \cdots + \frac{m_{p_n}(y)v_{p_n}}{m_1(y) + \cdots + m_{p_n}(y)}.$$

Set

$$m_i(y)/[m_1(y) + \cdots + m_{p_n}(y)] = M_i.$$

Therefore, since M is convex, $F_{\varepsilon_n}(y) \in M$, which implies that $T_n(M) \subset M$.

Let H_n be the finite-dimensional subspace of H which is spanned by $\{v_1, \ldots, v_{p_n}\}$. Let $M_n = M \cap H_n$. Now M is closed and H_n is closed since it is a finite-dimensional space. This implies that M_n is closed since the intersection of two closed sets is closed. Also M is convex, and H_n is convex since for $x_1 \in H_n$,

$$x_1 = \sum_{i=1}^{p_n} \alpha_i v_i,$$

and for $x_2 \in H_n$,

$$x_2 = \sum_{i=1}^{p_n} \beta_i v_i,$$

we have that

$$qx_1 + (1-q)x_2 = q \sum_{i=1}^{p_n} \alpha_i v_i + (1-q) \sum_{i=1}^{p_n} \beta_i v_i = \sum_{i=1}^{p_n} [q\alpha_i + (1-q)\beta_i] v_i$$

$$= \sum_{i=1}^{p_n} \gamma_i v_i \in H_n,$$

where $0 < q < 1$. Hence M_n is convex (the intersection of two convex sets is convex). Therefore M_n is a closed convex subset of H_n. The transformation T_n is defined on M and $M_n \subset M$, which implies that T_n is defined on M_n. Also, $T_n(M_n) \subset M_n$, for if $x \in M_n$, then $x = \sum_{i=1}^{p_n} \alpha_i v_i$, and

$$T_n(x) = F_{\varepsilon_n}(T(x)) = \sum_{i=1}^{p_n} m_i[T(x)] v_i \bigg/ \sum_{i=1}^{p_n} m_i[T(x)] = \sum_{i=1}^{p_n} \beta_i v_i \in H_n.$$

Also, $T_n(x) \in M$, since if $T_n(M) \subset M$ and $M_n \subset M$, we have that $x \in M_n \subset M$ and $T_n(x) \in M$. Thus

$$T_n(x) \in M \cap H_n = M_n,$$

which means that $T_n(M_n) \subset M_n$. Since F_{ε_n} is continuous and T is continuous, we have that T_n is continuous, and by using Brouwer's fixed-point theorem (Theorem 1.A.2), there is a point $x_n \in M_n$ such that

$$T_n(x_n) = x_n.$$

The set $\{T(x_n)\}$ is contained in the closed compact set $\overline{T(M)}$.

REMARK The set $\overline{T(M)}$ is compact since the ε_n-net gives a finite covering.

Now $\overline{T(M)} \subset M$ implies $\{T(x_n)\} \subset M$. Thus $T(x_n)$ has a limit point x_0 and $x_0 \in M$ since M is closed. Either the sequence $\{T(x_n)\}$ converges to x_0 or there is a subsequence of $\{T(x_n)\}$ which converges to x_0. For simplicity of notation, assume that $\{T(x_n)\}$ converges to x_0. Then

$$\|T(x_n) - x_0\| < \varepsilon \quad \text{for} \quad n > n(\varepsilon). \tag{1.A.3}$$

Also from the definition of T_n we have

$$\|T_n(x_n) - T(x_n)\| < \varepsilon_n. \tag{1.A.4}$$

Then from (1.A.3) and (1.A.4) we have

$$\|T_n(x_n) - x_0\| \le \|T_n(x_n) - T(x_n)\| + \|T(x_n) - x_0\| < \varepsilon_n + \varepsilon.$$

Since $T_n(x_n) = x_n$, we obtain

$$\|x_n - x_0\| < \varepsilon_n + \varepsilon.$$

Let ε' be given. T is continuous, and we have that there exists a $\delta(\varepsilon') > 0$ such that

$$\|T(x_n) - T(x_0)\| < \varepsilon'$$

whenever $\|x_n - x_0\| < \delta(\varepsilon')$. To make $\|x_n - x_0\| < \delta(\varepsilon')$, choose n large enough so that $\varepsilon_n + \varepsilon \le \delta(\varepsilon')$. We can do this since $\lim_{n \to \infty} \varepsilon_n = 0$. Hence we have shown that

$$\|T(x_n) - T(x_0)\| < \varepsilon$$

whenever n is large enough, which means that $T(x_n) \to T(x_0)$ as $n \to \infty$. Since the limit of the sequence $\{T(x_n)\}$ is unique, then we must have

$$T(x_0) = x_0,$$

completing the proof.

Finally, we present the proof of Barbalat's lemma which was stated in Section 1.

Lemma 1.A.6 (*Barbalat*) If
 (i) $f(t)$ is a continuous function, and its derivatives $f'(t)$ are bounded for $t \ge 0$;
 (ii) $G(x)$ is a continuous function, $G(x) > 0$ for $x \ne 0$, and $G(0) = 0$;
 (iii) $\int_0^\infty G[f(t)]\,dt < \infty$;

then

$$\lim_{t \to \infty} f(t) = 0.$$

PROOF We shall prove this lemma by contradiction. From the hypothesis of the lemma, we have for every $t \ge 0$

$$|f'(t)| < b < \infty \quad \text{and} \quad \int_0^\infty G(f(t))\,dt = c < \infty.$$

Let us assume that $\lim_{t \to \infty} f(t) \ne 0$. This implies that there exists a sequence, say $\{t_k\}$, $t_k > 0$ for $k = 1, 2, \ldots$, and some $\varepsilon > 0$, such that

$$|f(t_k)| \ge \varepsilon > 0.$$

We can further assume that for all k

$$t_{k+1} - t_k \geq m > 0 \tag{1.A.5}$$

that is, the elements in sequence (1.A.5) are distinct, $t_1 < t_2 < \cdots < t_k$. If this condition does not hold, then we can choose a new sequence which would satisfy (1.A.5). Since $f'(t)$ is bounded for $t \geq 0$, then using the mean value theorem, we can write

$$|f(t) - f(t_k)| \leq b|t - t_k| \qquad \text{for all} \quad k.$$

It is given that $G(x) > 0$ for $x \neq 0$, so we have

$$\int_0^\infty G[f(t)]\,dt \geq \sum_{k=1}^\infty \left\{ \int_{t_k - (m/2)}^{t_k + (m/2)} G[f(t)]\,dt \right\}.$$

For the length of the interval m we can write

$$t_k - \tfrac{1}{2}m \leq t \leq t_k + \tfrac{1}{2}m \qquad \text{for all} \quad k. \tag{1.A.6}$$

On this interval (1.A.6) we can construct the following inequality:

$$|f(t)| = |f(t_k) + [f(t) - f(t_k)]| \geq |f(t_k)| - |f(t) - f(t_k)|$$
$$\geq \varepsilon - b|t - t_k| \geq \varepsilon - \tfrac{1}{2}bm \geq \varepsilon_0.$$

Since $G(x)$ is continuous, we can define some

$$r = \min_{\varepsilon \leq |x| \leq \alpha} G(x) > 0;$$

then we can write

$$\int_{t_k - (m/2)}^{t_k + (m/2)} G[f(t)]\,dt \geq r \int_{t_k - (m/2)}^{t_k + (m/2)} dt \geq rm. \tag{1.A.7}$$

Taking the sum of both sides of (1.A.7), that is,

$$\sum_{k=1}^\infty \int_{t_k - (m/2)}^{t_k + (m/2)} G[f(t)]\,dt \geq \sum_{k=1}^\infty rm = \infty,$$

this implies that

$$\int_0^\infty G[f(t)]\,dt = \infty,$$

but by hypothesis

$$\int_0^\infty G[f(t)]\,dt = c < \infty,$$

hence, a contradiction. Therefore we conclude that

$$\lim_{t \to \infty} f(t) = 0.$$

CHAPTER II

Some Random Integral Equations of the Volterra Type with Applications

2.0 Introduction

In this chapter we shall present some of the most general results which have been obtained to date concerning random integral equations of the Volterra type. In Section 2.1 some results of Tsokos [4] will be given for the random integral equation

$$x(t;\omega) = h(t;\omega) + \int_0^t k(t,\tau;\omega)f(\tau,x(\tau;\omega))\,d\tau. \qquad (2.0.1)$$

We shall investigate the *existence* and *uniqueness* of a random solution of Eq. (2.0.1). The *asymptotic behavior* of the random solution and its *stability properties* also will be considered. In Section 2.2 several applications of Eq. (2.0.1) will be presented in the areas of telephone traffic theory, hereditary mechanics, and a generalization of the classical Poincaré–Lyapunov theorem (Tsokos [3]).

In Section 2.3 some recent results of Bharucha-Reid, Mukherjea, and Tserpes [1] will be given along with theorems of the authors concerning the existence of a random solution of the random Volterra integral equation of the form

$$x(t;\omega) = h(t;\omega) + \int_0^t K(u, x(u;\omega); \omega)\, du. \qquad (2.0.2)$$

Then in Section 2.4 applications of (2.0.2) in the theory of turbulence and in the theory of chemotherapy will be given.

In later chapters we shall consider some stochastic differential systems which reduce to stochastic integral equations of the form just given.

2.1 The Random Integral Equation

$$x(t;\omega) = h(t;\omega) + \int_0^t k(t,\tau;\omega) f(\tau, x(\tau;\omega))\, d\tau$$

This equation seems to be more general than any random Volterra integral equation which has been studied to date. The generality consists primarily in the choice of the stochastic kernel. The origin and the importance of this random integral equation have already been discussed. In this section we shall investigate the *existence* of a random solution and its *uniqueness* and *asymptotic* behavior, and shall consider a number of special cases as corollaries of the main theorems. Finally, the *stability* properties of the random solution will be investigated.

To accomplish our objectives here, we employ certain aspects of the methods of "admissibility theory," which has been utilized quite recently in the theory of deterministic integral equations by Corduneanu [4] as presented in Chapter I.

2.1.1 Existence and Uniqueness of a Random Solution

Let B and D be a pair of Banach spaces and T a linear operator. With respect to the study of this section, we state the following lemma, which will be used in the main theorems.

Lemma 2.1.1 Let T be a continuous operator from $C_c(R_+, L_2(\Omega, \mathcal{A}, \mathcal{P}))$ into itself. If B and D are Banach spaces stronger than C_c and the pair (B, D) is admissible with respect to T, then T is a continuous operator from B to D.

2.1 THE RANDOM INTEGRAL EQUATION

PROOF First we will prove that the operator T is closed from B to D. Let us consider the sequence $x_n(t;\omega) \in B$ such that $x_n(t;\omega) \xrightarrow{B} x(t;\omega)$ as $n \to \infty$. Let us assume $(Tx_n)(t;\omega) \xrightarrow{D} y(t;\omega)$ as $n \to \infty$. Now we must show that $(Tx)(t;\omega) = y(t;\omega)$. Since $x_n(t;\omega) \to x(t;\omega)$ in B, $x_n(t;\omega) \to x(t;\omega)$ in C_c. But since $T: C_c \to C_c$ is continuous we have $(Tx_n)(t;\omega) \to (Tx)(t;\omega)$ in C_c. On the other hand, $(Tx_n)(t;\omega) \to y(t;\omega)$ in D, which implies that $(Tx_n)(t;\omega) \to y(t;\omega)$ in C_c. Hence $(Tx)(t;\omega) = y(t;\omega)$, because the limit is unique in C_c. Therefore the operator T is closed. Then by the closed-graph theorem (Theorem 1.1.3) it follows that T is a continuous operator from B to D.

REMARK Since T is a closed and continuous linear operator, it is also bounded (Yosida [1, pp. 10–11]). Then it follows that we can find a constant $K > 0$ such that

$$\|(Tx)(t;\omega)\|_D \leqslant K\|x(t;\omega)\|_B$$

(see Definition 1.1.13).

With respect to our aims here, we state and prove the following theorems.

Theorem 2.1.2 Let us consider Eq. (2.0.1) under the following conditions:

(i) B and D are Banach spaces stronger than $C_c(R_+, L_2(\Omega, \mathscr{A}, \mathscr{P}))$ such that (B, D) is admissible with respect to the operator

$$(Tx)(t;\omega) = \int_0^t k(t,\tau;\omega)x(\tau;\omega)\,d\tau,$$

where $k(t,\tau;\omega)$ behaves as in Chapter I.

(ii) $x(t;\omega) \to f(t, x(t;\omega))$ is a continuous operator on

$$S = \{x(t;\omega) : x(t;\omega) \in D, \quad \|x(t;\omega)\|_D \leqslant \rho\}$$

with values in B, also satisfying

$$\|f(t, x(t;\omega)) - f(t, y(t;\omega))\|_B \leqslant \lambda \|x(t;\omega) - y(t;\omega)\|_D$$

with $x(t;\omega), y(t;\omega) \in S$ and λ a constant.

(iii) $\|h(t;\omega)\| \in D$.

Then there exists a *unique* random solution of the random integral equation (2.0.1), provided that

$$\lambda < K^{-1}, \quad \|h(t;\omega)\|_D + K\|f(t,0)\|_B \leqslant \rho(1 - \lambda K),$$

where K is the norm of the operator T (see Remark to Lemma 2.1.1).

PROOF Let us define an operator U on S into D as follows:

$$(Ux)(t;\omega) = h(t;\omega) + \int_0^t k(t,\tau;\omega) f(\tau, x(\tau;\omega))\, d\tau. \qquad (2.1.1)$$

Now we must show that U is a contracting operator and $U(S) \subset S$. Consider a function $y(t;\omega)$ in S. We can write

$$(Uy)(t;\omega) = h(t;\omega) + \int_0^t k(t,\tau;\omega) f(\tau, y(\tau;\omega))\, d\tau. \qquad (2.1.2)$$

Subtracting Eq. (2.1.2) from Eq. (2.1.1), we have

$$(Ux)(t;\omega) - (Uy)(t;\omega) = \int_0^t k(t,\tau;\omega)[f(\tau, x(\tau;\omega)) - f(\tau, y(\tau;\omega))]\, d\tau.$$

Since $U(S) \subset D$ and D is a Banach space, then

$$(Ux)(t;\omega) - (Uy)(t;\omega) \in D.$$

By assumptions (i) and (ii), $[f(\tau, x(\tau;\omega)) - f(\tau, y(\tau;\omega))] \in B$. From Lemma 2.1.1 we have seen that T is a continuous operator from the Banach space B into D, which implies that we can find a constant $K > 0$ such that

$$\|(Tx)(t;\omega)\|_D \leqslant K\|x(t;\omega)\|_B.$$

That is,

$$\|(Ux)(t;\omega) - (Uy)(t;\omega)\|_D \leqslant K\|f(t, x(t;\omega)) - f(t, y(t;\omega))\|_B.$$

Now, applying Lipschitz's condition given in (ii), we have

$$\|(Ux)(t;\omega) - (Uy)(t;\omega)\|_D \leqslant \lambda K\|x(t;\omega) - y(t;\omega)\|_D.$$

Using the condition that $\lambda K < 1$, the operator U is a contracting operator. It now remains to be shown that $U(S) \subset S$. For every function $x(t;\omega) \in S$ we have

$$(Ux)(t;\omega) = h(t;\omega) + \int_0^t k(t,\tau;\omega) f(\tau, x(\tau;\omega))\, d\tau. \qquad (2.1.3)$$

Applying Condition (iii) and Lemma 2.1.1, we can write expression (2.1.3) as follows:

$$\|(Ux)(t;\omega)\|_D \leqslant \|h(t;\omega)\|_D + K\|f(t, x(t;\omega))\|_B. \qquad (2.1.4)$$

In (2.1.4), $\|f(t, x(t;\omega))\|_B$ can be written as

$$\|f(t, x(t;\omega))\|_B = \|f(t, x(t;\omega)) - f(t,0) + f(t,0)\|_B$$
$$\leqslant \|f(t, x(t;\omega)) - f(t,0)\|_B + \|f(t,0)\|_B.$$

2.1 THE RANDOM INTEGRAL EQUATION

Using Lipschitz's condition, we have

$$\|f(t, x(t;\omega))\|_B \leq \lambda \|x(t;\omega) - 0\|_D + \|f(t,0)\|_B.$$

We can now write expression (2.1.4) as follows:

$$\|(Ux)(t;\omega)\|_D \leq \|h(t;\omega)\|_D + K\lambda \|x(t;\omega)\|_D + K\|f(t,0)\|_B. \quad (2.1.5)$$

Since $x(t;\omega) \in S$ and $\|x(t;\omega)\|_D \leq \rho$, (2.1.5) can be written as

$$\|(Ux)(t;\omega)\|_D \leq \|h(t;\omega)\|_D + K\lambda\rho + K\|f(t,0)\|_B. \quad (2.1.6)$$

Applying the condition of the theorem that

$$\|h(t;\omega)\|_D + K\|f(t,0)\|_B \leq \rho(1 - \lambda K),$$

(2.1.6) becomes

$$\|(Ux)(t;\omega)\|_D \leq \rho(1 - \lambda K) + K\rho\lambda \quad \text{or} \quad \|(Ux)(t;\omega)\|_D \leq \rho,$$

which implies that $(Ux)(t;\omega) \in S$ for all random variables $x(t;\omega) \in S$ or $U(S) \subset S$. Therefore, since U is a contracting operator and $U(S) \subset S$ (inclusion property), applying Banach's fixed-point theorem (Theorem 1.1.2), there *exists* a unique random solution $x(t;\omega) \in S$ such that

$$(Ux)(t;\omega) = h(t;\omega) + \int_0^t k(t,\tau;\omega) f(\tau, x(\tau;\omega)) \, d\tau = x(t;\omega).$$

2.1.2 Some Special Cases

Now we shall derive some particular cases of Theorem 2.1.2 choosing in a convenient manner the spaces B and D. Recall that a C_g space is a space of all continuous functions from $R_+ \to L_2(\Omega, \mathscr{A}, \mathscr{P})$ such that

$$\|x(t;\omega)\| = \left\{ \int_\Omega |x(t;\omega)|^2 \, d\mathscr{P}(\omega) \right\}^{\frac{1}{2}} \leq Zg(t),$$

where $t \in R_+$, Z is a number greater than zero, and $g(t)$ is a continuous function greater than zero. Also, $C_c(R_+, L_2(\Omega, \mathscr{A}, \mathscr{P}))$ is the space of all continuous functions from R_+ into $L_2(\Omega, \mathscr{A}, \mathscr{P})$, with the topology of uniform convergence on the interval $[0, T]$ for any $T > 0$, and the norm of the stochastic kernel of the integral equation can be defined as follows:

$$K(t, \tau) = \|k(t, \tau; \omega)\| = \mathscr{P}\text{-ess sup } |k(t, \tau; \omega)|$$

with respect to $\omega \, \varepsilon \, \Omega$. That is,

$$\|k(t, \tau; \omega)\| = \inf_{\Omega_0} \left\{ \sup_{\Omega - \Omega_0} |k(t, \tau; \omega)| \right\}$$

with $\mathscr{P}(\Omega_0) = 0$.

II SOME VOLTERRA TYPE EQUATIONS WITH APPLICATIONS

Theorem 2.1.3 Let us consider the random integral equation (2.0.1) under the following conditions:

(i) There exists a number $A > 0$ and a continuous function (on R_+) $g(t) > 0$ such that

$$\int_0^t \|k(t, \tau; \omega)\| g(\tau) \, d\tau \leq A, \qquad t \in R_+.$$

(ii) $f(t, x)$ is a continuous vector-valued function for $t \in R_+$, $\|x(t; \omega)\| \leq \rho$, such that

$$f(t, 0) \in C_g, \qquad \|f(t, x(t; \omega)) - f(t, y(t; \omega))\| \leq \lambda g(t) \|x(t; \omega) - y(t; \omega)\|.$$

(iii) $h(t; \omega)$ is a continuous bounded function on R_+ whose values are in $L_2(\Omega, \mathscr{A}, \mathscr{P})$. Then there exists a *unique* random solution $x(t; \omega) \in C$ of the random integral equation (2.0.1) such that

$$\|x(t; \omega)\|_C = \sup_{0 \leq t} \left\{ \int_\Omega |x(t; \omega)|^2 \, d\mathscr{P}(\omega) \right\}^{\frac{1}{2}} \leq \rho$$

for $t \in R_+$, as long as $\|h(t; \omega)\|$, λ, and $\|f(t, 0)\|_{C_g}$ are small enough.

PROOF We must show that under Condition (i) of the theorem the pair of Banach spaces (C_g, C) is admissible. That is, (C_g, C) is admissible with respect to the integral operator

$$(Tx)(t; \omega) = \int_0^t k(t, \tau; \omega) x(\tau; \omega) \, d\tau.$$

For $x(t; \omega) \in C_g$ we have

$$(Tx)(t; \omega) = \int_0^t k(t, \tau; \omega) x(\tau; \omega) \, d\tau$$

or

$$\|(Tx)(t; \omega)\| \leq \int_0^t \|k(t, \tau; \omega) x(\tau; \omega)\| \, d\tau$$

$$\leq \int_0^t \|k(t, \tau; \omega)\| [\|x(\tau; \omega)\|/g(\tau)] g(\tau) \, d\tau, \qquad (2.1.7)$$

where $\|k(t, \tau; \omega)\| = \mathscr{P}\text{-ess sup}_\omega |k(t, \tau; \omega)|$ is a function only of (t, τ). Using the definition of the norm in C_g, that is,

$$\sup_{0 \leq t} \left\{ \left\{ \int_\Omega |x(t; \omega)|^2 \, d\mathscr{P}(\omega) \right\}^{\frac{1}{2}} [1/g(t)] \right\} = \|x(t; \omega)\|_{C_g}$$

2.1 THE RANDOM INTEGRAL EQUATION

for $g(t) > 0$, we can write (2.1.7) as follows:

$$\|(Tx)(t;\omega)\| \leq \int_0^t \|k(t,\tau;\omega)\| \|x(\tau;\omega)\|_{C_g} g(\tau) \, d\tau$$

$$\leq \|x(t;\omega)\|_{C_g} \int_0^t \|k(t,\tau;\omega)\| g(\tau) \, d\tau$$

$$\leq A \|x(t;\omega)\|_{C_g}.$$

Therefore $\|(Tx)(t;\omega)\|$ is bounded and hence $(Tx)(t;\omega) \in C$ for all $x(t;\omega) \in C_g$. Hence $TC_g \subset C$, which implies that the pair of Banach spaces (C_g, C) is admissible with respect to the integral operator as defined here.

The remainder of the proof is analogous to that of Theorem 2.1.2 and is omitted.

For the special case where $g(t) = 1$ we state and prove the following corollary.

Corollary 2.1.4 Let us assume that the random integral equation (2.0.1) satisfies the following conditions:

(i) $\int_0^t \|k(t,\tau;\omega)\| \, d\tau \leq A$, $t \in R_+$, where A is some constant greater than zero.

(ii) $f(t, x)$ is a continuous function from R_+ into R uniformly in x such that
$$|f(t, x) - f(t, y)| \leq \lambda |x - y|.$$

(iii) $h(t;\omega)$ is a continuous bounded function from R_+ into $L_2(\Omega, \mathscr{A}, \mathscr{P})$. Then there exists a *unique* bounded random solution on R_+ of the random integral equation (2.0.1) if λ is small enough.

PROOF We must show that under Condition (i) of the corollary the pair of Banach spaces (C, C) is admissible. For a function $x(t;\omega) \in C$ we have

$$(Tx)(t;\omega) = \int_0^t k(t,\tau;\omega) x(\tau;\omega) \, d\tau$$

or

$$\|(Tx)(t;\omega)\| \leq \int_0^t \|k(t,\tau;\omega)\| \|x(\tau;\omega)\| \, d\tau. \tag{2.1.8}$$

Applying the definition of the norm as used in Theorem 2.1.3, inequality (2.1.8) can be written as follows:

$$\|(Tx)(t;\omega)\| \leq \|x(t;\omega)\| \int_0^t \|k(t,\tau;\omega)\| \, d\tau$$

$$\leq A \|x(t;\omega)\|, \quad t \in R_+.$$

Therefore $(Tx)(t;\omega) \in C$ for every random variable $x(t;\omega)$ or $TC \subset C$, which implies that (C, C) is admissible. The remainder of the proof follows from Theorem 2.1.2.

The following two corollaries are particular cases of Theorem 2.1.3.

Corollary 2.1.5 Assume that the random integral equation (2.0.1) satisfies the following conditions:

(i) $\|k(t,\tau;\omega)\| \leq \Lambda_1$, for $0 \leq \tau \leq t < \infty$, and $\int_0^\infty g(t)\,dt < \infty$.
(ii) Same condition as in Theorem 2.1.3, Condition (ii).
(iii) Same condition as in Theorem 2.1.3, Condition (iii).

Then there exists a *unique* random solution of Eq. (2.0.1) bounded on R_+ if $\|h(t;\omega)\|$, λ, and $\|f(t,0)\|$ are sufficiently small.

PROOF It is only necessary to show that the pair of Banach spaces (C_g, C) is admissible with respect to the integral operator

$$(Tx)(t;\omega) = \int_0^t k(t,\tau;\omega)x(\tau;\omega)\,d\tau, \qquad (2.1.9)$$

along with Condition (i) of the corollary. For a function $x(t;\omega) \in C_g$ expression (2.1.9) implies that

$$\|(Tx)(t;\omega)\| \leq \int_0^t \|k(t,\tau;\omega)\|\,\|x(\tau;\omega)\|\,d\tau.$$

Applying hypothesis (i) of the corollary, we have

$$\|(Tx)(t;\omega)\| \leq \Lambda_1 \int_0^t \|x(\tau;\omega)\|\,d\tau \leq \Lambda_1 \int_0^t [\|x(\tau;\omega)\|/g(\tau)]g(\tau)\,d\tau. \qquad (2.1.10)$$

Utilizing the definition of norm as applied in Theorem 2.1.3 to inequality (2.1.10), we have

$$\|(Tx)(t;\omega)\| \leq \Lambda_1 \|x(t;\omega)\|_{C_g} \int_0^t g(\tau)\,d\tau, \qquad (2.1.11)$$

but, applying Condition (i) of the corollary, (2.1.11) is written as

$$\|(Tx)(t;\omega)\| \leq M \qquad \text{for all} \quad t \geq 0.$$

Therefore the function $x(t;\omega) \in C_g$ implies that $(Tx)(t;\omega) \in C$, or $TC_g \subset C$. Hence the pair (C_g, C) is admissible, and, since Conditions (ii) and (iii) are the same as in Theorem 2.1.3, the proof is complete.

2.1 THE RANDOM INTEGRAL EQUATION

Corollary 2.1.6 Let us consider the random integral equation (2.0.1) under the following conditions:

(i) $\|k(t, \tau; \omega)\| \leq \Lambda_2 e^{-\alpha(t-\tau)}$, for $0 \leq \tau \leq t < +\infty$, and

$$\sup_{t \in R_+} \left\{ \int_t^{t+1} g(\tau) \, d\tau \right\} < \infty,$$

where Λ_2 and α are positive numbers.

(ii) Same as Condition (ii) of Theorem 2.1.3.
(iii) Same as Condition (iii) of Theorem 2.1.3.

Then there exists a *unique* random solution of the random integral equation (2.0.1) bounded on R_+ if $\|h(t; \omega)\|$, λ, and $\|f(t, 0)\|$ are small enough.

PROOF We must show that the pair of Banach spaces (C_g, C) is admissible with respect to the integral operator

$$(Tx)(t; \omega) = \int_0^t k(t, \tau; \omega) x(\tau; \omega) \, d\tau \tag{2.1.12}$$

along with Condition (i) of the corollary. Taking the norms of both sides of (2.1.12), we have

$$\|(Tx)(t; \omega)\| \leq \int_0^t \|k(t, \tau; \omega)\| \|x(\tau; \omega)\| \, d\tau, \tag{2.1.13}$$

but $\|k(t, \tau; \omega)\| \leq \Lambda_2 e^{-\alpha(t-\tau)}$, which implies that (2.1.13) can be written as

$$\|(Tx)(t; \omega)\| \leq \Lambda_2 \int_0^t e^{-\alpha(t-\tau)} \|x(\tau; \omega)\| \, d\tau$$

$$\leq \Lambda_2 \int_0^t e^{-\alpha(t-\tau)} [\|x(\tau; \omega)\|/g(\tau)] g(\tau) \, d\tau. \tag{2.1.14}$$

From the definition of the norm as used in Corollary 2.1.5, inequality (2.1.14) can be written as follows:

$$\|(Tx)(t; \omega)\| \leq \|x(t; \omega)\|_{C_g} \Lambda_2 \int_0^t e^{-\alpha(t-\tau)} g(\tau) \, d\tau. \tag{2.1.15}$$

But

$$\sup_{0 \leq t} \left\{ \int_t^{t+1} g(\tau) \, d\tau \right\} < \infty$$

implies that

$$\int_0^t e^{-\alpha(t-\tau)} g(\tau) \, d\tau \leq M < +\infty,$$

and (2.1.15) can be written as

$$\|(Tx)(t;\omega)\| \leq \|x(t;\omega)\|_{C_g} M\Lambda_2.$$

Therefore $(Tx)(t;\omega) \in C$ or $TC_g \subset C$, which implies that the pair of Banach spaces (C_g, C) is admissible. Since Conditions (ii) and (iii) are the same as in Theorem 2.1.3 and admissibility has been shown, the proof is complete.

2.1.3 Asymptotic Stability of the Random Solution

With respect to the asymptotic behavior of the random solution of the stochastic integral equation (2.0.1), we state and prove the following theorem, the objective of which is to investigate the possibility of the random solution being asymptotically exponentially stable.

Theorem 2.1.7 Let us consider the stochastic integral equation (2.0.1) under the following conditions:

(i) $\|k(t,\tau;\omega)\| \leq \Lambda_2 e^{-\alpha(t-\tau)}$, for $0 \leq \tau \leq t < +\infty$, $\Lambda_2 > 0$, and $\alpha > 0$.

(ii) $f(t,x)$ is a continuous function from $R_+ \times R$ into R such that $f(t,0) = 0$, and

$$|f(t,x) - f(t,y)| \leq \lambda|x - y|.$$

(iii) $\|h(t;\omega)\| \leq \rho e^{-\beta t}$, where ρ and β are positive numbers such that $0 < \beta < \alpha$.

Then there exists a unique random solution of Eq. (2.0.1) such that

$$\left\{\int_\Omega |x(t;\omega)|^2 \, d\mathscr{P}(\omega)\right\}^{\frac{1}{2}} \leq \rho e^{-\beta t}, \qquad t \in R_+,$$

as long as λ is small enough.

PROOF We must show that the pair of Banach spaces (C_g, C_g), with $g(t) = e^{-\beta t}$, is admissible under Conditions (i) and (iii) of the Theorem. That is, (C_g, C_g) is admissible with respect to the operator defined by

$$(Tx)(t;\omega) = \int_0^t k(t,\tau;\omega)x(\tau;\omega) \, d\tau. \qquad (2.1.16)$$

The norm of expression (2.1.16) can be written as

$$\|(Tx)(t;\omega)\| \leq \int_0^t \|k(t,\tau;\omega)\| \, \|x(\tau;\omega)\| \, d\tau. \qquad (2.1.17)$$

Applying Condition (i) of the theorem, we have

$$\|(Tx)(t;\omega)\| \leq \Lambda_2 \int_0^t e^{-\alpha(t-\tau)} \|x(\tau;\omega)\| \, d\tau. \qquad (2.1.18)$$

Hence

$$\|(Tx)(t;\omega)\| \leq \Lambda_2 \int_0^t e^{-\alpha(t-\tau)}[\|x(\tau;\omega)\|/g(\tau)]g(\tau)\,d\tau. \quad (2.1.19)$$

Using the definition of the norm on the C_g space, inequality (2.1.19) can be written as

$$\|(Tx)(t;\omega)\| \leq M \int_0^t e^{-\alpha(t-\tau)}g(\tau)\,d\tau = M \int_0^t e^{-\alpha(t-\tau)} e^{-\beta\tau}\,d\tau$$

$$= M e^{-\alpha t}[1/(\alpha-\beta)](e^{(\alpha-\beta)t} - 1)$$

$$= M(\alpha-\beta)^{-1}(e^{-\beta t} - e^{-\alpha t}). \quad (2.1.20)$$

Since $0 < \beta < \alpha$, we can majorize inequality (2.1.20) as follows:

$$\|(Tx)(t;\omega)\| \leq M(\alpha-\beta)^{-1}(e^{-\beta t} - e^{-\alpha t}) < M(\alpha-\beta)^{-1}e^{-\beta t}, \qquad t \in R_+,$$

which implies that $(Tx)(t;\omega) \in C_g$ for a function $x(t;\omega) \in C_g$. Therefore the pair of Banach spaces (C_g, C_g) is admissible with respect to the operator T, where $g(t) = e^{-\beta t}$. Condition (iii) of the theorem means that $h(t;\omega) \in C_g$. Applying Condition (ii), we have

$$\|f(t, x(t;\omega)) - f(t, y(t;\omega))\|_{C_g} \leq \lambda \|x(t;\omega) - y(t;\omega)\|_{C_g}.$$

Hence, all the conditions of Theorem 2.1.3 have been satisfied, which implies that there exists a unique random solution of the integral equation (2.0.1) such that

$$\left\{\int_\Omega |x(t;\omega)|^2\,d\mathscr{P}(\omega)\right\}^{\frac{1}{2}} \leq \rho\, e^{-\beta t}.$$

REMARK It is now clear that under these conditions there exists a random solution of the random integral equation (2.0.1) which is exponentially asymptotically stable, that is,

$$\lim_{t\to\infty} \left\{\int_\Omega |x(t;\omega)|^2\,d\mathscr{P}(\omega)\right\}^{\frac{1}{2}} = 0.$$

2.2 Some Applications of the Equation

$$x(t;\omega) = h(t;\omega) + \int_0^t k(t, \tau;\omega)f(\tau, x(\tau;\omega))\,d\tau$$

In this section we shall present some applications of the results of the previous section. We shall first consider a generalization of the classical stability theorem of Poincaré and Lyapunov. We shall then study a stochastic integral equation arising in the theory of telephone traffic, a related study of which was done by Fortet [1]. A problem in hereditary mechanics also will be considered which results in a pair of stochastic integral equations of the Volterra type, for which Distefano [1] showed existence of a solution by the method of successive approximations. In each case we shall briefly describe the problem and indicate that a random solution exists by applying the results of Section 2.1.

2.2.1 Generalization of Poincaré–Lyapunov Stability Theorem

As an example to illustrate our results, we shall generalize the classical stability theorem of Poincaré and Lyapunov (Tsokos [3]). That is, consider the following random differential system:

$$\dot{x}(t;\omega) = A(\omega)x(t;\omega) + f(t, x(t;\omega)), \qquad t \geq 0, \qquad (2.2.1)$$

where (i) $x(t;\omega)$ is the unknown $n \times 1$ random vector; (ii) $A(\omega)$ is an $n \times n$ matrix whose elements are measurable functions; and (iii) $f(t, x)$ is, for $t \in R_+$ and $x \in R$, an $n \times 1$ vector-valued function.

Now we shall reduce the random differential system (2.2.1) to a stochastic integral equation which will be a special case of the stochastic integral equation (2.0.1). Multiplying the random system (2.2.1) by $e^{-A(\omega)t}$, we have

$$e^{-A(\omega)t}\dot{x}(t;\omega) - A(\omega)e^{-A(\omega)t}x(t;\omega) = e^{-A(\omega)t}f(t, x(t;\omega)).$$

But

$$(d/dt)\{e^{-A(\omega)t}x(t;\omega)\} = e^{-A(\omega)t}(d/dt)x(t;\omega) - A(\omega)e^{-A(\omega)t}x(t;\omega).$$

Therefore

$$(d/dt)\{e^{-A(\omega)t}x(t;\omega)\} = e^{-A(\omega)t}f(t, x(t;\omega)). \qquad (2.2.2)$$

Integrating both sides of Eq. (2.2.2) from t_0 to t, we have

$$e^{-A(\omega)t}x(t;\omega) - e^{-A(\omega)t_0}x(t_0;\omega) = \int_{t_0}^{t} e^{-A(\omega)\tau}f(\tau, x(\tau;\omega))\, d\tau. \qquad (2.2.3)$$

Multiplying Eq. (2.2.3) by $e^{A(\omega)t}$ and letting $t_0 = 0$, it reduces to

$$x(t;\omega) = e^{A(\omega)t}x_0(\omega) + \int_0^t e^{A(\omega)(t-\tau)}f(\tau, x(\tau;\omega))\, d\tau, \qquad (2.2.4)$$

where $x_0(\omega) = x(0;\omega)$. Hence, if we let

$$h(t;\omega) = e^{A(\omega)t}x_0(\omega) \quad \text{and} \quad k(t, \tau;\omega) = e^{A(\omega)(t-\tau)}, \qquad 0 \leq \tau \leq t < \infty,$$

Eq. (2.2.4) can be written as

$$x(t;\omega) = h(t;\omega) + \int_0^t k(t, \tau;\omega)f(\tau, x(\tau;\omega))\, d\tau.$$

2.2 SOME APPLICATIONS OF THE EQUATION

Hence, the stochastic differential system (2.2.1) reduces to the stochastic integral equation (2.2.4), which is a special form of Eq. (2.0.1).

Now we state the following theorem.

Theorem 2.2.1 Let us assume that the following conditions hold with respect to the stochastic integral equation (2.2.4):

(i) The matrix $A(\omega)$ is stochastically stable, that is, there exists an $\alpha > 0$ such that

$$\mathscr{P}\{\omega; \operatorname{Re}\psi_k(\omega) < -\alpha, \quad k = 1, 2, \ldots, n\} = 1,$$

where $\psi_k(\omega)$, $k = 1, 2, \ldots, n$ are the characteristic roots of the matrix.

(ii) $f(t, x)$ is a continuous function from $R_+ \times R^n \to R^n$ such that

$$|f(t, x) - f(t, y)| \leq \lambda |x - y|$$

with $f(t, 0) = 0$ and λ sufficiently small.

Then there exists a *unique* random solution of the stochastic integral equation (2.2.4) such that

$$\lim_{t \to \infty} \left\{ \int_\Omega |x(t;\omega)|^2 \, d\mathscr{P}(\omega) \right\}^{\frac{1}{2}} = 0.$$

PROOF To prove this result, we want to prove that the pair of Banach spaces (C_g, C_g) is admissible under Conditions (i) and (ii) with $g(t) = e^{-\beta t}$, and then apply Theorem 2.1.3.

Recall that the norm in the space $C_g(R_+, L_2(\Omega, \mathscr{A}, \mathscr{P}))$ is defined by

$$\|x(t;\omega)\|_{C_g} = \sup_{t \in R_+} [1/g(t)] \left\{ \int_\Omega |x(t;\omega)|^2 \, d\mathscr{P}(\omega) \right\}^{\frac{1}{2}},$$

and for any function $x(t;\omega) \in C_g(R_+, L_2(\Omega, \mathscr{A}, \mathscr{P}))$, let us define the following integral operator:

$$(Tx)(t;\omega) = \int_0^t k(t, \tau; \omega) x(\tau; \omega) \, d\tau. \tag{2.2.5}$$

Since $k(t, \tau; \omega) = e^{A(\omega)(t-\tau)}$, $0 \leq \tau \leq t < \infty$, Eq. (2.2.5) becomes

$$(Tx)(t;\omega) = \int_0^t e^{A(\omega)(t-\tau)} x(\tau; \omega) \, d\tau$$

or

$$\|(Tx)(t;\omega)\| \leq \int_0^t \|e^{A(\omega)(t-\tau)}\| \, \|x(\tau;\omega)\| \, d\tau. \tag{2.2.6}$$

It has been shown by Morozan [2, 3] that there exists a subset D of Ω such that $\mathscr{P}(D) = 1$ and

$$\|e^{A(\omega)(t-\tau)}\| \leq K e^{-\alpha(t-\tau)} \tag{2.2.7}$$

for $\omega \in D$, $K > 0$, and α as defined previously. Now, putting (2.2.7) into inequality (2.2.6), we have

$$\|(Tx)(t;\omega)\| \leq K \int_0^t e^{-\alpha(t-\tau)} \|x(\tau;\omega)\| \, d\tau$$

$$\leq K \int_0^t e^{-\alpha(t-\tau)} [\|x(\tau;\omega)\|/g(\tau)] g(\tau) \, d\tau. \tag{2.2.8}$$

Since $g(t) = e^{-\beta t}$, $0 < \beta < \alpha$, inequality (2.2.8) becomes

$$\|(Tx)(t;\omega)\| \leq K \int_0^t e^{-\alpha(t-\tau)} (1/e^{-\beta \tau}) \|x(\tau;\omega)\| \, e^{-\beta \tau} \, d\tau$$

$$\leq K \|x(t;\omega)\|_{C_g} e^{-\alpha t} \int_0^t e^{\tau(\alpha - \beta)} \, d\tau$$

$$\leq K \|x(t;\omega)\|_{C_g} (\alpha - \beta)^{-1} (e^{-\beta t} - e^{-\alpha t}), \qquad t \geq 0. \tag{2.2.9}$$

Inequality (2.2.9) can be majorized as follows:

$$\|(Tx)(t;\omega)\| \leq K \|x(t;\omega)\|_{C_g} (\alpha - \beta)^{-1} e^{-\beta t} \tag{2.2.10}$$

because $0 < \beta < \alpha$. Dividing inequality (2.2.10) by $e^{-\beta t}$, we have

$$\|(Tx)(t;\omega)\|_{C_g} \leq K(\alpha - \beta)^{-1} \|x(t;\omega)\|_{C_g}.$$

Hence for $x(t;\omega) \in C_g(R_+, L_2(\Omega, \mathscr{A}, \mathscr{P}))$ we have $TC_g(R_+, L_2(\Omega, \mathscr{A}, \mathscr{P})) \subset C_g(R_+, L_2(\Omega, \mathscr{A}, \mathscr{P}))$, and the pair of Banach spaces (C_g, C_g) is admissible. The rest of the proof is due to Theorem 2.1.3.

2.2.2 A Problem in Telephone Traffic Theory

In this subsection we shall examine a stochastic integral equation arising in the study of telephone traffic. We shall describe the problem in detail and then apply Corollary 2.1.5 to show existence of a unique random solution.

Consider a telephone exchange, and suppose that calls arrive at the exchange at time instants $t_1, t_2, \ldots, t_n, \ldots$, where $0 < t_1 < t_2 < \cdots < t_n < \infty$. These arrival times must be considered as random instants, so we denote the distribution function by $A(t)$ on the time axis. For a call arriving at time t let the random variable $H(t;\omega)$ denote the holding time, that is, the length of time that a "conversation" is held for a call arriving at the exchange at time t. The $H(t_1;\omega)$, $H(t_2;\omega), \ldots$ are considered as being mutually independent for different times $t_1, t_2, \ldots$ and as being independent of the state of the exchange, where the state of the exchange is the number of busy channels.

2.2 SOME APPLICATIONS OF THE EQUATION

The number m of trunks or channels of the exchange is assumed to be finite and large, so that we approximate a continuous process. It is also assumed that any channel not being used may be utilized by an incoming call and that the holding time for a channel begins at the time instant that the call arrives at the exchange.

A conversation (or connection) is realized if a channel is not busy at the time a call arrives. If all channels are busy at the time t that a call arrives, then either the call is lost or a queueing problem develops. Only the first case will be considered here. Various problems have been studied in this situation. For example, the probability $P_k(t)$ that at time t, k of the m channels are busy has been examined in detail (Fortet [1]). We are concerned with the total number of "conversations" held (the number of busy channels) at time t, which for each t is a random variable and may be described by a stochastic integral equation.

Let $x(t;\omega)$ be the total number of conversations held at time t. That is, $x(t;\omega)$ is a random variable for each $t \in R_+$, and $x(0;\omega) = 0$. Let $J(t;\omega)$ be a random function with value one if a call arising at time $t > 0$ is not lost and value zero if the call is lost.

Let

$$K(t,\tau;\omega) = \begin{cases} 1 & \text{if } t - \tau \in [0, H(\tau;\omega)], \\ 0 & \text{if } t - \tau \notin [0, H(\tau;\omega)], \end{cases}$$

such that $K(t,\tau;\omega)$ is equal to one if a conversation from a call arising at time τ is still being held at time $t \geq \tau$ and is equal to zero otherwise. Thus we may write

$$x(t;\omega) = \int_0^t J(\tau;\omega) K(t,\tau;\omega) \, dA(\tau). \tag{2.2.11}$$

Equation (2.2.11) is interpreted as the total number of telephone calls arising at times τ, $0 \leq \tau \leq t$, that were not lost such that the conversation is still being held at time t.

Suppose that $V(k)$ is any function such that

$$V(k) = \begin{cases} 1 & \text{if } k = 0, 1, \ldots, m-1, \\ 0 & \text{otherwise}. \end{cases}$$

Clearly, $x(t;\omega) \leq m$ for all $t \in R_+$ and $\omega \in \Omega$. Hence we may write

$$V[x(t;\omega)] = \begin{cases} 1 & \text{if } x(t;\omega) = 0, 1, \ldots, m-1, \\ 0 & \text{otherwise}, \end{cases}$$

which means that $V[x(t;\omega)]$ has value one if a call arising at time t is not lost and value zero otherwise. Then Eq. (2.2.11) may be written as the nonlinear stochastic integral equation

$$x(t;\omega) = \int_0^t K(t,\tau;\omega)V[x(\tau;\omega)]\,dA(\tau), \qquad (2.2.12)$$

which Bharucha-Reid [7] refers to as the *Fortet integral equation*.

Suppose that the distribution $A(t)$ of arrival times has a density function $a(t)$. Then Eq. (2.2.12) reduces to a stochastic integral equation of the Volterra type

$$x(t;\omega) = \int_0^t K(t,\tau;\omega)V[x(\tau;\omega)]a(\tau)\,d\tau.$$

If we let

$$f(\tau, x(\tau;\omega)) = V[x(\tau;\omega)]a(\tau) = \begin{cases} a(\tau) & \text{if } x(\tau;\omega) = 0, 1, \ldots, m-1, \\ 0 & \text{otherwise,} \end{cases}$$

then we obtain a stochastic Volterra integral equation of the form of (2.0.1), with the stochastic free term identically zero,

$$x(t;\omega) = \int_0^t K(t,\tau;\omega)f(\tau, x(\tau;\omega))\,d\tau. \qquad (2.2.13)$$

$K(t,\tau;\omega)$ is the stochastic kernel defined for $0 \leq \tau \leq t < \infty$ and taking only the value zero or one.

Before showing that (2.2.13) possesses a unique random solution, we observe that the above description applies to many systems. If we replace the words "telephone exchange" with "serving mechanism," and the words "channel," "call," and "conversation" with the words "server," "customer," and "service," respectively, then we are dealing with a general system in which "customers" are being "served" by $m < \infty$ "servers." If we assume that a customer does not wait when he finds all m servers busy so that no queue develops, then the random solution of the stochastic integral equation (2.2.13) gives the total number of "services" being performed at time t. Also, the functions in (2.2.13) may be any functions which behave as given in Chapter I and describe the physical situation. For example, the stochastic kernel may be of the form

$$K(t,\tau;\omega) = I_{X(\omega)}(t-\tau)e^{-(t-\tau)},$$

where $I_{X(\omega)}(\cdot)$ is the indicator function of a random set $X(\omega)$, which means that solutions at earlier times $\tau \leq t$ have a decaying effect on the system.

We now show that the stochastic integral equation (2.2.13) satisfies the conditions of Corollary 2.1.5. We first show that

$$f(t, x(t;\omega)) \in L_2(\Omega, \mathscr{A}, \mathscr{P}) \quad \text{and} \quad K(t,\tau;\omega) \in L_\infty(\Omega, \mathscr{A}, \mathscr{P}).$$

2.2 SOME APPLICATIONS OF THE EQUATION

Let $t \geq 0$ be fixed. Since $a(t)$ is a density function, it may be assumed to be bounded for all t except on a set of measure zero. Hence for some $M > 0$ and all t, $0 \leq a(t) \leq M < \infty$, and we have

$$\int_\Omega |f(t, x(t;\omega))|^2 \, d\mathscr{P}(\omega) \leq \int_\Omega |a(t)|^2 \, d\mathscr{P}(\omega) \leq M^2 < \infty$$

by definition of $f(t, x(t;\omega))$, so that $f(t, x(t;\omega)) \in L_2(\Omega, \mathscr{A}, \mathscr{P})$ for each $t \in R_+$. By definition of $K(t, \tau; \omega)$, $0 \leq \tau \leq t < \infty$, we obviously have that the $\mathscr{P}$-measure of

$$\{\omega : |K(t, \tau; \omega)| > 1\}, \quad 0 \leq \tau \leq t < \infty,$$

is zero, that is, a $\mathscr{P}$-null set. Hence $K(t, \tau; \omega)$ is bounded $\mathscr{P}$-a.e. and is in $L_\infty(\Omega, \mathscr{A}, \mathscr{P})$. Also, if $(t_n, \tau_n) \to (t, \tau)$ as $n \to \infty$, we have

$$\mathscr{P}\{\omega : |K(t_n, \tau_n; \omega) - K(t, \tau; \omega)| > 0\} \to 0$$

as $n \to \infty$ since $K(t, \tau; \omega)$ has value zero or one only.

That $x(t; \omega) \in C$, a continuous bounded function for each ω, is easily shown. Let $\omega \in \Omega$ and choose $t \in R_+$. For $\varepsilon > 0$ and $h > 0$,

$$|x(t+h;\omega) - x(t;\omega)| = \left| \int_t^{t+h} K(t, \tau; \omega) f(\tau, x(\tau; \omega)) \, d\tau \right| \leq \left| \int_t^{t+h} a(\tau) \, d\tau \right|.$$

Thus for $\varepsilon > 0$ there is a $\delta > 0$ such that when $|(t+h) - t| < \delta$

$$\int_t^{t+h} a(\tau) \, d\tau < \varepsilon.$$

Therefore $x(t; \omega)$ is continuous on R_+ for each $\omega \in \Omega$ and bounded by m. Also,

$$\int_\Omega |x(t;\omega)|^2 \, d\mathscr{P}(\omega) \leq m^2 < \infty, \quad t \in R_+,$$

which means that $x(t; \omega) \in L_2(\Omega, \mathscr{A}, \mathscr{P})$ for each t.

We note that $\Lambda_1 = 1$ in Corollary 2.1.5, since

$$\|K(t, \tau; \omega)\| = \mathscr{P}\text{-ess sup}_\omega |K(t, \tau; \omega)| \leq 1.$$

Since $h(t; \omega) \equiv 0$ $\mathscr{P}$-a.e., trivially, it belongs to C. Since $a(t)$ is a density function, $f(t, 0) = a(t)$ is continuous, bounded, and nonnegative, and $\int_{-\infty}^{\infty} a(t) \, dt < \infty$. Suppose $g(t)$ is any function satisfying Condition (i) of

Corollary 2.1.5. Then we must have

$$|f(t,x) - f(t,y)| = \begin{cases} 0 \leq \lambda g(t) \cdot 0 & \text{if } x = y, \\ 0 \leq \lambda g(t)(m-1) & \text{if } x \neq y; \ x, y \in \{0, 1, \ldots, m-1\}, \\ a(t) \leq \lambda g(t)m & \text{if } x \neq m \text{ and } y = m \text{ or } \\ & x = m \text{ and } y \neq m. \end{cases}$$

If we choose $\lambda = 1/m$, then in order for Condition (ii) of Corollary 2.1.5 to be satisfied, there must exist a positive continuous function $g(t)$ on R_+ such that

$$\int_0^\infty g(t)\,dt < \infty \quad \text{and} \quad a(t) \leq g(t), \quad t \in R_+.$$

This restriction is not too severe, since $a(t)$ is a density function. We may take $g(t) = e^{-qt}$, $q > 0$, for example. Therefore, since $\|h(t;\omega)\|_C = 0$, $\|f(t,0)\| = a(t) \leq M$, and $\lambda = 1/m$ is small for large m, there exists a unique random solution of equation (2.2.13) if $a(t) \leq g(t)$, where $g(t)$ satisfies the given conditions.

2.2.3 A Stochastic Integral Equation in Hereditary Mechanics

A simple example of a stochastic integral equation of the Volterra type comes from the field of hereditary mechanics where the "forcing term" depends on the deviation of the system from a natural position of equilibrium as well as on an external source of excitation (Distefano [1]). When forces α and η are applied to two hinged bars, deflection of the bars is prevented by a viscoelastic spring reacting with an upward force S. Here η is the axial load and α is the resulting downward force on the spring. The bars are deflected a certain amount s, a nonlinear function of η, in general (see Figure 2.2.1). The forces and deflection are functions of time $t \geq 0$, and the displacement at time $t > 0$, $s(t)$, is some functional form of the force $S(\tau)$ exerted by the spring for $\tau \leq t$. However, $S(\tau)$ is a function of $\alpha(\tau)$, $\eta(\tau)$, and $s(\tau)$ for each $\tau \leq t$, which accounts for the hereditary effects of the system.

Distefano [1] considered the stochastic version of a system of two Volterra integral equations which arises in this problem. We are interested in $s(t)$, the displacement at time $t \geq 0$, of the linear hereditary phenomenon.

The linearized version of this problem leads to the integral equation

$$[1 - \eta(t)]s(t) - \int_0^t f(t,\tau)\eta(\tau)s(\tau)\,d\tau = g(t), \tag{2.2.14}$$

2.2 SOME APPLICATIONS OF THE EQUATION

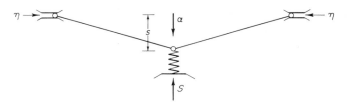

Figure 2.2.1 From Distefano[1].

where $f(t, \tau)$ is the "memory function" for the hereditary phenomenon and

$$g(t) = \alpha(t) + \int_0^t f(t, \tau)\alpha(\tau)\,d\tau,$$

assuming initial straightness of the bars, that is, $s(t) = 0$ at $t = 0$.

When the axial load has a small, randomly fluctuating component, that is, $\eta(t; \omega)$, $\omega \in \Omega$, is a random variable for each fixed $t > 0$, then Eq. (2.2.14) reduces to the two stochastic Volterra integral equations

$$u(t; \omega) = g(t; \omega) + \int_0^t \phi(\tau; \omega)f(t, \tau)u(\tau; \omega)\,d\tau,$$

$$v(t; \omega) = G(t; \omega) + \phi(t; \omega)\int_0^t f(t, \tau)v(\tau; \omega)\,d\tau,$$

where it is assumed that $0 \leq \eta(t; \omega) \leq \beta < 1$ $\mathscr{P}$-a.e. for all $t \in R_+$. Also,

$$G(t; \omega) = g(t; \omega)\phi(t; \omega), \qquad \phi(t; \omega) = \eta(t; \omega)/[1 - \eta(t; \omega)],$$

$$u(t; \omega) = [1 - \eta(t; \omega)]s(t; \omega), \quad \text{and} \quad v(t; \omega) = \eta(t; \omega)s(t; \omega)$$

are random variables for each $t \in R_+$. Then the quantity of interest $s(t; \omega)$ is the sum of the two random solutions $u(t; \omega) + v(t; \omega)$. Let

$$k_1(t, \tau; \omega) = \begin{cases} \phi(\tau; \omega)f(t, \tau), & 0 \leq \tau \leq t < \infty, \\ 0 & \text{otherwise,} \end{cases}$$

and

$$k_2(t, \tau; \omega) = \begin{cases} \phi(t; \omega)f(t, \tau), & 0 \leq \tau \leq t < \infty, \\ 0 & \text{otherwise.} \end{cases}$$

Then the random integral equations reduce to two stochastic integral equations each of the form (2.0.1) with $f(t, x(t; \omega)) = x(t; \omega)$, that is, linear stochastic integral equations,

$$u(t;\omega) = g(t;\omega) + \int_0^t k_1(t,\tau;\omega)u(\tau;\omega)\,d\tau,$$

$$v(t;\omega) = G(t;\omega) + \int_0^t k_2(t,\tau;\omega)v(\tau;\omega)\,d\tau. \tag{2.2.15}$$

Suppose the "memory function" is of the form

$$f(t,\tau) = Qe^{-Q(t-\tau)}, \qquad 0 \leq \tau \leq t < \infty, \qquad Q > 0.$$

Since $0 \leq \eta(t;\omega) \leq \beta$ $\mathscr{P}$-a.e. for all $t \in R_+$, $\phi(t;\omega)$ is bounded $\mathscr{P}$-a.e. on R_+, and we may assume $\alpha(t;\omega)$ and $s(t;\omega)$ are bounded $\mathscr{P}$-a.e. and continuous on R_+ by the nature of the physical situation. Thus there is some $M > 0$ such that

$$\phi(t;\omega) = \eta(t;\omega)/[1 - \eta(t;\omega)] \leq M \quad \mathscr{P}\text{-a.e.} \qquad \text{for all} \quad t \in R_+.$$

Hence

$$\|k_1(t,\tau;\omega)\| = \|\phi(\tau;\omega)f(t,\tau)\| \leq MQ\,e^{-Q(t-\tau)} \leq MQ < \infty$$

and

$$\|k_2(t,\tau;\omega)\| = \|\phi(t;\omega)f(t,\tau)\| \leq MQ < \infty.$$

Also, $f(t,0) = 0$, since $f(t,u(t;\omega)) = u(t;\omega)$ and $f(t,v(t;\omega)) = v(t;\omega)$ by comparison of (2.0.1) and (2.2.15), and $G(t;\omega)$ and $g(t;\omega)$ are continuous and bounded $\mathscr{P}$-a.e. on R_+ since $\phi(t;\omega)$ and $\alpha(t;\omega)$ are assumed to have these properties.

If in Corollary 2.1.6 we take the function $g(t) \equiv 1$ for all $t \in R_+$, then

$$\sup_{t \in R_+} \left\{ \int_t^{t+1} g(\tau)\,d\tau \right\} = 1 < \infty.$$

Also,

$$|f(t,u) - f(t,v)| = |u - v|$$

so that $\lambda = 1$. We may take $\Lambda_2 = MQ > 0$ and $\alpha = Q > 0$, and then all conditions of the corollary are satisfied. Therefore there exists a unique random solution of each of the equations (2.2.15), provided that $\|g(t;\omega)\|$ and $\|G(t;\omega)\|$ are small enough.

2.3 The Random Integral Equation

$$x(t;\omega) = h(t;\omega) + \int_0^t K(u, x(u;\omega);\omega)\,du$$

This equation occurs in many important situations, as has already been stated. To complete the objectives of this part of the book, we shall investigate the *existence* and *uniqueness* of a random solution of the stochastic integral equation (2.0.2). Some results of Bharucha-Reid, Mukherjea, and Tserpes [1] will be given which use methods of probabilistic functional analysis from Chapter I. Also a theorem of the authors will be presented which utilizes some of the concepts of "admissibility theory" as was discussed in Chapter I and employed in Section 2.1.

We shall first assume that the functions $x(t;\omega)$, $h(t;\omega)$, and $K(u,x;\omega)$ in Eq. (2.0.2) are real-valued, x and h are product-measurable on $R_+ \times \Omega$, and $K(u,x;\omega)$ is product-measurable on $R_+ \times \Omega$ for each x.

Theorem 2.3.1 Equation (2.0.2) with $t \in R_+$ has a random solution if the following conditions are satisfied:

(i) The kernel $K(u,x;\omega)$ is a continuous function of u and x, $u \in R_+$ and $x \in R$ for each $\omega \in \Omega$, and $K(u,x;\omega)$ is a random variable for each u and x.

(ii) $\sup_x |K(u,x;\omega)| \leqslant g(\omega) \in L_p(\Omega, \mathscr{A}, \mathscr{P})$, $1 \leqslant p < \infty$.

(iii) Let $h(t;\omega)$ be a continuous map from R_+ into $L_p(\Omega, \mathscr{A}, \mathscr{P})$, $1 \leqslant p < \infty$, and

$$|h(t;\omega)| + \int_0^t \sup_x |K(u,x;\omega)|\,du \leqslant N < \infty$$

for each $t \in R_+$ and $\omega \in \Omega$.

(iv) The kernel satisfies the following weak continuity property: If $x_n(\omega)$ converges weakly to $x(\omega)$ in $L_p(\Omega, \mathscr{A}, \mathscr{P})$, then $K(u, x_n(\omega);\omega)$ converges weakly to $K(u, x(\omega);\omega)$ whenever they are in $L_p(\Omega, \mathscr{A}, \mathscr{P})$ for each u.

PROOF We first consider $t \in [0, M]$. Define the sequence $x_0(t;\omega) = h(t;\omega)$,

$$x_n(t;\omega) = \begin{cases} x_0(t;\omega) & \text{if } 0 \leqslant t \leqslant 1/n, \\ x_0(t;\omega) + \displaystyle\int_0^{t-(1/n)} K(u, x_{n-1}(u;\omega);\omega)\,du \end{cases}$$

if $m/n \leqslant t \leqslant (m+1)/n$,

$$1 \leqslant m \leqslant Mn - 1, \quad n \geqslant 1.$$

Let $1 \leqslant p < \infty$ throughout the proof.

Since $h(t;\omega) \in L_p(\Omega, \mathscr{A}, \mathscr{P})$, then $x_0(t;\omega)$ is in $L_p(\Omega, \mathscr{A}, \mathscr{P})$. Thus $x_n(t;\omega)$, $n \geq 1$, is in $L_p(\Omega, \mathscr{A}, \mathscr{P})$, since by Condition (ii) and an extension of Minkowski's inequality (Beckenbach and Bellman [1, p. 22]), for $t \in [0, M]$ we have

$$\left\{ \int_\Omega \left| \int_0^t K(u, x_n(u;\omega);\omega) \, du \right|^p d\mathscr{P}(\omega) \right\}^{1/p}$$
$$\leq \int_0^t \left\{ \int_\Omega |K(u, x_n(u;\omega);\omega)|^p \, d\mathscr{P}(\omega) \right\}^{1/p} du$$
$$\leq \int_0^t \left\{ \int_\Omega |g(\omega)|^p \, d\mathscr{P}(\omega) \right\}^{1/p} du < \infty. \tag{2.3.0}$$

Then

$$\left\{ \int_\Omega |x_n(t;\omega)|^p \, d\mathscr{P}(\omega) \right\}^{1/p}$$
$$= \left\{ \int_\Omega \left| h(t;\omega) + \int_0^t K(u, x_{n-1}(u;\omega);\omega) \, du \right|^p d\mathscr{P}(\omega) \right\}^{1/p}$$
$$\leq \left\{ \int_\Omega |h(t;\omega)|^p \, d\mathscr{P}(\omega) \right\}^{1/p}$$
$$+ \left\{ \int_\Omega \left| \int_0^t K(u, x_{n-1}(u;\omega);\omega) \, du \right|^p d\mathscr{P}(\omega) \right\}^{1/p}$$
$$< \infty$$

by Minkowski's inequality, (2.3.0), and the fact that $h(t;\omega) \in L_p(\Omega, \mathscr{A}, \mathscr{P})$ for each $t \in [0, \infty)$.

Since $h(t;\omega)$ is continuous from R_+ into $L_p(\Omega, \mathscr{A}, \mathscr{P})$ we have that $\{x_n(t;\omega)\}_{n=0}^\infty$ is an equicontinuous family of functions from $[0, M]$ into $L_p(\Omega, \mathscr{A}, \mathscr{P})$. To show this, let $\varepsilon > 0$ and $n \geq 0$ be fixed but arbitrary. Then for $t_1, t_2 \in [0, M]$,

$$|x_n(t_1;\omega) - x_n(t_2;\omega)| \leq |h(t_1;\omega) - h(t_2;\omega)|$$
$$+ \left| \int_0^{t_1 - (1/n)} K(u, x_{n-1}(u;\omega);\omega) \, du \right.$$
$$\left. - \int_0^{t_2 - (1/n)} K(u, x_{n-1}(u;\omega);\omega) \, du \right|$$

for $\omega \in \Omega$. Thus we can find $\delta > 0$ such that for $|t_1 - t_2| < \delta$

$$|h(t_1;\omega) - h(t_2;\omega)| < \varepsilon/2$$

and

$$\left| \int_0^{t_1-(1/n)} K(u, x_{n-1}(u;\omega); \omega) \, du - \int_0^{t_2-(1/n)} K(u, x_{n-1}(u;\omega); \omega) \, du \right| < \varepsilon/2.$$

Since n is arbitrary, the functions $x_n(t;\omega)$ are equicontinuous from $[0, M]$ into $L_p(\Omega, \mathscr{A}, \mathscr{P})$.

Also, since $h(t;\omega)$ is product-measurable on $[0, \infty) \times \Omega$ and $K(u, x; \omega)$ is product-measurable on $[0, \infty) \times \Omega$ for each x, then $x_n(t;\omega)$ is product-measurable on $[0, M] \times \Omega$, $n \geqslant 0$.

Now, for $t \in [0, M]$ and any $x(t;\omega)$, we have by Condition (iii)

$$|x(t;\omega)| \leqslant |h(t;\omega)| + \int_0^t |K(u, x(t;\omega); \omega)| \, du$$

$$\leqslant |h(t;\omega)| + \int_0^t \sup_x |K(u, x; \omega)| \, du \leqslant N < \infty.$$

Then the sequence $\{x_n(t;\omega)\}$ satisfies the conditions of the Arzela-Ascoli-type Theorem 1.2.1, and there exists a subsequence $\{x_{n_i}(t;\omega)\}$ that converges weakly to a product-measurable function $x(t;\omega)$ in $L_p(\Omega, \mathscr{A}, \mathscr{P})$, for $t \in [0, M]$, that satisfies

$$x(t;\omega) = h(t;\omega) + \int_0^t K(u, x(u;\omega); \omega) \, du, \quad \mathscr{P}\text{-a.e.}$$

Thus for each $M > 0$ we have a random solution $x_M(t;\omega)$ of (2.0.2) such that for each $t \in [0, M]$

$$x_M(t;\omega) = h(t;\omega) + \int_0^t K(u, x_M(u;\omega); \omega) \, du, \quad \mathscr{P}\text{-a.e.}$$

We then define $z(t;\omega) \equiv x_M(t;\omega)$ for $M - 1 \leqslant t < M$ so that $z(t;\omega)$ is product-measurable and is a random solution of Eq. (2.0.2).

The proof of the following theorem is similar to that of Theorem 2.3.1 and is therefore omitted.

Theorem 2.3.2 Equation (2.0.2) with $t \in [0, M]$, for $0 < M < \infty$, has a random solution if the following conditions are satisfied:

(i) The kernel $K(u, x(u;\omega); \omega)$ is a continuous function of u and x, $u \in R_+$, $x \in R$, for each $\omega \in \Omega$, and $K(u, x; \omega) \in L_p(\Omega, \mathscr{A}, \mathscr{P})$ for each u and x, $1 \leqslant p < \infty$.

(ii) For all $t \in [0, M]$, $\sup_x |K(t, x; \omega)| \leqslant g(\omega) \in L_p(\Omega, \mathscr{A}, \mathscr{P})$, $1 \leqslant p < \infty$.

(iii) $h(t;\omega)$ is a continuous function from $[0, M]$ into $L_p(\Omega, \mathcal{A}, \mathcal{P})$, and for each $t \in [0, M]$, $0 < M < \infty$, there exists a constant $N_M > 0$ such that

$$|h(t;\omega)| + \int_0^t \sup_x |K(u, x;\omega)| \, du \leq N_M < \infty, \quad \omega \in \Omega.$$

(iv) Same as Condition (iv) in Theorem 2.3.1.

The following theorem presents conditions under which a random solution exists and is unique.

Theorem 2.3.3 Equation (2.0.2) with $t \in [0, \infty)$ has a *unique* random solution if the following conditions are satisfied:

(i) For each $\omega \in \Omega$ and $u \in [0, \infty)$ the kernel $K(u, x;\omega)$ is a continuous function of x in $(-\infty, \infty)$, and for each x, $K(u, x;\omega)$ is product-measurable on $[0, \infty) \times \Omega$ and $\int_0^t \sup_x |K(u, x;\omega)| \, du < \infty$.

(ii) The kernel $K(u; x;\omega)$ satisfies the following Lipschitz condition: For each u

$$|K(u, x;\omega) - K(u, y;\omega)| < a(\omega)|x - y|,$$

where $a(\omega)$ is a nonnegative, real-valued function on Ω.

(iii) $h(t;\omega)$ is continuous in t for each $\omega \in \Omega$.

PROOF We shall obtain a solution first on an arbitrary closed, bounded interval $[0, M]$. Let $C[0, M]$ be the space of all continuous functions on $[0, M]$, where for each $\omega \in \Omega$ we introduce a norm $\|\cdot\|_\omega$ in the following way:

$$\|x(t)\|_\omega \equiv \sup_{t \geq 0} \{e^{-b(\omega)t}|x(t)|\},$$

where $b(\omega) > a(\omega)$. Under this norm $C[0, M]$ is a Banach space. Now let E be the set of all mappings $x(t;\omega)$ from Ω into $C[0, M]$ such that $x(t;\omega)$ is a random variable for every t. Then $x(t;\omega) \in E$ is product-measurable on $[0, M] \times \Omega$, and so $\int_0^t K(u, x(u;\omega);\omega) \, du$ is defined and is a random variable for every t. Hence if we define

$$T[x(t;\omega)] = h(t;\omega) + \int_0^t K(u, x(u;\omega);\omega) \, du,$$

then clearly $T[x(t;\omega)]$ is continuous in t for each $\omega \in \Omega$, and hence product-measurable on $[0, M] \times \Omega$. If $z(t;\omega) \in E$, then one can check easily that

$$\|T[x(t;\omega)] - T[z(t;\omega)]\|_\omega < C(\omega)\|x(t;\omega) - z(t;\omega)\|_\omega$$

2.4 APPLICATIONS OF THE INTEGRAL EQUATION

2.4.2 Stochastic Models for Chemotherapy

We shall now consider two models for chemotherapy which were developed deterministically by Bellman, Jacquez, and Kalaba [2]. We will present stochastic versions of these models that describe the distribution of a drug in one-organ and two-organ biological systems and which lead to *semistochastic* integral equations of the form (2.0.2) (Padgett and Tsokos [1, 10]). That is, the resulting integral equations have deterministic solutions for $0 \leqslant t < \tau$, for some fixed $\tau > 0$, and random solutions for $\tau \leqslant t$. Solutions of this type are called *semirandom solutions*.

2.4.2.a A Stochastic Model for Drug Distribution in a One-Organ Biological System

Consider a closed system with a simplified heart, one organ or capillary bed, and recirculation of the blood with a constant rate of flow, where the heart is considered as a mixing chamber of constant volume. The flow of blood is assumed to be "slug" flow, that is, no mixing occurs in the vessels. It is assumed that an injection of drug is given directly at the entrance of the heart, producing a known concentration in the blood plasma. Also, as the blood passes through the capillary bed or organ, the particles of drug are assumed to enter the extracellular space only by the process of diffusion through the capillary walls.

Since it is impossible to know the concentration of drug at every point in the plasma at any given time after injection, for a given experiment measurements of drug concentration in the plasma should be made at several points in a particular area of the system at the same instant of time and the *mean value* of these measurements should be used as the drug concentration in that area of the system. For example, several measurements may be taken at points between the heart exit and the entrance to the capillary bed at time $t > 0$, since the concentration is considered fairly uniform in certain areas of the system, as assumed by Bellman, Jacquez, and Kalaba [1, 2]. It is realistic to assume that this mean value estimates the true state of nature at time $t > 0$, and if another initial injection is given in the same system and the experiment is repeated under the same conditions, then a different mean value would result. Thus the concentration of drug in the plasma in given areas of the system is more realistically considered as a random function of time rather than a deterministic one.

We use the following notation for $t \geqslant 0$: $u(s, t; \omega)$ is the concentration in moles per unit volume at point s in the capillary at time t; $\omega \in \Omega$ (a random variable for each t); c is the constant volume flow rate of plasma in the capillary bed; and l is the mean length of capillary in the organ. We assume that all capillaries in the capillary bed or organ are lumped together into one capillary of length l, total volume flow rate c, and total surface area equal to that of all of the capillaries combined, and that the blood enters at the zero end and exits at the l end.

For a one-organ system maintained by a simplified heart (see Fig. 2.4.1) let the heart be considered as a mixing chamber of constant volume given by

$$V^* = V_e/[\ln(1 + V_e/V_r)],$$

where V_r is the residual volume of the heart and V_e is the ejection volume, and let the heart have constant entering and leaving flow rates c. This is obtained by representing the concentration $\gamma(t)$ in plasma leaving the heart at time $t > 0$ as a function of V_e, V_r, and the initial concentration of drug at time zero, γ_0, in the residual blood of the heart,

$$\gamma(t) = \gamma_0 \exp(-ct/V^*), \qquad t \geq 0.$$

We assume that an initial injection is given at the entrance of the heart resulting in a concentration $u_1(t)$, $0 \leq t \leq t_1$, of drug in plasma entering the heart, where t_1 is the duration of injection. Let the time required for the blood to flow from the heart exit to the entrance of the organ be $\tau > 0$, and also let τ be the time required for blood to flow from the exit of the organ to the heart entrance. Then plasma containing drug particles reaches the organ τ time units after injection, and while flowing through the organ, diffusion of drug through the capillary walls into the organ tissue occurs. Therefore after time $\tau > 0$ the concentration of drug in blood plasma in the system is a random variable due to the random nature of the diffusion process and recirculation of the blood. Hence the concentration of drug in plasma entering the heart at time $t > \tau$, $u_R(t; \omega)$, is a random variable and is given by

$$u_R(t; \omega) = \begin{cases} 0, & t < 0, \\ u_1(t), & 0 \leq t < \tau, \\ u_1(t) + u(l, t - \tau; \omega), & t \geq \tau, \end{cases} \qquad (2.4.2)$$

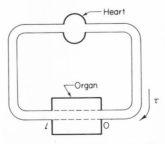

Figure 2.4.1 One-organ model.

2.4 APPLICATIONS OF THE INTEGRAL EQUATION

where $u_I(t) = 0$ for $t > t_1$, and $u(l, t; \omega)$ is the concentration of drug in plasma leaving the organ at time t. The concentration of drug in plasma leaving the heart, $u_L(t; \omega)$, satisfies the integral equation (see Bellman, Jacquez, Kalaba, and Kotkin [1])

$$u_L(t; \omega) = (c/V^*) \int_0^t [u_R(s; \omega) - u_L(s; \omega)] \, ds, \qquad t \geq 0. \qquad (2.4.3)$$

Then the concentration of drug entering the organ at time t is given by

$$u(0, t; \omega) = \begin{cases} 0, & 0 \leq t < \tau, \\ u_L(t - \tau; \omega), & t \geq \tau. \end{cases}$$

Equation (2.4.3) is a *semistochastic* integral equation as stated. The solution is deterministic for $0 \leq t < \tau$, that is, $u_L(t; \omega) \equiv \xi(t), 0 \leq t < \tau$, and $u_L(t; \omega)$ is a random function of t for $t \geq \tau$. This is due to the diffusion process in the organ and the recirculation of the blood.

Substituting Eq. (2.4.2) into (2.4.3), we have

$$u_L(t; \omega) = \int_0^t (c/V^*)[u_I(s) + u(l, s - \tau; \omega) - u_L(s; \omega)] \, ds$$

$$= \int_0^{T(t)} (c/V^*) u_I(s) \, ds - (c/V^*) \int_0^t [u_L(s; \omega) - u(l, s - \tau; \omega)] \, ds$$

$$= G(t) + \int_0^t k(s, u_L(s; \omega); \omega) \, ds, \qquad (2.4.4)$$

where

$$G(t) = \int_0^{T(t)} (c/V^*) u_I(s) \, ds, \qquad T(t) = \begin{cases} t, & 0 \leq t < t_1, \\ t_1, & t \geq t_1, \end{cases}$$

$$k(s, u_L(s; \omega); \omega) = (-c/V^*)[u_L(s; \omega) - u(l, s - \tau; \omega)],$$

and $u(l, s; \omega) = 0$ if $s < 0$. Let the initial concentration be given by

$$u_I(t) = \begin{cases} u^*, & 0 \leq t \leq t_1, \\ 0, & \text{otherwise.} \end{cases}$$

We shall now show that a *semirandom* solution of (2.4.4) exists, that is, a deterministic solution $\xi(t)$ for $0 \leq t < \tau$ and a random solution $u_L(t; \omega)$ for $\tau \leq t \leq M$, where $\tau < M < \infty$, $t_1 < M$. To show that a deterministic solution $\xi(t), 0 \leq t < \tau$, exists, we shall use the ordinary method of successive approximations for deterministic integral equations (Mikhlin, [1]). To indicate that a random solution $u_L(t; \omega), \tau \leq t \leq M < \infty$, exists, we employ Theorem 2.3.2.

II SOME VOLTERRA TYPE EQUATIONS WITH APPLICATIONS

Suppose that $0 \leq t < \tau$. Since $u(l, t; \omega) = 0$ for $t < \tau$, Eq. (2.4.4) becomes a linear (deterministic) Volterra integral equation with $u_L(t; \omega) \equiv \xi(t)$,

$$\xi(t) = G(t) + \int_0^t (-c/V^*)\xi(s)\,ds. \tag{2.4.5}$$

Let the kernel be given by

$$K(t, y) = \begin{cases} 1, & 0 \leq y \leq t, \\ 0, & \text{otherwise.} \end{cases}$$

Then by the method of successive approximations define the sequence

$$\xi_0(t) = G(t),$$

$$\xi_1(t) = G(t) + (-c/V^*)\int_0^t K(t, y)\xi_0(y)\,dy,$$

$$\xi_n(t) = G(t) + (-c/V^*)\int_0^t K(t, y)\xi_{n-1}(y)\,dy, \qquad n > 1.$$

Let

$$K_1(t, y) = K(t, y), \qquad K_m(t, y) = \int_y^t K(t, s)K_{m-1}(s, y)\,ds, \qquad m \geq 2,$$

since

$$K(t, s)K(s, y) = \begin{cases} 0 & \text{if } s < y \text{ or } s > t, \\ 1 & \text{if } y \leq s \leq t. \end{cases}$$

Then the nth approximation is given by

$$\xi_n(t) = G(t) + \sum_{m=1}^n (-c/V^*)^m \int_0^t K_m(t, y)G(y)\,dy$$

by successive substitution in the sequence and then interchanging the order of summation and integration. As $n \to \infty$, the sequence $\xi_n(t)$ converges to a solution of (2.4.5) if the series on the right converges, that is,

$$\xi(t) = G(t) + \sum_{m=1}^\infty (-c/V^*)^m \int_0^t K_m(t, y)G(y)\,dy. \tag{2.4.6}$$

To see that the series in (2.4.6) converges, we note that since $K(t, y) \leq 1$ for all t and y, there exists a $Q > 0$ such that $|K(t, y)| < Q$. By definition, we have

$$|K_1(t, y)| = |K(t, y)| < Q$$

and

$$|K_2(t, y)| = \left|\int_y^t K(t, s)K(s, y)\,ds\right| < Q^2(t - y).$$

2.4 APPLICATIONS OF THE INTEGRAL EQUATION

Assume that

$$|K_m(t, y)| < Q^m(t - y)^{m-1}/(m - 1)! \tag{2.4.7}$$

Then

$$|K_{m+1}(t, y)| \leq \int_y^t |K(t, s)||K_{m-1}(s, y)| \, ds < \frac{Q^{m+1}}{(m-1)!} \int_y^t (s - y)^{m-1} \, ds$$

$$= \frac{Q^{m+1}(t - y)^m}{m!}.$$

Therefore, by induction, inequality (2.4.7) holds for all $m \geq 1$, and we have from (2.4.6)

$$|\xi(t)| \leq G(t) + \sum_{m=1}^{\infty} \left|(-c/V^*)^m \int_0^t K_m(t, y)G(y) \, dy\right|$$

$$< G(t) + \sum_{m=1}^{\infty} (-c/V^*)^m (Q^m t_1^{m+1} u^*/m!),$$

where $t_1 = \min(t, t_1)$, $0 \leq t < \tau$. The series on the right converges by the ratio test. Hence there exists a solution of (2.4.4) for $0 \leq t < \tau$.

Suppose that $\tau \leq t \leq M$, $\tau < M < \infty$. Then Eq. (2.4.4) is

$$u_L(t; \omega) = G(t) + \int_0^t k(s, u_L(s; \omega); \omega) \, ds.$$

For $\omega \in \Omega$ the concentrations $u_L(t; \omega)$ and $u(l, t; \omega)$ can be considered as continuous functions of time, and hence $k(t, u_L(t; \omega); \omega)$ is continuous in t and u_L. For each t and u_L, $k(t, u_L; \omega)$ is a constant. Hence $k(t, u_L; \omega) \in L_p(\Omega, \mathscr{A}, \mathscr{P})$, $1 \leq p < \infty$, trivially, and Condition (i) of Theorem 2.3.2 is satisfied.

By the nature of the physical situation, the maximum value of $u_L(t; \omega)$ and $u(l, t - \tau; \omega)$ is u^*, and the minimum value is zero for all $t \in [\tau, M]$. Thus

$$\sup_{u_L}|k(t, u_L; \omega)| = \sup_{u_L}\{(c/V^*)|u_L(t; \omega) - u(l, t - \tau; \omega)|\} = (c/V^*)u^*.$$

Define

$$g(\omega) = cu^*/V^*, \qquad \omega \in \Omega,$$

and then $g(\omega) \in L_p(\Omega, \mathscr{A}, \mathscr{P})$, since

$$\left\{\int_\Omega |g(\omega)|^p d\mathscr{P}(\omega)\right\}^{1/p} = cu^*/V^* < \infty.$$

Therefore Condition (ii) of Theorem 2.3.2 is satisfied.

The stochastic free term $h(t;\omega)$ is the deterministic function $G(t)$, which is bounded,

$$|G(t)| \leq \int_0^{T(t)} (c/V^*)u_1(s)\,ds \leq t_1 cu^*/V^* < \infty$$

for all $t \in [0, M]$. Hence $h(t;\omega)$ is in $L_p(\Omega, \mathscr{A}, \mathscr{P})$. Obviously, $h(t;\omega)$ is continuous in t since $G(t)$ is an integral. Let $t \in [\tau, M] \subset [0, M]$. Then

$$|h(t;\omega)| + \int_0^t \sup|k(s, u_L;\omega)|\,ds \leq t_1(cu^*/V^*) + t(cu^*/V^*)$$
$$\leq (cu^*/V^*)(t_1 + M).$$

Take

$$N_M = [(cu^*/V^*)(t_1 + M)] + 1 < \infty,$$

where $[\,\cdot\,]$ is the greatest integer function. Then Condition (iii) of Theorem 2.3.2 holds.

Suppose the sequence $\{u_{L,n}(t,\omega)\}$ converges to $u_L(t;\omega)$ in $L_p(\Omega, \mathscr{A}, \mathscr{P})$ for each $t \in [\tau, M]$. Then for every $\varepsilon_1 > 0$ there exists an $N > 0$ such that whenever $n > N$,

$$\left\{\int_\Omega |u_{L,n}(t;\omega) - u_L(t;\omega)|^p\,d\mathscr{P}(\omega)\right\}^{1/p} < \varepsilon_1.$$

Therefore, letting $\varepsilon = (c/V^*)\varepsilon_1 > 0$, we have for $n > N$

$$\left\{\int_\Omega |k(t, u_{L,n}(t;\omega);\omega) - k(t, u_L(t;\omega);\omega)|^p\,d\mathscr{P}(\omega)\right\}^{1/p}$$
$$= (c/V^*)\left\{\int_\Omega |u_{L,n}(t;\omega) - u_L(t;\omega)|^p\,d\mathscr{P}(\omega)\right\}^{1/p} < (c/V^*)\varepsilon_1 = \varepsilon.$$

That is, $k(t, u_{L,n}(t;\omega);\omega)$ converges to $k(t;u_L(t;\omega);\omega)$ in $L_p(\Omega, \mathscr{A}, \mathscr{P})$. Hence Condition (iv) of Theorem 2.3.2 is satisfied.

Therefore for $\tau \leq t \leq M < \infty$, $t_1 < M$, there exists a random solution of Eq. (2.4.4).

Thus we have shown that the semistochastic integral equation (2.4.4) possesses a semirandom solution for all $t \in [0, M]$, where $0 < M < \infty$, $\tau < M$, and $t_1 < M$.

2.4.2.b Two-Organ Biological Systems

In this section we will extend the results of the previous section to biological systems which consist of two organs.

2.4 APPLICATIONS OF THE INTEGRAL EQUATION

Consider a closed system with a simplified heart, two organs or capillary beds, and recirculation of the blood with a constant rate of flow assumed, where the heart is considered as a mixing chamber of constant volume. With respect to the flow of blood, we assume that there is no mixing in the vessels as before. This system is not as simplified as it may appear at a casual glance, since we may consider one organ as a certain organ of the body, such as the lungs, and the other organ as being a collection of the remaining capillary beds or organs of the body. See Fig. 2.4.2 for a schematic description of the system.

As in Section 2.4.2.a, it is assumed that an injection of drug, or a chemical agent, is given to the system directly at the entrance of the heart, producing a known concentration of drug in the blood plasma entering the heart; that is, drug particles are dissolved in the blood plasma. Also, as the blood passes through each of the organs, the particles of drug are assumed to enter the extracellular space only by the process of diffusion through the capillary walls. Due to the nondeterministic nature of a diffusion process, a random amount of drug is removed from the blood plasma in the organs, and hence the concentration of drug in the blood is reduced by a random amount. Therefore it is impossible to know exactly the point-by-point concentration of the drug in the blood plasma and in the extracellular space of the organs at any given time after the blood containing the drug passes through the organs.

In addition to the notation already introduced in the previous section, the following will be used:

$u_j(s, t; \omega)$ is the random concentration of drug in organ j at point s in the capillary at time t, for $j = 1, 2$.

c_j is the constant-volume flow rate of blood in organ j, $j = 1, 2$, and $c = c_1 + c_2$ is the total constant-volume flow rate of blood in the system.

Again, suppose that the injection of drug produces an initial concentration $u_1(t)$, $0 \leq t \leq t_1$, entering the heart, and let τ denote the time lag due to the

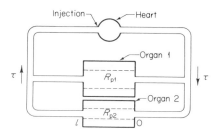

Figure 2.4.2 The two-organ system.

circulation of the blood as described in Section 2.4.2.a. Thus the drug enters each organ after a time τ, and due to the diffusion process in the organs, the concentration of drug entering the heart, $u_R(t;\omega)$, after time τ is a random variable for each $t \geqslant \tau$ and is given by

$$u_R(t;\omega) = \begin{cases} 0, & t < 0, \\ u_1(t), & 0 \leqslant t, \\ u_1(t) + \{[c_1 u_1(l, t-\tau;\omega) + c_2 u_2(l, t-\tau;\omega)]/c\} & \tau \leqslant t \leqslant M, \end{cases}$$

where $u_j(l, t;\omega)$ is the concentration of drug in plasma leaving organ j (at the l end) at time t, $j = 1, 2$, and $u_1(t) = 0$ for $t > t_1$. Also, the concentration of drug in plasma leaving the heart at time $t \geqslant 0$ is given by the semirandom solution $u_L(t;\omega)$ of the semistochastic integral equation (2.4.3). We showed in the previous section that this semistochastic integral equation has a semirandom solution for each t, $0 \leqslant t \leqslant M$, where $M < \infty$ is a constant. Then the concentration of drug in plasma entering organ j at time t is given by

$$u_j(0, t;\omega) = \begin{cases} 0, & 0 \leqslant t < \tau, \\ u_L(t-\tau;\omega), & \tau \leqslant t \leqslant M, \end{cases}$$

where $j = 1, 2$.

These models for chemotherapy are realistic in that they retain such properties as recirculation of the blood, mixing in the heart, the presence of more than one organ, and randomness in the diffusion of the drug into the organ tissues, even though several simplifying assumptions were made in order to deal with the mathematics involved. Such assumptions seem necessary in obtaining mathematical descriptions of biological systems since they are in general very complex systems.

CHAPTER III

Approximate Solution of the Random Volterra Integral Equation and an Application to Population Growth Modeling

3.0 Introduction

In this chapter we shall present some methods of approximating the unique random solution $x(t;\omega)$ of a stochastic integral equation of the Volterra type of the form

$$x(t;\omega) = h(t;\omega) + \int_0^t k(t,\tau;\omega) f(\tau, x(\tau;\omega)) \, d\tau, \qquad (3.0.1)$$

where $x(t;\omega)$, $h(t;\omega)$, $k(t,\tau;\omega)$, and $f(t,x)$ behave as described in Chapters I and II. We shall consider the problem of obtaining an approximation to a realization of $x(t;\omega)$ by the method of successive approximations at each $t \in R_+$ and also by applying some of the theory of stochastic approximation.

A new stochastic formulation of a population growth model will also be given along with a numerical example.

In the method of successive approximations we shall investigate the convergence of a generated sequence of random variables to the unique random solution $x(t; \omega)$ at each $t \in R_+$. Also, the rate of convergence, the maximum error of approximation, and the combined error of approximation with the error of numerically evaluating the integral are considered.

A general theorem of Burkholder [1] in the theory of stochastic approximation is also applied to Eq. (3.0.1) resulting in conditions under which a sequence of approximations converges with probability one to the unique random solution $x(t; \omega)$ at each $t \in R_+$.

3.1 The Method of Successive Approximations

Let $C_c(R_+, L_2(\Omega, \mathcal{A}, \mathcal{P}))$ be the space described in Chapter I, and let $B, D \subset C_c(R_+, L_2(\Omega, \mathcal{A}, \mathcal{P}))$ be Banach spaces with the norm in D defined such that

$$\|x(t;\omega)\|_D \geq \sup_{t \in R_+} \left\{ \int_\Omega |x(t;\omega)|^2 \, d\mathcal{P}(\omega) \right\}^{\frac{1}{2}}.$$

Let

$$S = \{x(t;\omega) : x(t;\omega) \in D, \quad \|x(t;\omega)\|_D \leq \rho\},$$

as in Theorem 2.1.2, with $x(t; \omega)$, $h(t; \omega)$, $f(t, x(t; \omega))$, and $k(t, \tau; \omega)$, for $0 \leq \tau \leq t < \infty$, behaving as described previously. As in the existence proof of Theorem 2.1.2, let U be the contraction mapping from S into S defined by

$$(Ux)(t;\omega) = h(t;\omega) + \int_0^t k(t,\tau;\omega) f(\tau, x(\tau;\omega)) \, d\tau,$$

which has the unique fixed point $x(t; \omega)$.

It is assumed here that the distribution of the random variable $h(t; \omega)$ is known at each $t \in R_+$, or that a value of $h(t; \omega)$ can be observed at each $t \in R_+$. Define the sequence of successive approximations $\{x_n(t; \omega)\}$ by

$$x_0(t;\omega) = h(t;\omega), \quad x_{n+1}(t;\omega) = (Ux_n)(t;\omega), \quad n \geq 0. \quad (3.1.1)$$

The sequence defined recursively here is contained in the set S, which is a result of the following lemma.

Lemma 3.1.1 $h(t; \omega) \in S$ and hence $x_n(t; \omega) \in S$, $n = 0, 1, 2, \ldots$. Also,

$$x_n(t;\omega) \xrightarrow{D} x(t;\omega) \in S.$$

3.1 THE METHOD OF SUCCESSIVE APPROXIMATIONS

PROOF From the condition of Theorem 2.1.2

$$\|h(t;\omega)\|_D \leq \rho(1 - \lambda K) - K\|f(t,0)\|_B \leq \rho(1 - \lambda K) < \rho$$

since $\lambda K < 1$. Hence $h(t;\omega) \in S$.

For an arbitrary integer $r > 0$ consider

$$\|x_{n+r}(t;\omega) - x_n(t;\omega)\|_D = \|(Ux_{n+r-1})(t;\omega) - (Ux_{n-1})(t;\omega)\|_D$$

$$= \left\| h(t;\omega) + \int_0^t k(t,\tau;\omega) f(\tau, x_{n+r-1}(\tau;\omega))\, d\tau \right.$$

$$\left. - h(t;\omega) - \int_0^t k(t,\tau;\omega) f(\tau, x_{n-1}(\tau;\omega))\, d\tau \right\|_D$$

$$= \left\| \int_0^t k(t,\tau;\omega)[f(\tau, x_{n+r-1}(\tau;\omega)) \right.$$

$$\left. - f(\tau, x_{n-1}(\tau;\omega))]\, d\tau \right\|_D$$

$$\leq K\|f(t, x_{n+r-1}(t;\omega)) - f(t, x_{n-1}(t;\omega))\|_B$$

$$\leq \lambda K\|x_{n+r-1}(t;\omega) - x_{n-1}(t;\omega)\|_D$$

by Condition (ii) of Theorem 2.1.2 and the boundedness of the operator T. Repeating the same argument $n - 1$ times gives

$$\|x_{n+r}(t;\omega) - x_n(t;\omega)\|_D \leq (\lambda K)^2 \|x_{n+r-2}(t;\omega) - x_{n-2}(t;\omega)\|_D$$

$$\leq \cdots \leq (\lambda K)^n \|x_r(t;\omega) - x_0(t;\omega)\|_D$$

$$\leq (\lambda K)^n [\|x_r(t;\omega)\|_D + \|h(t;\omega)\|_D]$$

$$\leq (\lambda K)^n 2\rho.$$

But as $n \to \infty$, $(\lambda K)^n \to 0$ since $\lambda K < 1$, and hence

$$\lim_{n \to \infty} \|x_{n+r}(t;\omega) - x_n(t;\omega)\|_D = 0.$$

Since $r > 0$ is arbitrary, $\{x_n(t;\omega)\}$ is a Cauchy sequence in $S \subset D$. Thus, since D is complete, $\{x_n(t;\omega)\}$ converges to a point in D. But the unique solution $x(t;\omega)$ is in $S \subset D$, and since

$$\|x_n(t;\omega) - x(t;\omega)\|_D = \|(Ux_{n-1})(t;\omega) - (Ux)(t;\omega)\|_D$$

$$\leq \lambda K\|x_{n-1}(t;\omega) - x(t;\omega)\|_D$$

$$\leq \cdots \leq (\lambda K)^n \|x_0(t;\omega) - x(t;\omega)\|_D \to 0$$

as $n \to \infty$, then $x_n(t;\omega) = (Ux_{n-1})(t;\omega) \to x(t;\omega) = (Ux)(t;\omega) \in S$.

3.1.1 Convergence of the Successive Approximations†

We shall now investigate the mode of convergence of the successive approximations defined by Eq. (3.1.1). We shall use the definition of almost sure convergence (or convergence with probability one) and the Markov inequality as given by Loève [1].

Definition 3.1.1 Let $X_n(\omega)$ be a sequence of random variables defined on the probability measure space $(\Omega, \mathscr{A}, \mathscr{P})$ and let $X(\omega)$ be a random variable defined on $(\Omega, \mathscr{A}, \mathscr{P})$. The sequence $X_n(\omega)$ *converges almost surely* (a.s.) *to* $X(\omega)$, $X_n(\omega) \to^{\text{a.s.}} X(\omega)$, if $X_n(\omega) \to X(\omega)$, except perhaps on a set of probability zero. Equivalently, $X_n(\omega) \to^{\text{a.s.}} X(\omega)$ if for every $\varepsilon > 0$

$$\mathscr{P}\{\bigcup_{k \geq n} [\omega : |X_k(\omega) - X(\omega)| \geq \varepsilon]\} \to 0 \quad \text{as} \quad n \to \infty.$$

Markov Inequality For $a \geq 0$, $r > 0$, we have

$$\mathscr{P}\{\omega : |X(\omega)| \geq a\} \leq E|X(\omega)|^r / a^r,$$

if $E|X(\omega)|^r$ exists.

Theorem 3.1.2 (Loève [1], p. 173) If for some $r > 0$

$$\sum_{n=1}^{\infty} E|X_n(\omega) - X(\omega)|^r < \infty,$$

then

$$X_n(\omega) \stackrel{\text{a.s.}}{\to} X(\omega).$$

PROOF By the Markov inequality, for every $\varepsilon > 0$

$$\mathscr{P}\{\omega : |X_n(\omega) - X(\omega)| \geq \varepsilon\} \leq E|X_n(\omega) - X(\omega)|^r / \varepsilon^r$$

for every $n \geq 1$, $r > 0$. Hence

$$\sum_{n=1}^{\infty} \mathscr{P}\{\omega : |X_n(\omega) - X(\omega)| \geq \varepsilon\} \leq \sum_{n=1}^{\infty} [E|X_n(\omega) - X(\omega)|^r / \varepsilon^r] < \infty$$

for some $r > 0$, by hypothesis. But

$$\mathscr{P}\{\bigcup_{k \geq n} [\omega : |X_k(\omega) - X(\omega)| \geq \varepsilon]\} \leq \sum_{k \geq n} \mathscr{P}\{\omega : |X_k(\omega) - X(\omega)| \geq \varepsilon\}, \quad (3.1.2)$$

and since

$$\sum_{n=1}^{\infty} \mathscr{P}\{\omega : |X_n(\omega) - X(\omega)| \geq \varepsilon\} < \infty$$

for every $\varepsilon > 0$, the sum on the right in (3.1.2) must tend to zero as $n \to \infty$. Hence,

$$\mathscr{P}\{\bigcup_{k \geq n} [\omega : |X_k(\omega) - X(\omega)| \geq \varepsilon]\} \to 0 \quad \text{as} \quad n \to \infty,$$

† Sections 3.1.1–3.1.3 adapted from Padgett and Tsokos [5] with permission of Taylor and Francis, Ltd.

3.1 THE METHOD OF SUCCESSIVE APPROXIMATIONS

and $X_n(\omega) \to^{a.s.} X(\omega)$, by definition.

We shall now show that the sequence of successive approximations (3.1.1) converges a.s. to the unique random solution of (3.0.1).

Theorem 3.1.3 For each $t \in R_+$, $x_n(t;\omega) \to^{a.s.} x(t;\omega)$ under the conditions of Theorem 2.1.2.

PROOF By definition, for $t \in R_+$,

$$\sum_{n=0}^{\infty} \left\{ \int_\Omega |x_n(t;\omega) - x(t;\omega)|^2 \, d\mathscr{P}(\omega) \right\}^{\frac{1}{2}}$$

$$\leq \sum_{n=0}^{\infty} \sup_{t \geq 0} \left\{ \int_\Omega |x_n(t;\omega) - x(t;\omega)|^2 \, d\mathscr{P}(\omega) \right\}^{\frac{1}{2}}$$

$$\leq \sum_{n=0}^{\infty} \|x_n(t;\omega) - x(t;\omega)\|_D.$$

However, from the conditions of Theorem 2.1.2, we obtain

$$\|x_n(t;\omega) - x(t;\omega)\|_D = \|(Ux_{n-1})(t;\omega) - (Ux)(t;\omega)\|_D$$

$$= \left\| h(t;\omega) + \int_0^t k(t,\tau;\omega) f(\tau, x_{n-1}(\tau;\omega)) \, d\tau \right.$$

$$\left. - h(t;\omega) - \int_0^t k(t,\tau;\omega) f(\tau, x(\tau;\omega)) \, d\tau \right\|_D$$

$$= \left\| \int_0^t k(t,\tau;\omega) [f(\tau, x_{n-1}(\tau;\omega)) - f(\tau, x(\tau;\omega))] \, d\tau \right\|_D$$

$$\leq K \|f(t, x_{n-1}(t;\omega)) - f(t, x(t;\omega))\|_B$$

$$\leq (\lambda K) \|x_{n-1}(t;\omega) - x(t;\omega)\|_D.$$

Repeating the argument $n - 1$ times, we have

$$\|x_n(t;\omega) - x(t;\omega)\|_D \leq (\lambda K)^n \|x_0(t;\omega) - x(t;\omega)\|_D$$

$$= (\lambda K)^n \|h(t;\omega) - x(t;\omega)\|_D \leq (\lambda K)^n 2\rho$$

since $h(t;\omega), x(t;\omega) \in S$ by Lemma 3.1.1. Thus

$$\sum_{n=0}^{\infty} \left\{ \int |x_n(t;\omega) - x(t;\omega)|^2 \, d\mathscr{P}(\omega) \right\}^{\frac{1}{2}} \leq \sum_{n=0}^{\infty} \|x_n(t;\omega) - x(t;\omega)\|_D$$

$$\leq \sum_{n=0}^{\infty} (\lambda K)^n 2\rho = 2\rho/(1 - \lambda K) < \infty$$

since $\lambda K < 1$. Hence the nth term of the series converges to zero as $n \to \infty$, and there exists an $N > 0$ such that for $k > N$

$$\left\{\int_\Omega |x_k(t;\omega) - x(t;\omega)|^2 \, d\mathcal{P}(\omega)\right\}^{\frac{1}{2}} < 1,$$

so that for $k > N$ we have

$$\left\{\int_\Omega |x_k(t;\omega) - x(t;\omega)|^2 \, d\mathcal{P}(\omega)\right\} \leq \left\{\int_\Omega |x_k(t;\omega) - x(t;\omega)|^2 \, d\mathcal{P}(\omega)\right\}^{\frac{1}{2}}.$$

Therefore

$$\sum_{n=0}^{N} \left\{\int_\Omega |x_n(t;\omega) - x(t;\omega)|^2 \, d\mathcal{P}(\omega)\right\}^{\frac{1}{2}}$$

$$+ \sum_{n=N+1}^{\infty} \left\{\int_\Omega |x_n(t;\omega) - x(t;\omega)|^2 \, d\mathcal{P}(\omega)\right\}$$

$$\leq \sum_{n=0}^{\infty} \left\{\int_\Omega |x_n(t;\omega) - x(t;\omega)|^2 \, d\mathcal{P}(\omega)\right\}^{\frac{1}{2}} \leq 2\rho/(1-\lambda K) < \infty.$$

Hence

$$\sum_{n=0}^{N} \{E|x_n(t;\omega) - x(t;\omega)|^2\}^{\frac{1}{2}} + \sum_{n=N+1}^{\infty} E|x_n(t;\omega) - x(t;\omega)|^2 \leq 2\rho/(1-\lambda K).$$

Since $x_n(t;\omega), x(t;\omega) \in L_2(\Omega, \mathcal{A}, \mathcal{P})$, so is $x_n(t;\omega) - x(t;\omega)$, and for each $n = 0, \ldots, N$ we have

$$\{E|x_n(t;\omega) - x(t;\omega)|^2\}^{\frac{1}{2}} = \left\{\int_\Omega |x_n(t;\omega) - x(t;\omega)|^2 \, d\mathcal{P}(\omega)\right\}^{\frac{1}{2}} \leq M_n < \infty$$

for some constant $M_n > 0$. Thus

$$\sum_{n=0}^{\infty} E|x_n(t;\omega) - x(t;\omega)|^2 = \sum_{n=0}^{N} E|x_n(t;\omega) - x(t;\omega)|^2$$

$$- \sum_{n=0}^{N} \{E|x_n(t;\omega) - x(t;\omega)|^2\}^{\frac{1}{2}}$$

$$+ \left[\sum_{n=0}^{N} \{E|x_n(t;\omega) - x(t;\omega)|^2\}^{\frac{1}{2}}\right.$$

$$\left. + \sum_{n=N+1}^{\infty} E|x_n(t;\omega) - x(t;\omega)|^2\right]$$

$$\leq \sum_{n=0}^{N} M_n^2 - \sum_{n=0}^{N} M_n + [2\rho/(1-\lambda K)]$$

$$= \sum_{n=0}^{N} M_n(M_n - 1) + [2\rho/(1-\lambda K)]$$

$$< \infty, \quad \text{for} \quad t \in R_+.$$

3.1 THE METHOD OF SUCCESSIVE APPROXIMATIONS

Therefore, by Theorem 3.1.2, for each $t \in R_+$

$$x_n(t;\omega) \stackrel{\text{a.s.}}{\to} x(t;\omega).$$

Thus the sequence of successive approximations converges to the unique random solution $x(t;\omega)$ with probability one for each $t \in R_+$. Therefore the sequence $\{x_n(t;\omega)\}$ converges to $x(t;\omega)$ in probability and in distribution for each $t \in R_+$. As a by-product of this theorem, we also obtain that $x_n(t;\omega)$ converges to $x(t;\omega)$ in mean-square (quadratic mean) for each $t \in R_+$, since

$$\sum_{n=0}^{\infty} E|x_n(t;\omega) - x(t;\omega)|^2 < \infty.$$

3.1.2 Rate of Convergence and Error of Approximation

We now consider the rate of convergence of the sequence of successive approximations given by (3.1.1) and obtain the maximum error of approximating the true solution $x(t;\omega)$ by the nth successive approximation, $x_n(t;\omega)$, at each $t \in R_+$.

For the investigation of the rate of convergence, let $t \in R_+$ be fixed. We now obtain a bound on the norm in $L_2(\Omega, \mathscr{A}, \mathscr{P})$ of the difference between the nth and $(n+1)$th successive approximations, giving the speed of convergence of (3.1.1) for each $t \in R_+$. We have

$$\|x_{n+1}(t;\omega) - x_n(t;\omega)\|_{L_2(\Omega, \mathscr{A}, \mathscr{P})}$$

$$= \left\{\int_\Omega |x_{n+1}(t;\omega) - x_n(t;\omega)|^2 d\mathscr{P}(\omega)\right\}^{\frac{1}{2}}$$

$$\leq \sup_{t \geq 0} \left\{\int_\Omega |x_{n+1}(t;\omega) - x_n(t;\omega)|^2 d\mathscr{P}(\omega)\right\}^{\frac{1}{2}}$$

$$\leq \|x_{n+1}(t;\omega) - x_n(t;\omega)\|_D$$

$$= \left\|h(t;\omega) + \int_0^t k(t,\tau;\omega)f(\tau, x_n(\tau;\omega))\, d\tau - h(t;\omega)\right.$$

$$\left. - \int_0^t k(t,\tau;\omega)f(\tau, x_{n-1}(\tau;\omega))\, d\tau\right\|_D$$

$$= \left\|\int_0^t k(t,\tau;\omega)[f(\tau, x_n(\tau;\omega)) - f(\tau, x_{n-1}(\tau;\omega))]\, d\tau\right\|$$

$$\leq K\|f(t, x_n(t;\omega)) - f(t, x_{n-1}(t;\omega))\|_B$$

$$\leq \lambda K\|x_n(t;\omega) - x_{n-1}(t;\omega)\|_D$$

as in the previous section. Repeating the argument $n - 1$ times, we obtain

$$\|x_{n+1}(t;\omega) - x_n(t;\omega)\|_{L_2(\Omega,\mathscr{A},\mathscr{P})} \leq \|x_{n+1}(t;\omega) - x_n(t;\omega)\|_D$$
$$\leq \cdots \leq (\lambda K)^n \|x_1(t;\omega) - x_0(t;\omega)\|_D.$$

But since $x_0(t;\omega) = h(t;\omega)$, we have $x_1(t;\omega) = (Ux_0)(t;\omega) = (Uh)(t;\omega)$ and

$$(\lambda K)^n \|x_1(t;\omega) - x_0(t;\omega)\|_D$$

$$= (\lambda K)^n \left\| h(t;\omega) + \int_0^t k(t,\tau;\omega) f(\tau, h(\tau;\omega)) \, d\tau - h(t;\omega) \right\|_D$$

$$= (\lambda K)^n \left\| \int_0^t k(t,\tau;\omega) f(\tau, h(\tau;\omega)) \, d\tau \right\|_D$$

$$\leq (\lambda K)^n K \|f(t, h(t;\omega))\|_B$$

$$\leq (\lambda K)^n K [\|f(t, h(t;\omega)) - f(t,0)\|_B + \|f(t,0)\|_B]$$

$$\leq (\lambda K)^n K [\lambda \|h(t;\omega)\|_D + \|f(t,0)\|_B]$$

$$= (\lambda K)^n [\lambda K \|h(t;\omega)\|_D + K \|f(t,0)\|_B]$$

$$< (\lambda K)^n [\|h(t;\omega)\|_D + K \|f(t,0)\|_B] \leq (\lambda K)^n \rho (1 - \lambda K)$$

from the assumption of Theorem 2.1.2 that $\lambda K < 1$,

$$\|h(t;\omega)\|_D + K \|f(t,0)\|_B \leq \rho (1 - \lambda K),$$

and $f(t,x)$ satisfies a Lipschitz condition. Therefore for each $n \geq 0$ and $t \in R_+$ we have that

$$\|x_{n+1}(t;\omega) - x_n(t;\omega)\|_{L_2(\Omega,\mathscr{A},\mathscr{P})} < (\lambda K)^n \rho (1 - \lambda K), \qquad (3.1.3)$$

where $\lambda K < 1$ and $\rho > 0$.

Now, to find a bound on the error of approximating the random solution $x(t;\omega)$ at $t \in R_+$ by the nth successive approximation given by (3.1.1), we use a technique similar to that used by Rall [1] in the nonstochastic case. As before,

$$\|x(t;\omega) - x_n(t;\omega)\|_{L_2(\Omega,\mathscr{A},\mathscr{P})} \leq \|x(t;\omega) - x_n(t;\omega)\|_D.$$

An upper bound on the quantity on the right-hand side is found as follows. Since $x_n(t;\omega) \xrightarrow{D} x(t;\omega)$ from Lemma 3.1.1, let $p > 0$ and note that, as $p \to \infty$, for every $n \geq 0$,

$$\|x_{n+p}(t;\omega) - x_n(t;\omega)\|_D \to \|x(t;\omega) - x_n(t;\omega)\|_D.$$

3.1 THE METHOD OF SUCCESSIVE APPROXIMATIONS

Now,

$$\|x_n(t;\omega) - x_0(t;\omega)\|_D$$
$$= \|[x_n(t;\omega) - x_{n-1}(t;\omega)]$$
$$\quad + [x_{n-1}(t;\omega) - x_{n-2}(t;\omega)] + \cdots + [x_1(t;\omega) - x_0(t;\omega)]\|_D$$
$$\leq \sum_{k=0}^{n-1} \|x_{k+1}(t;\omega) - x_k(t;\omega)\|_D$$
$$\leq \sum_{k=0}^{n-1} (\lambda K)^k \|x_1(t;\omega) - x_0(t;\omega)\|_D$$
$$\leq \sum_{k=0}^{n-1} (\lambda K)^k [\lambda K \|h(t;\omega)\|_D + K \|f(t,0)\|_B]$$

from the previous results. Similarly,

$$\|x_{n+p}(t;\omega) - x_n(t;\omega)\|_D$$
$$= \|[x_{n+p}(t;\omega) - x_{n+p-1}(t;\omega)] + \cdots + [x_{n+1}(t;\omega) - x_n(t;\omega)]\|_D$$
$$\leq \sum_{k=n}^{n+p-1} \|x_{k+1}(t;\omega) - x_k(t;\omega)\|_D$$
$$\leq \sum_{k=n}^{n+p-1} (\lambda K)^k [\lambda K \|h(t;\omega)\|_D + K \|f(t,0)\|_B].$$

Then as $p \to \infty$ we obtain

$$\|x(t;\omega) - x_n(t;\omega)\|_D \leq \sum_{k=n}^{\infty} (\lambda K)^k [\lambda K \|h(t;\omega)\|_D + K \|f(t,0)\|_B]$$
$$= (\lambda K)^n [\lambda K \|h(t;\omega)\|_D + K \|f(t,0)\|_B] \sum_{l=0}^{\infty} (\lambda K)^l$$
$$< (\lambda K)^n [\|h(t;\omega)\|_D + K \|f(t,0)\|_B][1/(1 - \lambda K)]$$
$$\leq (\lambda K)^n \rho (1 - \lambda K)[1/(1 - \lambda K)] = (\lambda K)^n \rho$$

since $\lambda K < 1$. Therefore the error of approximation of $x(t;\omega)$ by $x_n(t;\omega)$ for each $t \geq 0$ is less than $(\lambda K)^n \rho$, that is,

$$\|x(t;\omega) - x_n(t;\omega)\|_{L_2(\Omega,\mathscr{A},\mathscr{P})} < (\lambda K)^n \rho. \quad (3.1.4)$$

We may also remark that these results support the fact that $x_n(t;\omega)$ converges to $x(t;\omega)$ in mean-square, as was shown in Section 3.1.1, and we have a bound on $E\{|x(t;\omega) - x_n(t;\omega)|^2\}$ for each $t \in R_+$ and $n \geq 0$ from (3.1.4), which is given by

$$E\{|x(t;\omega) - x_n(t;\omega)|^2\} < [(\lambda K)^n \rho]^2.$$

3.1.3 Combined Error of Approximation and Numerical Integration

In this section we shall consider the error of approximation of a random solution $x(t;\omega)$ of (3.0.1) when the integral is evaluated numerically. We state a definition given by Rall [1].

Definition 3.1.2 The operator U on a Banach space X into itself such that for some $x \in X$, $U(x) = x$, is said to be an *arithmetic fixed-point problem* if the function $U(x)$ can be calculated to any desired accuracy by a finite number of arithmetic operations.

As before we write Eq. (3.0.1) as a fixed-point problem,

$$x(t;\omega) = (Ux)(t;\omega) = h(t;\omega) + \int_0^t k(t,\tau;\omega) f(\tau, x(\tau;\omega))\, d\tau, \qquad t \geq 0.$$

(3.1.5)

We shall consider the discrete approximation of (3.1.5). That is, we obtain a solution at each of the discrete points $0 = t_0 < t_1 < \cdots < t_n < \cdots < \infty$, where $t_i - t_{i-1} = r$, $i = 1, 2, \ldots$, and $t_n = t_0 + nr = nr$. For fixed $t = t_n$ in R_+ the interval from zero to t is divided into n subintervals, $0 = t_0 < t_1 < \cdots < t_n = t$. Note that as $r \to 0$, then for fixed t such that $t = t_n = nr$, we must have $n \to \infty$. This discrete version is equivalent to writing the integral in (3.1.5) as a finite sum which approaches the true value of the integral as $r \to 0$ for fixed $t = t_n$. Thus we transform (3.1.5) into an arithmetic fixed-point problem.

For $i = 0, 1, 2, \ldots$ we use the following notation throughout this section:

$$x_i(\omega) = x(t_i;\omega), \qquad f_i(x_i(\omega)) = f(t_i, x(t_i;\omega)),$$
$$h_i(\omega) = h(t_i;\omega), \qquad k_{n,i}(\omega) = k(t_n, t_i, \omega),$$

where $t_n = t = nr$ and the functions satisfy the conditions of Theorem 2.1.2 for each $t_i \in R_+$.

Now, for any $t = t_n = nr$ and each $\omega \in \Omega$ we can write the integral in (3.1.5) as

$$\int_0^t k(t_n, \tau;\omega) f(\tau, x(\tau;\omega))\, d\tau = \sum_{i=0}^n W_{n,i} k_{n,i}(\omega) f_i(x_i(\omega)) - \delta^{(n)}(\omega),$$

3.1 THE METHOD OF SUCCESSIVE APPROXIMATIONS

where $W_{n,i}$ are appropriate weights (such as the composite trapezoidal rule), and the error of approximation $\delta^{(n)}(\omega)$ can be made as small as desired by choosing $r = t_i - t_{i-1}$, $0 \leq t_i \leq t_n = t$, appropriately. Then we may write the discrete version of (3.1.5) as (exactly)

$$x_n(\omega) = h_n(\omega) + \sum_{i=0}^{n} W_{n,i} k_{n,i}(\omega) f_i(x_i(\omega)) - \delta^{(n)}(\omega) = (Ux_n)(\omega).$$

However, if we ignore the error of approximating the integral by the sum, we obtain an approximate value of $x_n(\omega)$, denoted by $\tilde{x}_n(\omega)$, for each $t = t_n$, where

$$x_0(\omega) = \tilde{x}_0(\omega) = h_0(\omega).$$

Then we have

$$\tilde{x}_n(\omega) = h_n(\omega) + \sum_{i=0}^{n} W_{n,i} k_{n,i}(\omega) f_i(\tilde{x}_i(\omega)). \quad (3.1.6)$$

Let

$$(F\tilde{x}_n)(\omega) = h_n(\omega) + \sum_{i=0}^{n} W_{n,i} k_{n,i}(\omega) f_i(\tilde{x}_i(\omega)).$$

Then

$$(F\tilde{x}_n)(\omega) = U(\tilde{x}_n)(\omega) + \delta^{(n)}(\omega).$$

Define the sequence of successive approximations to $x_n(\omega)$ for each $t = t_n$ by

$$\tilde{x}_n^{(0)}(\omega) = h_n(\omega) = x_n^{(0)}(\omega),$$
$$\tilde{x}_n^{(m+1)}(\omega) = (U\tilde{x}_n^{(m)})(\omega) + \delta^{(n)}(\omega) = (F\tilde{x}_n^{(m)})(\omega), \quad m \geq 0.$$

Suppose the set S is given by

$$S = \{x(t;\omega) : x(t;\omega) \in D, \quad \|x(t;\omega)\|_D \leq \rho\},$$

as in Theorem 2.1.2. The operator U is a contraction operator on S, and there exists a unique fixed point $x_n(\omega) = (Ux_n)(\omega)$ in S at $t = t_n \geq 0$, that is, a unique random solution of (3.1.5) exists at each $t = t_n$.

We assume that the error $\delta^{(n)}(\omega) \in L_2(\Omega, \mathscr{A}, \mathscr{P})$ for each t_n. We define the norm

$$\|\delta^{(n)}(\omega)\| \equiv \|\delta^{(n)}(\omega)\|_D$$

for n (the number of subintervals) fixed. Also, from Theorem 2.1.2 we have that

$$\|f_n(x_n(\omega)) - f_n(y_n(\omega))\|_B \leq \lambda \|x_n(\omega) - y_n(\omega)\|_D$$

for $x_n(\omega), y_n(\omega) \in S$ and $\lambda > 0$ a constant.

If we have $\|h_n(\omega)\|_D \leq \rho_1$ and

$$\rho \geq \rho_1 + [\|\delta^{(n)}(\omega)\|/(1 - \lambda K)]$$

and
$$\|h_n(\omega)\|_D + K\|f_n(0)\|_B \leq \rho_1(1 - \lambda K),$$
then the sequence $x_n^{(m)}(\omega) = x_m(t_n; \omega)$ defined in Section 3.1.2 is in the set
$$S_1 = \{x(t; \omega) : x(t; \omega) \in D, \quad \|x(t; \omega)\|_D \leq \rho_1\} \subset S.$$
This can be shown as follows:
$$x_0(t_n; \omega) = x_n^{(0)}(\omega) = h(t_n; \omega) = h_n(\omega) \in S_1$$
by assumption. Then the successive approximations defined by $x_n^{(m)}(\omega)$ $(Ux_n^{(m-1)})(\omega)$ are in S_1, since
$$\|x_n^{(1)}(\omega)\|_D = \|(Ux_n^{(0)})(\omega)\|_D$$
$$= \left\| h_n(\omega) + \int_0^{t_n} k(t_n, \tau; \omega) f(\tau, h(\tau; \omega))\, d\tau \right\|_D$$
$$\leq \|h_n(\omega)\|_D + K\lambda \|h_n(\omega)\|_D + K\|f_n(0)\|_B$$
$$\leq \rho_1(1 - \lambda K) + \lambda K \rho_1 = \rho_1,$$
$$\vdots$$
$$\|x_n^{(m)}(\omega)\|_D \leq \|h_n(\omega)\|_D + \lambda K \|x_n^{(m-1)}(\omega)\|_D + K\|f_n(0)\|_B$$
$$\leq \rho_1(1 - \lambda K) + \rho_1 \lambda K = \rho_1.$$

Hence all $x_n^{(m)}(\omega) \in S_1 \subset S$ and converge to the unique random solution $x_n(\omega) \in S_1$ as $m \to \infty$, as shown in the previous section.

We now show that the sequence $\tilde{x}_n^{(m)}(\omega)$ of approximations to $\tilde{x}_n(\omega)$ is in S for all $m \geq 0$ under the given conditions on $h_n(\omega)$ and f. We have
$$\tilde{x}_n^{(0)}(\omega) = h_n(\omega) \in S_1 \subset S.$$
Assume that $\tilde{x}_n^{(1)}(\omega), \ldots, \tilde{x}_n^{(m)}(\omega)$ are in S. Then we wish to show that $\tilde{x}_n^{(m+1)}(\omega) \in S$, and hence that all $\tilde{x}_n^{(m)}(\omega)$ are in S. We have
$$\|\tilde{x}_n^{(m+1)}(\omega) - x_n^{(m+1)}(\omega)\|_D = \|(U\tilde{x}_n^{(m)})(\omega) + \delta^{(n)}(\omega) - (Ux_n^{(m+1)})(\omega)\|_D$$
$$\leq \lambda K \|\tilde{x}_n^{(m)}(\omega) - x_n^{(m)}(\omega)\|_D + \|\delta^{(n)}(\omega)\|$$
since U is a contraction operator on S. Continuing in this manner, we obtain
$$\|\tilde{x}_n^{(m+1)}(\omega) - x_n^{(m+1)}(\omega)\|_D$$
$$\leq (\lambda K)^2 \|\tilde{x}_n^{(m-1)}(\omega) - x_n^{(m-1)}(\omega)\|_D + (\lambda K)\|\delta^{(n)}(\omega)\| + \|\delta^{(n)}(\omega)\|$$
$$\leq \cdots \leq (\lambda K)^{m+1} \|\tilde{x}_n^{(0)}(\omega) - x_n^{(0)}(\omega)\|_D + \|\delta^{(n)}(\omega)\|$$
$$\times [1 + \lambda K + \cdots + (\lambda K)^m]$$
$$= (\lambda K)^{m+1} \|h_n(\omega) - h_n(\omega)\|_D + \|\delta^{(n)}(\omega)\|[1 + \lambda K + \cdots + (\lambda K)^m]$$
$$= \|\delta^{(n)}(\omega)\|\{[1 - (\lambda K)^{m+1}]/(1 - \lambda K)\}. \tag{3.1.7}$$

3.1 THE METHOD OF SUCCESSIVE APPROXIMATIONS 77

Hence for an arbitrary $m \geq 0$

$$\|\tilde{x}_n^{(m+1)}(\omega)\|_D \leq \|\tilde{x}_n^{(m+1)}(\omega) - x_n^{(m+1)}(\omega)\|_D + \|x_n^{(m+1)}(\omega)\|_D$$
$$\leq \|\delta^{(n)}(\omega)\|\{[1 - (\lambda K)^{m+1}]/(1 - \lambda K)\} + \rho_1$$
$$\leq [\|\delta^{(n)}(\omega)\|/(1 - \lambda K)] + \rho_1 \leq \rho,$$

since $\lambda K < 1$. Hence for all m, $\tilde{x}_n^{(m)}(\omega) \in S$.

We shall now give the following theorem.

Theorem 3.1.4 If $\|h_n(\omega)\|_D \leq \rho_1$,

$$\|h_n(\omega)\|_D + K\|f_n(0)\|_B \leq \rho_1(1 - \lambda K), \qquad \lambda K < 1,$$

and

$$\rho \geq \rho_1 + [\|\delta^{(n)}(\omega)\|/(1 - \lambda K)],$$

then the sequence of successive approximations defined by

$$\tilde{x}_n^{(0)}(\omega) = h_n(\omega) = x_n^{(0)}(\omega),$$
$$\tilde{x}_n^{(m+1)}(\omega) = (U\tilde{x}_n^{(m)})(\omega) + \delta^{(n)}(\omega), \qquad m \geq 0,$$
$$= (F\tilde{x}_n^{(m)})(\omega),$$

converges in the Banach space D to within an error of $\|\delta^{(n)}(\omega)\|/(1 - \lambda K)$ to the true solution $x_n(\omega)$ at $t = t_n = nr$. That is,

$$\|\tilde{x}_n^{(m+1)}(\omega) - x_n(\omega)\|_D < (\lambda K)^{m+1}\rho + [\|\delta^{(n)}(\omega)\|/(1 - \lambda K)][1 - (\lambda K)^{m+1}]$$
$$\to \|\delta^{(n)}(\omega)\|/(1 - \lambda K)$$

as $m \to \infty$. Also, since we can choose r as small as desired to make $\|\delta^{(n)}(\omega)\|$ small, this error can be made to go to zero as $r \to 0$, $n \to \infty$ such that $t = nr$ remains fixed.

PROOF From inequality (3.1.7) we have

$$\|\tilde{x}_n^{(m+1)}(\omega) - x_n^{(m+1)}(\omega)\|_D \leq \|\delta^{(n)}(\omega)\|[1 - (\lambda K)^{m+1}]/(1 - \lambda K).$$

From Section 3.1.2 and the argument that $x_n^{(m)}(\omega) \in S_1 \subset S$, we also have that

$$\|x_n^{(m+1)}(\omega) - x_n(\omega)\|_D < (\lambda K)^{m+1}\rho.$$

Hence

$$\|\tilde{x}_n^{(m+1)}(\omega) - x_n(\omega)\|_D \leq \|\tilde{x}_n^{(m+1)}(\omega) - x_n^{(m+1)}(\omega)\|_D + \|x_n^{(m+1)}(\omega) - x_n(\omega)\|_D$$
$$< \|\delta^{(n)}(\omega)\|\{[1 - (\lambda K)^{m+1}]/(1 - \lambda K)\} + (\lambda K)^{m+1}\rho.$$

Thus as $m \to \infty$ we have that $\|\tilde{x}_n^{(m+1)}(\omega) - x_n(\omega)\|_D$ becomes smaller than $\|\delta^{(n)}(\omega)\|/(1 - \lambda K)$.

Since we may choose $r = t_i - t_{i-1}$, to calculate the integral as accurately as desired, for every $\varepsilon_1 > 0$ we may choose r so small (n large) that

$$\|\delta^{(n)}(\omega)\| < \varepsilon_1(1 - \lambda K).$$

Also for every $\varepsilon_2 > 0$ we may choose m so large that $(\lambda K)^{m+1}\rho < \varepsilon_2$. Thus for every $\varepsilon_1 > 0$ and $\varepsilon_2 > 0$ we may choose r so small (n large) and m so large that

$$\|\tilde{x}_n^{(m+1)}(\omega) - x_n(\omega)\|_D < \varepsilon_1 + \varepsilon_2 = \varepsilon.$$

As an example, suppose that we use a quadrature such as the composite trapezoidal rule (Kopal [1], pp. 397–410). The error of approximating the integral at each $\omega \in \Omega$ is then of the order of r^3. Hence, using r as small as possible gives a good approximation for sufficiently large m. In fact, if the interval $[0, t]$ is divided into n subintervals such that $r = t/n$, then the error of integration by the composite trapezoidal rule of a function $g(t; \omega)$ is of the form

$$\delta^{(n)}(\omega) = \tfrac{1}{12}nr^3 g''(\xi; \omega),$$

where ξ is some point in $[0, t]$ and the double prime indicates second derivative with respect to t. Hence here we must assume that $k(t, \tau; \omega) f(\tau, x(\tau; \omega))$ has a second derivative with respect to τ, and as $r \to 0$ ($n \to \infty$), $\|\delta^{(n)}(\omega)\| \to 0$.

3.2 A New Stochastic Formulation of a Population Growth Problem

In mathematical models for biological processes it is usually the case that a complex biological system is replaced by a simpler, idealized, hypothetical one (Bartlett [1, 2], Moran [1], Chiang [1], Bharucha-Reid [8]). Many simplifying assumptions on the actual biological system may still result in a very complicated mathematical or statistical model. In obtaining a mathematical model of a biological system, either the random or stochastic changes in the system may be ignored and a deterministic model reflecting "averages" of the random phenomena used, or the random variations may be accounted for, resulting in a stochastic model. The latter case is more realistic than the former and should be used even though in most situations stochastic models are much more difficult to work with mathematically.

In formulating a mathematical model for the growth of a biological population consisting of a single species, such complications as age structure and random changes must be dealt with. In much of the classical theory of population growth in which the age structure is considered the changes produced in the population growth rate by the phenomena of birth, aging, and death are assumed to be deterministic (Feller [1], Kendall [1], Bartlett [1, 2], Moran [1]). In this deterministic theory the *expected* birth rate satisfies an integral equation which Feller [1] calls the renewal equation. This integral equation involves the *expected* reproduction rate of female individuals of a given age and the *expected* rate of reproduction of females at a given time by members of the parent population.

In this section we shall present a formulation of a population growth model which results in a stochastic integral equation that is similar in form to the deterministic integral equation just mentioned. However, the solution to the stochastic integral equation is a stochastic process giving the *birth rate* instead of the *expected birth rate* in the population. We will also show that the expected value of the birth rate process satisfies the deterministic integral equation previously mentioned.

The stochastic integral equation obtained is of the Volterra type given by Eq. (3.0.1).

The theory of random or stochastic integral equations of this type given in Chapter II and Section 3.1 permits one to show existence and uniqueness of a random solution $x(t;\omega)$, and to obtain an approximation to a realization of the process $x(t;\omega)$ (also see Anderson [1], Tsokos [4], Padgett and Tsokos [5, 6, 11], Milton and Tsokos [2, 3], Hardiman and Tsokos [2], A. N. V. Rao and Tsokos [2]). This theory allows one to obtain the existence and uniqueness of a random solution $x(t;\omega)$ without specifying the exact distributions of the stochastic processes at each time t. That is, the stochastic processes which constitute Eq. (3.0.1) may be very general processes.

In Section 3.2.1 we shall discuss the deterministic model, and we give the stochastic formulation of the population growth problem in Section 3.2.2. We shall show that it completely specifies the state of the population at each time t and indicate the connection between the present formulation and the deterministic formulations mentioned earlier. In Section 3.2.3 we will show that the stochastic integral equation obtained in Section 3.2.2 has a unique random solution under certain conditions. Finally, we will give an example to indicate the fruitfulness of the stochastic integral equation formulation.

3.2.1 The Deterministic Model

Consider the effect of age structure on the growth of a population consisting of a single species, in the absence of external influences such as emigration or immigration, dependence on the density of the population, and disease. In the classical theory the size of the population is treated as a continuous variable and it is supposed that the modifications produced in the population due to the phenomena of birth, aging, and death are deterministic (Kendall [1]).

The variables such as the birth rate and number of individuals in the population are usually interpreted only for the female or reproducing portion of the biological population with the possibility of unequal sex ratio being ignored.

Let $\lambda(t)\,dt$ be the *expected* number of females born during time $(t, t+dt)$ to females aged t as in Feller [1], Bartlett [1, 2], and Kendall [1]. Also, $b(t)\,dt$ is the *expected* number of female births occurring in the time interval $(t, t+dt)$ (the expected birth rate), and $n(s,t)\,ds$ is the *expected* number of females in the population at time t in the age group $(s, s+ds)$. Then the total number of females in the population, that is, the size of the female population at time t, is the quantity

$$n(t) = \int_0^t n(s,t)\,ds = \int_0^t l(s)b(t-s)\,ds, \qquad (3.2.1)$$

where $l(t)$ is the probability that an individual born at time zero is alive at time $t > 0$ and is given by

$$l(t) = \exp\left[-\int_0^t \mu(s)\,ds\right], \qquad (3.2.2)$$

where $\mu(s)$ is the death rate of individuals aged s, $0 \leqslant s \leqslant t$. Then we have that (Kendall [1])

$$n(s,t) = l(s)b(t-s) \qquad (3.2.3)$$

and that

$$b(t) = g(t) + \int_0^t \lambda(y)l(y)b(t-y)\,dy, \qquad (3.2.4)$$

where

$$g(t) = l(t)\int_0^\infty \lambda(t+u)n(u,0)\,du. \qquad (3.2.5)$$

The function $g(t)$ is completely specified if the age structure and population size are known at the time observation of the population begins, that is, at epoch $t = 0$.

The integral equation (3.2.4) was studied in detail by Feller [1]. He showed that under certain conditions Eq. (3.2.4) possessed a *unique solution*, and also the *asymptotic behavior* of the solution was investigated. He suggested the method of *successive approximations* in order to approximate the expected birth rate $b(t)$ at each time $t > 0$, after presenting a treatment of the equation by Lotka [1]. The paper by Feller [1] contains an extensive list of the research papers concerning Eq. (3.2.4) up to 1941.

3.2.2 The Stochastic Model

It should be evident that the deterministic theory of population growth does not provide an adequate and realistic description of the processes involved, because it does not take into account the random fluctuations which occur as the process develops. Using the theory of stochastic integral equations which was given in Chapter II, we will formulate stochastic versions of Eqs. (3.2.3)–(3.2.5). It will not be required to specify the exact behavior of the processes because of the very general theory involved. However, we will show that the stochastic formulation is related to the "expected value" (deterministic) formulation under the usual kind of assumptions that are made in the stochastic population models.

Let $m(t;\omega)$ be a stochastic process which enumerates the number of offspring (female) produced by individuals (females) aged t at epoch $t > 0$. Then $m(t;\omega) dt$ is the number of offspring produced by individuals aged t in the interval $(t, t + dt)$. It is clear that m should be treated as a stochastic process, since at any epoch t the number of offspring produced by individuals aged t is random. We will consider m to be a continuous function of t for almost all ω, which approximates the number of offspring at each $t \geq 0$. That is, for almost all ω we assume that the stochastic process $m(t;\omega)$ has continuous sample functions. Hence for each $t \geq 0$ the number of female births at that epoch in the population is a random variable $b(t;\omega)$. That is to say, $b(t;\omega)$ is a stochastic process, and the number of individuals aged s at epoch t in the population is given by

$$n(s, t;\omega) = l(s)b(t - s;\omega) \qquad (3.2.6)$$

and the process $b(t;\omega)$ must satisfy the stochastic integral equation of the type (3.0.1) given by

$$b(t;\omega) = g(t;\omega) + \int_0^t l(s)m(s;\omega)b(t - s;\omega) \, ds, \qquad (3.2.7)$$

where

$$g(t;\omega) = l(t) \int_0^\infty m(t + u;\omega)n(u, 0;\omega) \, du. \qquad (3.2.8)$$

The process $g(t;\omega)$ is completely specified if the distribution of the population size at time zero and the behavior of the process $m(t;\omega)$ are known. Thus the size of the population at time $t \geq 0$ is a random variable and is given by

$$n(t;\omega) = \int_0^t n(s, t;\omega) \, ds = \int_0^t l(s)b(t - s;\omega) \, ds, \qquad (3.2.9)$$

the stochastic version of Eq. (3.2.1).

III SOLUTION OF THE RANDOM VOLTERRA INTEGRAL EQUATION

Before investigating the stochastic integral equation with respect to the conditions which guarantee a unique random solution we will consider its relation to the classical stochastic theory of population growth. We assume that the process $N(t;\omega)$ is a discrete valued stochastic process giving the number of individuals in the population at time t.

Assume the following:

(i) The subpopulations which are generated by two coexisting individuals develop independently of each other.

(ii) An individual (female) age x existing at time t has a chance

$$\lambda(x)\,dt + o(dt)$$

of producing a single offspring (female) during any time interval of length dt, where $\lambda(x)$ is the same function for all individuals in the population.

Let $dN(x, t; \omega)$ denote the number of individuals in the small age interval $(x, x + dx)$ at time t, and

$$E[dN(x, t; \omega)] = \alpha(x, t)\,dx + o(dx).$$

This is also the variance of $dN(x, t; \omega)$ to the first order. [That is, $dN(x, t; \omega)$ is of Poisson character.] It is assumed that

$$
\begin{aligned}
dN(x, t; \omega) &= 0 &&\text{with probability} &&1 - \alpha(x, t)\,dx + o(dx) \\
&= 1 &&\text{with probability} &&\alpha(x, t)\,dx + o(dx) \\
&\geqslant 2 &&\text{with probability} &&o(dx).
\end{aligned}
$$

Also, let $dM(x; \omega)$ denote the number of offspring produced by individuals aged x in the time interval $(x, x + dx)$. It is assumed that

$$
\begin{aligned}
dM(x; \omega) &= 0 &&\text{with probability} &&1 - \lambda(x)\,dx + o(dx) \\
&= 1 &&\text{with probability} &&\lambda(x)\,dx + o(dx) \\
&\geqslant 2 &&\text{with probability} &&o(dx),
\end{aligned}
$$

where

$$E[dM(x; \omega)] = \lambda(x)\,dx + o(dx).$$

If $\lambda(x) \equiv \lambda$, a constant, then $dM(x; \omega)$ has the same probability distribution at each x. If the process $dB(t; \omega)$ denotes the number of new births (integer-valued) in the population in the time interval $(t, t + dt)$, then

$$dN(x, t; \omega) = l(x)\,dB(t - x; \omega).$$

Assume that the death rate $\mu(s)$ is the same for all ages s.

3.2 A NEW STOCHASTIC FORMULATION

Then from Eq. (3.2.2) we have $l(t) = e^{-\mu t}$. Hence

$$dN(x, t; \omega) = e^{-\mu x} dB(t - x; \omega). \qquad (3.2.10)$$

The joint probability distribution of $dN(x, t; \omega)$ and $dM(x; \omega)$ is given to the first order by

		$dN(x, t; \omega)$		
		0	1	Total
$dM(x; \omega)$	0	$1 - \alpha(x, t) dx$	$[1 - \lambda(x) dx]\alpha(x, t) dx$	$1 - \lambda(x) dx \cdot \alpha(x, t) dx$
	1	0	$\lambda(x) dx \cdot \alpha(x, t) dx$	$\lambda(x) dx \cdot \alpha(x, t) dx$
	Total	$1 - \alpha(x, t) dx$	$\alpha(x, t) dx$	1

Thus

$$E[dM(x; \omega) \, dN(x, t; \omega)] = \lambda(x) \, dx \cdot \alpha(x, t) \, dx$$

and hence

$$\text{cov}[dM(x; \omega), dN(x, t; \omega)] = 0;$$

that is, $dM(x; \omega)$ and $dN(x, t; \omega)$ are uncorrelated. Then

$$\text{cov}[dM(x; \omega)l(x), dB(t - x; \omega)] = \text{cov}[dM(x; \omega) \, e^{-\mu x}, dN(x, t; \omega) \, e^{\mu x}]$$
$$= 0$$

from Eq. (3.2.10) and the given covariance.

For the continuous case, therefore, we make the similar assumption that the processes $m(x; \omega)$ and $n(x, t; \omega)$ are uncorrelated. Hence if we take the expectation of the stochastic integral equation (3.2.7), we obtain

$$E[b(t; \omega)] = E[g(t; \omega)] + \int_0^t E[l(s)m(s; \omega)b(t - s; \omega)] \, ds$$

or

$$b(t) = g(t) + \int_0^t l(s)E[m(s; \omega)] \cdot E[b(t - s; \omega)] \, ds$$

$$= g(t) + \int_0^t l(s)\lambda(s)b(t - s) \, ds, \qquad (3.2.11)$$

since

$$E[g(t; \omega)] = l(t) \int_0^\infty E[m(t + u; \omega)n(u, 0; \omega)] \, du$$

$$= l(t) \int_0^\infty \lambda(t + u)n(u, 0) \, du = g(t).$$

Equation (3.2.11) is the deterministic integral equation (3.2.4) given by Feller [1], Bartlett [1, 2], Moran [1], and Kendall [1].

We make the assumption on the correlation of m and n in order to obtain (3.2.11). This assumption is not necessary in the general theory of stochastic integral equations of the form (3.0.1) as given in Chapter II.

3.2.3 Existence and Uniqueness of a Random Solution

As Feller [1] did for the deterministic integral equation (3.2.4), we will now obtain conditions under which the equation (3.2.7) possesses a unique random solution $b(t;\omega)$, the population birth rate, bounded for $t \geq 0$.

By changing the variable of integration we may rewrite the stochastic integral equation (3.2.7) as

$$b(t;\omega) = g(t;\omega) + \int_0^t l(t-s)m(t-s;\omega)b(s;\omega)\,ds. \qquad (3.2.12)$$

Then the stochastic kernel in Eq. (3.2.12) is, for $0 \leq \tau \leq t < \infty$,

$$k(t,\tau;\omega) = l(t-\tau)m(t-\tau;\omega). \qquad (3.2.13)$$

If $m(t;\omega)$ is assumed to be bounded for almost all ω by some positive constant Λ, that is, there is an upper bound on the number of offspring produced at epoch t by females aged $t \geq 0$, then since

$$l(t) = e^{-\mu t}$$

for a constant death rate μ, we have from (3.2.13)

$$|||k(t,\tau;\omega)||| = \mathscr{P} - \operatorname*{ess\,sup}_{\omega} |l(t-\tau)m(t-\tau;\omega)| \leq \Lambda |l(t-\tau)|$$

$$= \Lambda e^{-\mu(t-\tau)}.$$

We will now state and prove the following theorem, which gives conditions that guarantee the existence of a unique random solution of Eq. (3.2.12).

Theorem 3.2.1 Consider the stochastic integral equation (3.2.12) subject to the following conditions:

(i) $|||l(t-\tau)m(t-\tau;\omega)||| \leq \Lambda e^{-\mu(t-\tau)}$ for $0 \leq \tau \leq t < \infty$, where μ and Λ are positive constants.

(ii) $g(t;\omega)$ is in the space C.

Then there exists a unique random solution $b(t;\omega) \in C$ such that $\|b(t;\omega)\|_C \leq \rho$ for some $\rho > 0$, provided that

$$\|g(t;\omega)\|_C \leq \rho[1 - (\Lambda^*/\mu)] \qquad \text{and} \qquad \Lambda^* < \mu,$$

where Λ^* is the infimum of the set of all constants Λ satisfying (i).

3.2 A NEW STOCHASTIC FORMULATION

PROOF   It has been shown in Chapter I that under conditions similar to (i) and (ii) there exists a unique random solution of Eq. (3.0.1) if the Banach spaces involved are admissible with respect to the integral operator

$$(Tx)(t;\omega) = \int_0^t k(t,\tau;\omega)x(\tau;\omega)\,d\tau$$

and if the function f in Eq. (3.0.1) satisfies a Lipschitz condition on the set

$$S = \{x(t;\omega): x(t;\omega) \in C, \quad \|x(t;\omega)\|_C \leq \rho\}$$

for some constant $\rho > 0$.

In Eq. (3.2.12) f is the identity function with respect to $x(t;\omega)$, and hence

$$|f(t,x) - f(t,y)| = |x - y| \tag{3.2.14}$$

for $x, y \in S$. That is, f satisfies a Lipschitz condition on S with Lipschitz constant equal to one.

Therefore if we show that the pair of spaces (C, C) is admissible with respect to the operator T given by

$$(Tx)(t;\omega) = \int_0^t l(t-\tau)m(t-\tau;\omega)x(\tau;\omega)\,d\tau \tag{3.2.15}$$

for $x(t;\omega) \in S$, and verify that $g(t;\omega)$ given in Section 3.2.2 is in the space C, then the existence of a unique random solution of Eq. (3.2.10) follows from the result of Theorem 2.1.2 under Conditions (i) and (ii) of the theorem.

Taking the norm in $L_2(\Omega, \mathscr{A}, \mathscr{P})$ of both sides of Eq. (3.2.15), we have

$$\|(Tx)(t;\omega)\|_{L_2(\Omega,\mathscr{A},\mathscr{P})}$$

$$\leq \int_0^t \|l(t-\tau)m(t-\tau;\omega)x(\tau;\omega)\|_{L_2(\Omega,\mathscr{A},\mathscr{P})}\,d\tau$$

$$\leq \int_0^t \||l(t-\tau)m(t-\tau;\omega)\|| \cdot \|x(\tau;\omega)\|_{L_2(\Omega,\mathscr{A},\mathscr{P})}\,d\tau$$

$$\leq \sup_{t \geq 0} \{\|x(t;\omega)\|_{L_2(\Omega,\mathscr{A},\mathscr{P})}\} \int_0^t \||l(t-\tau)m(t-\tau;\omega)\||\,d\tau$$

$$\leq \|x(t;\omega)\|_C \Lambda \int_0^t e^{-\mu(t-\tau)}\,d\tau$$

by definition of the norm in C and inequality (3.2.14). Thus

$$\|(Tx)(t;\omega)\|_{L_2(\Omega,\mathscr{A},\mathscr{P})} \leq \|x(t;\omega)\|_C (\Lambda/\mu)(1 - e^{-\mu t}), \qquad t \geq 0,$$

since

$$\int_0^t e^{-\mu(t-\tau)}\,d\tau = (1/\mu)(1 - e^{-\mu t}).$$

Therefore $(Tx)(t;\omega)$ is bounded in mean square and is in C by definition. Hence the pair (C, C) is admissible with respect to the operator T given by Eq. (3.2.15).

Since
$$\sup_{t \geq 0} \|(Tx)(t;\omega)\|_{L_2(\Omega,\mathscr{A},\mathscr{P})} = \|(Tx)(t;\omega)\|_C$$
$$\leq \|x(t;\omega)\|_C (\Lambda/\mu) \sup_{t \geq 0} (1 - e^{-\mu t})$$
$$= (\Lambda/\mu)\|x(t;\omega)\|_C,$$

we see that the norm of T is the quantity Λ^*/μ, where Λ^* is the least upper bound of $m(t;\omega)$ for almost all ω.

Now, we must verify that $g(t;\omega)$ is in C. Note that $\int_0^\infty n(u,0;\omega)\,du$ is the number of female individuals in the population at time zero, that is, the size of the initial population. We may assume without losing generality that there is a finite number $N_0 > 0$ such that
$$\int_0^\infty n(u,0;\omega)\,du \leq N_0 \qquad \text{for almost all} \quad \omega \in \Omega.$$

Then
$$g(t;\omega) = \int_0^\infty l(t)m(t+u;\omega)n(u,0;\omega)\,du$$
$$\leq e^{-\mu t}\Lambda^* N_0 < \infty \qquad \text{for almost all} \quad \omega$$

by the assumption that $m(t;\omega)$ is bounded by Λ^* and $l(t) = e^{-\mu t}$. Therefore
$$\int_\Omega |g(t;\omega)|^2 \, d\mathscr{P}(\omega) \leq \int_\Omega |e^{-\mu t}\Lambda^* N_0|^2 \, d\mathscr{P}(\omega)$$
$$= (e^{-\mu t}\Lambda^* N_0)^2 < \infty, \qquad t \in R_+,$$

and so by definition $g(t;\omega)$ is in the space $L_2(\Omega, \mathscr{A}, \mathscr{P})$. Also,
$$\|g(t;\omega)\|_{L_2(\Omega,\mathscr{A},\mathscr{P})} = \left\{ \int_\Omega |g(t;\omega)|^2 \, dP(\omega) \right\}^{\frac{1}{2}} \leq e^{-\mu t}\Lambda^* N_0, \qquad t \in R_+,$$

which means that $g(t;\omega)$ is bounded in $L_2(\Omega, \mathscr{A}, \mathscr{P})$. Since $m(t;\omega)$ is a continuous function of t in $L_\infty(\Omega, \mathscr{A}, \mathscr{P})$, we have

$$\|g(t+s;\omega) - g(t;\omega)\|_{L_2(\Omega,\mathscr{A},\mathscr{P})}$$
$$\leq \int_0^\infty \|l(t+s)m(t+u+s;\omega)n(u,0;\omega)$$
$$- l(t)m(t+u;\omega)n(u,0;\omega)\|_{L_2(\Omega,\mathscr{A},\mathscr{P})}\,du$$
$$\leq \int_0^\infty \|n(u,0;\omega)\|_{L_2(\Omega,\mathscr{A},\mathscr{P})} \cdot \|l(t+s)m(t+u+s;\omega)$$
$$- l(t)m(t+u;\omega)\|_{L_2(\Omega,\mathscr{A},\mathscr{P})}\,du.$$

But as $s \to 0$

$$e^{-\mu(t+s)} \to e^{-\mu t},$$

and since $n(u, 0; \omega)$ is in $L_2(\Omega, \mathscr{A}, \mathscr{P})$ for $u \geq 0$, we have that the integrand approaches zero as $s \to 0$. Hence $g(t; \omega)$ is continuous in mean square. We then have that $g(t; \omega)$ is in C by definition.

Therefore there exists a unique random solution of the stochastic integral equation (3.2.12), $b(t; \omega) \in S$, provided

$$\Lambda^* < \mu \quad \text{and} \quad \|g(t; \omega)\|_C = \Lambda^* N_0^* \leq \rho[1 - (\Lambda^*/\mu)],$$

since we have

$$\|g(t; \omega)\|_C = \sup_{t \geq 0} \|g(t; \omega)\|_{L_2(\Omega, \mathscr{A}, \mathscr{P})} = \Lambda^* N_0^*,$$

where N_0^* is the infimum of $\{N_0\}$, completing the proof.

The assumption that $\Lambda^* < \mu$ means that the upper bound on the number of offspring of individuals of a given age is less than the death rate. This may be interpreted as meaning that the population will eventually die out as $t \to \infty$. Also the effect of $g(t; \omega)$ tends to be negligible as $t \to \infty$.

In Section 3.3.3 we shall give a numerical example for a hypothetical population model such as that given here.

3.3 Method of Stochastic Approximation

We shall now approach the problem of obtaining an approximation to the random solution of the stochastic integral equation (3.0.1) by applying the theory of stochastic approximation. We shall first discuss the technique of stochastic approximation that will be used and present a general theorem due to Burkholder [1] which describes the convergence of the approximations to the true solution. We then shall obtain the conditions which the functions in Eq. (3.0.1) must satisfy in order for the method of stochastic approximation to be applicable.

3.3.1 A Stochastic Approximation Procedure

Stochastic approximation was first introduced in 1951 in a paper by Robbins and Monro [1], in which they considered the problem of approximating the root of an unknown regression function $M(x) = \alpha$, where α is a known constant. Their results were generalized and extended by Wolfowitz [1], Blum [1, 2], Kallianpur [1], Kiefer and Wolfowitz [1], Burkholder [1], and others (see Wasan [1]). Morozan [4] has applied stochastic approximation techniques to the theory of Lyapunov functions.

Robbins and Monro showed that under certain conditions on $M(x)$, where

$$M(x) = E[Y(x(\omega))|x(\omega) = x],$$

the sequence defined by

$$x_{n+1}(\omega) = x_n(\omega) + a_n[\alpha - Y(x_n(\omega))], \qquad n \geq 1, \qquad (3.3.1)$$

where $x_1(\omega)$ is an arbitrary, real, random variable, converges to the root of $M(x) = \alpha$ in mean square and in probability. We shall discuss a theorem of Burkholder [1] which requires somewhat weaker conditions than those of Robbins and Monro but results in the sequence $\{x_n(\omega)\}$ converging with probability one.

It is assumed that a value of $x(\omega)$ can be fixed, x_1, say, and the value $Y(x_1)$ can be found or observed. Then the next value x_2 is found from (3.3.1), and so on. Under the conditions to be given, the sequence (3.3.1) converges with probability one to a real number θ such that $M(\theta) = 0 \, (= \alpha)$, that is, θ is a root of $M(x) = 0$.

Let $M(\cdot)$ be a function from R (the real numbers) into R. For each $x \in R$ let $Y(x)$ be a random variable with probability distribution function $G(\cdot|x)$ such that $E[Y(x)] = M(x)$. Let $\{a_n\}$ be a positive number sequence, and let $x_1(\omega)$ be a random variable. If n is a positive integer, let

$$x_{n+1}(\omega) = x_n(\omega) - a_n y_n(\omega),$$

where $y_n(\omega)$ is a random variable with conditional distribution function $G(\cdot|x_n)$, given $x_1, \ldots, x_n, y_1, \ldots, y_{n-1}$. The random variable sequence $\{x_n(\omega)\}$ so defined is a stochastic approximation process which Burkholder [1] refers to as type A_0.

We denote by $V(x)$ the function

$$V(x) = \text{var}[Y(x)]$$

for $x \in R$.

We shall now state the following theorem.

Theorem 3.3.1 (Burkholder [1]) Suppose $\{x_n(\omega)\}$ is a stochastic approximation process of the type A_0 and θ is a real number such that the following conditions hold:

(i) For every $\varepsilon > 0$, if $|x - \theta| > \varepsilon$, then $(x - \theta)M(x) > 0$.
(ii) $\sup_x \{|M(x)|/(1 + |x|)\} < \infty$.
(iii) $\sup_x V(x) < \infty$.

(iv) If $0 < \delta_1 < \delta_2 < \infty$, then
$$\inf_{\delta_1 \leq |x-\theta| \leq \delta_2} |M(x)| > 0.$$
(v) $\sum_{n=1}^{\infty} a_n = \infty$ and $\sum_{n=1}^{\infty} a_n^2 < \infty$.
(vi) $M(x)$ and $V(x)$ are Borel measurable.

Then $\mathscr{P}\{\omega : \lim_{n\to\infty} x_n(\omega) = 0\} = 1$.

The conditions in Theorem 3.3.1 seem to be more general than those given by Blum [1], although the proofs are similar. In the next section this theorem will be applied to the stochastic integral equation (3.0.1).

3.3.2 Solution of Eq. (3.0.1) by Stochastic Approximation

We have that for each t and τ satisfying $0 \leq \tau \leq t < \infty$ the variances exist for $h(t;\omega)$, $k(t,\tau;\omega)$, and $x(t;\omega)$, since for each $t \in R_+$, $h(t;\omega)$ and $x(t;\omega)$ are in $L_2(\Omega, \mathscr{A}, \mathscr{P})$ and for each t and τ, $0 \leq \tau \leq t < \infty$, $k(t,\tau;\omega)$ is in $L_\infty(\Omega, \mathscr{A}, \mathscr{P})$. Again we assume that the distribution function of $h(t;\omega)$ is known for each $t \in R_+$ or that a value of $h(t;\omega)$ can be observed for each t. It is also assumed that $x(t;\omega)$, $h(t;\omega)$, and $k(t,\tau;\omega)$ are mutually independent, real-valued random variables at each t and τ, $0 \leq \tau \leq t < \infty$.

As in Section 3.1, we write

$$(Ux)(t;\omega) = h(t;\omega) + \int_0^t k(t,\tau;\omega) f(\tau, x(\tau;\omega)) \, d\tau$$

for $t \in R_+$. By Theorem 2.1.2 there exists a unique random solution $x(t;\omega)$ of (3.0.1),

$$(Ux)(t;\omega) = x(t;\omega).$$

Let

$$(Yx)(t;\omega) = x(t;\omega) - (Ux)(t;\omega)$$

for x a continuous, real-valued function from R_+ into $L_2(\Omega, \mathscr{A}, \mathscr{P})$ contained in the set S of Theorem 2.1.2. For fixed $t \in R_+$ define

$$M[x(t)] = E[(Yx)(t;\omega)|x(t;\omega) = x(t), \quad x(\tau;\omega) = \theta(\tau) \quad \text{for} \quad \tau < t]$$
$$= E[x(t;\omega) - (Ux)(t;\omega)|x(t;\omega) = x(t), \quad x(\tau;\omega) = \theta(\tau) \quad \text{for} \quad \tau < t],$$

where $\theta(\tau)$ is a realization of the unique random solution of Eq. (3.0.1). That is, we have already obtained values $\theta(\tau)$ of the unique random solution for $\tau < t$. Hence we now wish to find the value $\theta(t)$ such that

$$M[\theta(t)] = 0.$$

90   III   SOLUTION OF THE RANDOM VOLTERRA INTEGRAL EQUATION

That is, a value of the unique random solution is now to be obtained at $t \in R_+$.

Suppose that at the fixed $t \in R_+$ we choose

$$x_1(t; \omega) = h(t; \omega), \qquad x_{n+1}(t; \omega) = x_n(t; \omega) - a_n y_n(t; \omega), \qquad n \geq 1, \quad (3.3.2)$$

where $y_n(t; \omega)$ is the random variable $(Yx_n)(t; \omega)$. We apply Theorem 3.3.1 to obtain conditions for which the sequence $\{x_n(t; \omega)\}$ defined by (3.3.2) converges to $\theta(t)$ with probability one.

We have

$$M[x(t)] = E[x(t; \omega) - (Ux)(t; \omega) | x(t; \omega) = x(t), \ x(\tau; \omega) = \theta(\tau) \text{ for } \tau < t]$$

$$= E[x(t; \omega) - h(t; \omega)$$

$$- \int_0^t k(t, \tau; \omega) f(\tau, x(\tau; \omega)) \, d\tau | x(t; \omega) = x(t), x(\tau; \omega) = \theta(\tau) \text{ if } \tau < t]$$

$$= x(t) - \mu_h(t) - \int_0^t \mu_k(t, \tau)$$

$$\times E[f(\tau, x(\tau; \omega)) | x(t; \omega) = x(t), \ x(\tau; \omega) = \theta(\tau) \text{ if } \tau < t] \, d\tau$$

$$= x(t) - \mu_h(t) - \int_0^t \mu_k(t, \tau) f(\tau, \theta(\tau)) \, d\tau, \qquad (3.3.3)$$

where

$$\mu_h(t) = E[h(t; \omega)] \qquad \text{and} \qquad \mu_k(t, \tau) = E[k(t, \tau; \omega)],$$

which exist by assumption.

We will now show that Condition (i) of Theorem 3.3.1 holds at fixed $t \in R_+$. Let $x(t) < \theta(t)$. Then from (3.3.3) we have

$$M[x(t)] = x(t) - \mu_h(t) - \int_0^t \mu_k(t, \tau) f(\tau, \theta(\tau)) \, d\tau$$

$$< \theta(t) - \mu_h(t) - \int_0^t \mu_k(t, \tau) f(\tau, \theta(\tau)) \, d\tau$$

$$= 0$$

since $\theta(t)$ is a value of the unique random solution of (3.0.1) at $t \in R_+$. Likewise, if $x(t) > \theta(t)$, then

$$M[x(t)] = x(t) - \mu_h(t) - \int_0^t \mu_k(t, \tau) f(\tau, \theta(\tau)) \, d\tau$$

$$> \theta(t) - \mu_h(t) - \int_0^t \mu_k(t, \tau) f(\tau, \theta(\tau)) \, d\tau$$

$$= 0.$$

Hence, Condition (i) of Burkholder's theorem holds.

3.3 METHOD OF STOCHASTIC APPROXIMATION

We now must show that Condition (ii) of the theorem is satisfied; that is,

$$\sup_{x(t)} \{|M[x(t)]|/[1 + |x(t)|]\} < \infty.$$

By definition, we have that

$$M[x(t)] = x(t) - \mu_h(t) - \int_0^t \mu_k(t, \tau)$$

$$\times E[f(\tau, x(\tau; \omega))|x(t; \omega) = x(t), \quad x(\tau; \omega) = \theta(\tau) \text{ if } \tau < t] \, d\tau.$$

But by assumption the means of $k(t, \tau; \omega)$, $h(t; \omega)$, and $f(\tau, x(\tau; \omega))$ exist, and hence

$$\frac{|M[x(t)]|}{1 + |x(t)|}$$

$$= \frac{|x(t) - \mu_h(t)|}{1 + |x(t)|}$$

$$- \frac{|\int_0^t \mu_k(t, \tau) E[f(\tau, x(\tau; \omega))|x(t; \omega) = x(t), \quad x(\tau; \omega) = \theta(\tau) \text{ if } \tau < t] \, d\tau|}{1 + |x(t)|}$$

$$< 1 + |\mu_h(t)|$$

$$+ \int_0^t |\mu_k(t, \tau) E[f(\tau, x(\tau; \omega))|x(t; \omega) = x(t), \quad x(\tau; \omega) = \theta(\tau) \text{ if } \tau < t]| \, d\tau$$

for all $x(t)$ since

$$|x(t)|/[1 + |x(t)|] < 1 \quad \text{and} \quad 1/[1 + |x(t)|] \leq 1.$$

Thus

$$\sup_{x(t)} \{|M[x(t)]|/[1 + |x(t)|]\} < \infty,$$

and Condition (ii) holds.

To show that Condition (iii) holds, we proceed as follows. Let

$$V[x(t)] = \text{var}[x(t; \omega) - (Ux)(t; \omega)|x(t; \omega) = x(t), \quad x(\tau; \omega) = \theta(\tau) \text{ if } \tau < t]$$

$$= \text{var}[(Ux)(t; \omega)|x(t; \omega) = x(t), \quad x(\tau; \omega) = \theta(\tau) \text{ if } \tau < t]$$

from the property that a constant $[x(t)]$ plus a random variable has the same variance as the random variable. But from the preceding

$$E[(Ux)(t; \omega)|x(t; \omega) = x(t), \quad x(\tau; \omega) = \theta(\tau) \text{ if } \tau < t]$$

$$= \mu_h(t) + \int_0^t \mu_k(t, \tau) f(\tau, \theta(\tau)) \, d\tau.$$

Therefore

$$V[x(t)] = E\left\{\left[h(t;\omega) + \int_0^t k(t,\tau;\omega)f(\tau,x(\tau;\omega))\,d\tau - \mu_h(t) \right.\right.$$
$$\left.\left. - \int_0^t \mu_k(t,\tau)f(\tau,\theta(\tau))\,d\tau\right]^2 \bigg| x(t;\omega) = x(t),\ x(\tau;\omega) = \theta(\tau)\text{ if }\tau < t\right\}$$

$$= E[h(t;\omega) - \mu_h(t)]^2 + 2E\left\{[h(t;\omega) - \mu_h(t)]\right.$$
$$\times \int_0^t [k(t,\tau;\omega)f(\tau,x(\tau;\omega)) - \mu_k(t,\tau)f(\tau,\theta(\tau))]\,d\tau$$
$$\bigg| x(t;\omega) = x(t),\ x(\tau;\omega) = \theta(\tau)\text{ if }\tau < t\right\}$$
$$+ E\left\{\left[\int_0^t [k(t,\tau;\omega)f(\tau,x(\tau;\omega)) - \mu_k(t,\tau)f(\tau,\theta(\tau))]\,d\tau\right]^2\right.$$
$$\bigg| x(t;\omega) = x(t),\ x(\tau;\omega) = \theta(\tau)\text{ if }\tau < t\right\}$$

$$= \text{var}[h(t;\omega)] + 0$$
$$+ E\left\{\left[\int_0^t [k(t,\tau;\omega)f(\tau,x(\tau;\omega)) - \mu_k(t,\tau)f(\tau,\theta(\tau))]\,d\tau\right]^2\right.$$
$$\bigg| x(t;\omega) = x(t),\ x(\tau;\omega) = \theta(\tau)\text{ if }\tau < t\right\},$$

where the second term is zero because of independence.
But the last expectation may be written as

$$E\left\{\int_0^t [k(t,\tau;\omega)f(\tau,x(\tau;\omega)) - \mu_k(t,\tau)f(\tau,\theta(\tau))]\,d\tau \right.$$
$$\times \int_0^t [k(t,s;\omega)f(s,x(s;\omega)) - \mu_k(t,s)f(s,\theta(s))]\,ds$$
$$\bigg| x(t;\omega) = x(t),\ x(u;\omega) = \theta(u)\text{ if }u < t\right\}$$
$$= \int_0^t \int_0^t E\{[k(t,\tau;\omega)f(\tau,x(\tau;\omega)) - \mu_k(t,\tau)f(\tau,\theta(\tau))]$$
$$\times [k(t,s;\omega)f(s,x(s;\omega)) - \mu_k(t,s)f(s,\theta(s))]$$
$$|x(t;\omega) = x(t),\ x(u;\omega) = \theta(u)\text{ if }u < t\}\,ds\,d\tau.$$

3.3 METHOD OF STOCHASTIC APPROXIMATION

However, since

$$E[k(t,s;\omega)f(s,x(s;\omega))|x(t;\omega) = x(t), \quad x(u;\omega) = \theta(u) \text{ if } u < t]$$
$$= \mu_k(t,s)f(s,\theta(s)), \quad s < t,$$

the integrand becomes

$$E\{[k(t,\tau;\omega)f(\tau,x(\tau;\omega))\cdot k(t,s;\omega)f(s,x(s;\omega))]|x(t;\omega)$$
$$= x(t), x(u;\omega) = \theta(u) \text{ if } u < t\}$$
$$- \mu_k(t,\tau)f(\tau,\theta(\tau))\cdot \mu_k(t\cdot s)f(s,\theta(s))$$
$$= \{E[k(t,\tau;\omega)k(t,s;\omega)]$$
$$- \mu_k(t,\tau)\mu_k(t,s)\}\cdot f(\tau,\theta(\tau))f(s,\tau(s)), \quad s<t, \quad \tau<t,$$
$$= \text{cov}[k(t,\tau;\omega),k(t,s;\omega)]f(\tau,\theta(\tau))f(s,\theta(s)), \quad s,\tau<t.$$

Hence the double integral becomes

$$\int_0^t \int_0^t \text{cov}[k(t,\tau;\omega),k(t,s;\omega)]f(\tau,\theta(\tau))f(s,\theta(s))\,ds\,d\tau.$$

Therefore if $f(\tau,x)$ is uniformly bounded for $x \in R$ and $\tau \in [0,t]$ for each fixed $t \in R_+$, and if

$$\int_0^t \int_0^t \text{cov}[k(t,\tau;\omega),k(t,s;\omega)]\,ds\,d\tau \leq \Gamma, \quad \text{some constant,}$$

then

$$V[x(t)] < \infty$$

for all $x(t)$ values, since $\text{var}[h(t;\omega)] < \infty$ by the assumption that $h(t;\omega) \in L_2(\Omega, \mathscr{A}, \mathscr{P})$. Hence

$$\sup_{x(t)} V[x(t)] < \infty$$

and Condition (iii) is satisfied. Note that the covariance exists for each fixed t, τ, and s, since $k(t,\tau;\omega)$ and $k(t,s;\omega)$ are in $L_\infty(\Omega, \mathscr{A}, \mathscr{P})$ for each (t,τ) and (t,s).

To show that Condition (iv) of Theorem 3.3.1 holds, let $0 < \delta_1 < \delta_2 < \infty$. By definition,

$$|M[x(t)]| = |x(t) - \mu_h(t) - \int_0^t \mu_k(t,\tau)f(\tau,\theta(\tau))\,d\tau|$$
$$= \left|[x(t) - \theta(t)] + \left[\theta(t) - \mu_h(t) - \int_0^t \mu_k(t,\theta)f(\tau,\theta(\tau))\,d\tau\right]\right|$$
$$= |x(t) - \theta(t)|$$

since $\theta(t)$ is a realization of the unique random solution of (3.0.1). Hence

$$\inf_{\delta_1 \leqslant |x(t)-\theta(t)| \leqslant \delta_2} |M[x(t)]| = \inf_{\delta_1 \leqslant |x(t)-\theta(t)| \leqslant \delta_2} |x(t) - \theta(t)| = \delta_1 > 0$$

and Condition (iv) holds.

Thus we have the following theorem.

Theorem 3.3.2 If $x(\tau;\omega)$, $h(t;\omega)$, and $k(t,\tau;\omega)$ are mutually independent, real-valued random variables for each t and τ, $0 \leqslant \tau \leqslant t < \infty$, and the conditions of Theorem 2.1.2 hold for a unique random solution of (3.0.1) to exist, and if for fixed $t \in R_+$:

(i) $f(\tau, x)$ is uniformly bounded for $x \in R$ and $\tau \in [0, t]$;
(ii) $\int_0^t \int_0^t \text{cov}[k(t,\tau;\omega), k(t,s;\omega)] \, ds \, d\tau \leqslant \Gamma$, some constant;
(iii) $\sum_{n=1}^\infty a_n = \infty$ and $\sum_{n=1}^\infty a_n^2 < \infty$;
(iv) $M(x)$ and $V(x)$ are Borel measurable;

then

$$\mathscr{P}\{\omega: \lim_{n\to\infty} x_n(t;\omega) = \theta(t)\} = 1,$$

where $\theta(t)$ is a value of the unique random solution $x(t;\omega)$ of the stochastic integral equation (3.0.1) and $\{x_n(t;\omega)\}$ is defined by (3.3.2).

Therefore the stochastic approximation procedure defined by (3.3.2) converges to a value of the unique random solution at each $t \in R_+$ with probability one if the conditions given are satisfied. This gives a very useful technique for numerically obtaining an approximation to a realization of the unique random solution of Eq. (3.0.1) in practical situations.

3.3.3 Numerical Solution for a Hypothetical Population

We will now illustrate the fruitfulness of the results obtained in the foregoing sections by using a hypothetical population and choosing certain distributions for the processes $m(t;\omega)$ and $n(u,0;\omega)$ given in Section 3.2. Then values will be obtained from these distributions and a realization of the unique random solution of (3.2.13), the birth rate $b(t;\omega)$, at each $t \geqslant 0$ will be found by the method of stochastic successive approximations presented in Section 3.1.

Feller [1] used successive approximations in solving the deterministic equation (3.2.4). Let

$$(Ub)(t;\omega) = g(t;\omega) + \int_0^t k(t,\tau;\omega)b(\tau;\omega)\,d\tau, \qquad t \geqslant 0. \qquad (3.3.4)$$

Define a sequence of random variables at each fixed value of $t \in R_+$ as follows:

$$b_1(t;\omega) = g(t;\omega), \qquad b_{n+1}(t;\omega) = (Ub_n)(t;\omega), \qquad n \geqslant 1. \qquad (3.3.5)$$

3.3 METHOD OF STOCHASTIC APPROXIMATION

The sequence $\{b_n(t;\omega)\}$ for each $t \geq 0$ can be shown to converge to the unique random solution of Eq. (3.2.13) with probability one and in mean square as $n \to \infty$ under the conditions of Theorem 3.2.1. We use Theorem 3.1.2.

Theorem 3.3.3 The sequence of random variables $b_n(t;\omega)$, $n = 1, 2, \ldots$, converges to $b(t;\omega)$ with probability one for each $t \geq 0$ under the conditions of Theorem 3.2.1.

PROOF The proof follows a similar argument to that for Theorem 3.1.3, and will be omitted.

We use a hypothetical population as an example. Suppose that the death rate, that is, the number of individuals dying per unit time, is $\mu = 2.1$. To obtain a *value* of the stochastic kernel $k(t,\tau;\omega)$, a *value* of the number of offspring produced at time $t - \tau$ by females aged $t - \tau \geq 0$, $m(t - \tau;\omega)$, is needed. We suppose that the biological population consists of organisms which produce offspring at the same rate at all ages from age zero until death. That is, for all t, $m(t;\omega)$ has the same distribution with mean $\lambda(t) \equiv \lambda$. It is assumed that $m(t;\omega)$ has the uniform distribution on the interval zero to two, so that the mean number of offspring produced per unit time by individuals aged t is 1.0 for all t. Therefore to find a value of the stochastic kernel, we may generate a value from this uniform distribution.

Also, in order to use the stochastic successive approximation method to approximate a realization of $b(t;\omega)$, we need a value of $g(t;\omega)$ at each t. The process $g(t;\omega)$ is given by Eq. (3.2.8), and thus we need values of $n(u, 0;\omega)$ in order to obtain values of $g(t;\omega)$. We choose as an approximate distribution of $n(u, 0;\omega)$, the number of organisms of age u in the population at time zero, a gamma distribution with mean

$$2/\beta(u) = \beta_0/e^u, \qquad u \geq 0,$$

which appropriately tends to zero as $u \to \infty$, where β_0 is the mean number of organisms in the population of age zero at time zero (that is, the average number of organisms aged zero at the time observation of the population is begun). Therefore we use values from the family of gamma densities

$$f(x;u) = \beta^2(u)x\, e^{-x\beta(u)}, \qquad x \geq 0, \quad u \geq 0,$$

as values of $n(u, 0;\omega)$, and then obtain values of $g(t;\omega)$ from Eq. (3.2.8) with $m(t;\omega)$ behaving as described and

$$l(t) = e^{-\mu t} = e^{-(2.1)t}.$$

We assume the mean number of organisms aged zero to be $\beta_0 = 50$ for convenience.

To evaluate the integrals in Eqs. (3.2.8) and (3.2.12), the composite trapezoidal rule was used (Kopal [1]) and an approximate realization of the birth rate $b(t;\omega)$ was obtained by the method of successive approximation with the aid of an electronic computer. Iteration of Eqs. (3.3.5) was continued until the absolute difference between the values of $b_n(t;\omega)$ and $b_{n-1}(t;\omega)$ at each t was less than the specified accuracy 0.001. The results of two simulations of this hypothetical population are shown in Fig. 3.3.1.

The hypothetical population serves to illustrate the usefulness of the formulations given here. In a more realistic situation the distributions of $n(u,0;\omega)$ and $m(t;\omega)$ may be known or approximated and the same techniques applied. Also, once these distributions are known and the values of $b(t;\omega)$ are found, then the number of individuals (female or reproducing organisms) in the population may be obtained from Eq. (3.2.9), and the growth process is completely specified. It is possible that more general growth processes than that given here may be formulated in terms of stochastic integral equations, and if so, the general theory discussed in Chapter II may apply.

The method of stochastic approximation given in section 3.3.2 may also be used to obtain the realizations shown in Fig. 3.3.1.

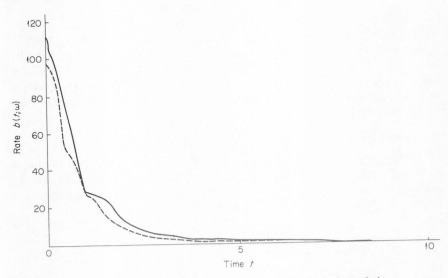

Figure 3.3.1 Two realizations of birth rate for hypothetical population.

CHAPTER IV

A Stochastic Integral Equation of the Fredholm Type and Some Applications

4.0 Introduction†

In this chapter we shall consider a stochastic integral equation of the Fredholm type of the form

$$x(t;\omega) = h(t;\omega) + \int_0^\infty k_0(t,\tau;\omega)\,e(\tau,x(\tau;\omega))\,d\tau, \qquad (4.0.1)$$

which was presented in Section 1.3.

We shall study the existence and uniqueness of a random solution of Eq. (4.0.1) using the concepts of admissibility introduced in Chapters I and II. We will also consider the asymptotic properties of the unique random

---

† Sections 4.0 and 4.1 adapted with permission of the publisher, The American Mathematical Society, from Padgett and Tsokos [11], *Proceedings of the American Mathematical Society.* Copyright © 1972, Volume 33, Number 2, pp. 534–542.

solution $x(t;\omega)$. In order to study the existence and uniqueness of a random solution of Eq. (4.0.1), we shall first consider the existence and uniqueness of a random solution of the stochastic integral equation of the mixed Volterra–Fredholm type of the form

$$x(t;\omega) = h(t;\omega) + \int_0^t k(t,\tau;\omega)f(\tau,x(\tau;\omega))\,d\tau + \int_0^\infty k_0(t,\tau;\omega)e(\tau,x(\tau;\omega))\,d\tau,$$
(4.0.2)

where the stochastic kernel $k(t,\tau;\omega)$ and the function $f(t,x)$ behave as in the first three chapters. Then Eq. (4.0.1) is a special case of Eq. (4.0.2), that is, we obtain (4.0.1) from (4.0.2) when $k(t,\tau;\omega)$ is identically equal to zero for almost all $\omega \in \Omega$. Likewise, for $k_0(t,\tau;\omega)$ equal to zero for all $t,\tau \in R_+$ and almost all $\omega \in \Omega$, we obtain the random Volterra integral equation of Chapters II and III.

A nonstochastic version of Eq. (4.0.2) has been studied by Miller, Nohel, and Wong [1], Petrovanu [1], and Corduneanu [5], among others.

An application of Eq. (4.0.1) will be presented in stochastic control theory. A linear system which was considered in the deterministic sense by Corduneanu [3] is generalized to the nonlinear stochastic case, and conditions are given that guarantee the existence of a unique random solution of the resulting stochastic control system such that the random solution is stochastically asymptotically exponentially stable.

4.1 Existence and Uniqueness of a Random Solution

We shall make the following assumptions throughout this chapter: The random solution $x(t;\omega)$ and $h(t;\omega)$ are functions of $t \in R_+$ with values in the space $L_2(\Omega, \mathscr{A}, \mathscr{P})$. The function $e(t,x(t;\omega))$ is a function of t with values in $L_2(\Omega, \mathscr{A}, \mathscr{P})$. For each t and τ such that $0 \leqslant t < \infty$ and $0 \leqslant \tau < \infty$, the stochastic kernel $k_0(t,\tau;\omega)$ will be an essentially bounded function with respect to $\mathscr{P}$-measure, that is, for each $t, \tau \in R_+$, $k_0(t,\tau;\omega)$ will be in $L_\infty(\Omega, \mathscr{A}, \mathscr{P})$. Then the product of $k_0(t,\tau;\omega)$ and $e(t,x(t;\omega))$ will always be in $L_2(\Omega, \mathscr{A}, \mathscr{P})$. Furthermore, with respect to the behavior of $k_0(t,\tau;\omega)$, we assume that the mapping $(t,\tau) \to k_0(t,\tau;\omega)$ from the set

$$\Delta_1 = \{(t,\tau): 0 \leqslant t < \infty,\ 0 \leqslant \tau < \infty\}$$

into $L_\infty(\Omega, \mathscr{A}, \mathscr{P})$ is continuous. That is,

$$\mathscr{P}\text{-ess}\sup_{\omega \in \Omega} |k_0(t_n,\tau_n;\omega) - k_0(t,\tau;\omega)| \to 0$$

4.1 EXISTENCE AND UNIQUENESS OF A RANDOM SOLUTION

as $n \to \infty$ whenever $(t_n, \tau_n) \to (t, \tau)$ as $n \to \infty$. Denote the norm of $k_0(t, \tau; \omega)$ in $L_\infty(\Omega, \mathscr{A}, \mathscr{P})$ by

$$|||k_0(t, \tau; \omega)||| = \mathscr{P}\text{-ess}\sup_{\omega \in \Omega} |k_0(t, \tau; \omega)| = \|k_0(t, \tau; \omega)\|_{L_\infty(\Omega, \mathscr{A}, \mathscr{P})}.$$

We also assume that for each $t \in R_+$, $|||k_0(t, \tau; \omega)|||$ and $|||k_0(t, \tau; \omega)||| \times \|x(\tau; \omega)\|_{L_2(\Omega, \mathscr{A}, \mathscr{P})}$ are integrable with respect to $\tau \in R_+$.

We let $C_c = C_c(R_+, L_2(\Omega, \mathscr{A}, \mathscr{P}))$, C_g, and C denote the various spaces of functions defined in Chapter I. Also, let $B, D \subset C_c$ be a pair of Banach spaces.

Define the operators $\mathscr{K}$ and $\mathscr{L}$ from B into C_c by

$$(\mathscr{K}x)(t; \omega) = \int_0^t k(t, \tau; \omega) x(\tau; \omega)\, d\tau,$$

$$(\mathscr{L}x)(t; \omega) = \int_0^\infty k_0(t, \tau; \omega) x(\tau; \omega)\, d\tau, \qquad t \in R_+.$$

If B and D are stronger than C_c and the pair (B, D) is admissible with respect to $\mathscr{K}$, then from Lemma 2.1.1 we have that $\mathscr{K}$ is continuous from B to D. Therefore there is a constant $K_1 > 0$ such that

$$\|(\mathscr{K}x)(t; \omega)\|_D \leqslant K_1 \|x(t; \omega)\|_B.$$

The following lemma shows that the operator $\mathscr{L}$ is continuous from $C_c(R_+, L_2(\Omega, \mathscr{A}, \mathscr{P}))$ into itself.

Lemma 4.1.1 The operator $\mathscr{L}$ defined here is a continuous linear operator from $C_c(R_+, L_2(\Omega, \mathscr{A}, \mathscr{P}))$ into itself.

PROOF As stated previously, the space $C_c(R_+, L_2(\Omega, \mathscr{A}, \mathscr{P}))$ is a Fréchet space with distance function given by

$$d(x, y) = \sum_{n=1}^\infty (1/2^n)[\|x - y\|_n / (1 + \|x - y\|_n)]$$

where $\|\cdot\|_n$ is the family of semi-norms defined in Chapter I.

By the assumption made on the stochastic kernel $k_0(t, \tau; \omega)$, the integral

$$(\mathscr{L}x)(t; \omega) = \int_0^\infty k_0(t, \tau; \omega) x(\tau; \omega)\, d\tau$$

exists for all $x(t; \omega) \in C_c(R_+, L_2(\Omega, \mathscr{A}, \mathscr{P}))$. Now, define the integral operator $\mathscr{L}_M$, $M = 1, 2, \ldots$, by

$$(\mathscr{L}_M x)(t; \omega) = \int_0^M k_0(t, \tau; \omega) x(\tau; \omega)\, d\tau, \qquad t \in R_+,$$

and hence
$$(\mathscr{L}_M x)(t;\omega) \to (\mathscr{L}x)(t;\omega) \quad \text{as} \quad M \to \infty.$$

We shall show that $\{\mathscr{L}_M\}$ is indeed a sequence of continuous linear operators from $C_c(R_+, L_2(\Omega, \mathscr{A}, \mathscr{P}))$ into itself. Obviously, $\mathscr{L}_M$ is linear, $M = 1, 2, \ldots$. Let $x_m(t;\omega) \to x(t;\omega)$ in $C_c(R_+, L_2(\Omega, \mathscr{A}, \mathscr{P}))$ as $m \to \infty$. Then by definition
$$\|x_m(t;\omega) - x(t;\omega)\|_{L_2(\Omega, \mathscr{A}, \mathscr{P})} \to 0$$
uniformly on every interval $[0, Q]$, $Q > 0$. Therefore for each $n = 1, 2, \ldots$ the semi-norms
$$\|x_m(t;\omega) - x(t;\omega)\|_n = \sup_{0 \leq t \leq n} \|x_m(t;\omega) - x(t;\omega)\|_{L_2(\Omega, \mathscr{A}, \mathscr{P})} \to 0$$
as $m \to \infty$ and $d(x_m, x) \to 0$ as $m \to \infty$. Hence for any $n = 1, 2, \ldots$ we have
$$\|(\mathscr{L}_M x_m)(t;\omega) - (\mathscr{L}_M x)(t;\omega)\|_n$$
$$= \left\| \int_0^M k_0(t, \tau; \omega)[x_m(\tau;\omega) - x(\tau;\omega)] \, d\tau \right\|_n$$
$$= \sup_{0 \leq t \leq n} \left\| \int_0^M k_0(t, \tau; \omega)[x_m(\tau;\omega) - x(\tau;\omega)] \, d\tau \right\|_{L_2(\Omega, \mathscr{A}, \mathscr{P})}$$
$$\leq \sup_{0 \leq t \leq n} \left\{ \int_0^M \|k_0(t, \tau; \omega)\| \cdot \|x_m(\tau;\omega) - x(\tau;\omega)\|_{L_2(\Omega, \mathscr{A}, \mathscr{P})} \, d\tau \right\}.$$

However, for every $\varepsilon > 0$ there exists an N_M such that $m > N_M$ implies that
$$\|x_m(\tau;\omega) - x(\tau;\omega)\|_{L_2(\Omega, \mathscr{A}, \mathscr{P})} < \varepsilon$$
uniformly in $\tau \in [0, M]$. Therefore we have
$$\|(\mathscr{L}_M x_m)(t;\omega) - (\mathscr{L}_M x)(t;\omega)\|_n < \varepsilon \sup_{0 \leq t \leq n} \left\{ \int_0^M \|k_0(t, \tau; \omega)\| \, d\tau \right\}.$$

Since $k_0(t, \tau; \omega)$ is continuous from Δ_1 into $L_2(\Omega, \mathscr{A}, \mathscr{P})$, the norm $\|k_0(t, \tau; \omega)\|$ is bounded on the set
$$\{(t, \tau) : 0 \leq t \leq n, \ 0 \leq \tau \leq M\}$$
by some number K_{nM}. Thus independent of $t \in [0, n]$ for $m > N_M$ we have
$$\|(\mathscr{L}_M x_m)(t;\omega) - (\mathscr{L}_M x)(t;\omega)\|_n < \varepsilon M K_{nM}.$$

Therefore for each M and $\varepsilon_n^* > 0$ there exists an N_M such that $m > N_M$ implies
$$\|(\mathscr{L}_M x_m)(t;\omega) - (\mathscr{L}_M x)(t;\omega)\|_n < \varepsilon_n^*,$$

4.1 EXISTENCE AND UNIQUENESS OF A RANDOM SOLUTION

which means that

$$(\mathscr{L}_M x_m)(t;\omega) \to (\mathscr{L}_M x)(t;\omega)$$

in $L_2(\Omega, \mathscr{A}, \mathscr{P})$-norm uniformly on $[0, n]$. Since $n = 1, 2, \ldots$ is arbitrary, we may take $Q < n$ so that

$$\sup_{0 \leq t \leq Q} \|(\mathscr{L}_M x_m)(t;\omega) - (\mathscr{L}_M x)(t;\omega)\|_{L_2(\Omega, \mathscr{A}, \mathscr{P})}$$

$$\leq \sup_{0 \leq t \leq n} \|(\mathscr{L}_M x_m)(t;\omega) - (\mathscr{L}_M x)(t;\omega)\|_{L_2(\Omega, \mathscr{A}, \mathscr{P})} < \varepsilon_n^*$$

for $m > N_M$, and thus we have uniform convergence on every set $[0, Q]$, $Q > 0$. Hence $(\mathscr{L}_M x_m)(t;\omega)$ converges to $(\mathscr{L}_M x)(t;\omega)$ in the metric d by definition, that is, $\mathscr{L}_M$ is a continuous mapping from the Fréchet space $C_c(R_+, L_2(\Omega, \mathscr{A}, \mathscr{P}))$ into itself for each $M = 1, 2, \ldots$.

Now, applying Theorem 1.1.10 to the sequence of continuous linear operators $\{\mathscr{L}_M\}$, we obtain the fact that $\mathscr{L}$ is a continuous linear operator from $C_c(R_+, L_2(\Omega, \mathscr{A}, \mathscr{P}))$ into itself, completing the proof.

Let $H_1, H_2 \subset H$, where H is the subset of C_c of all functions $x(t;\omega)$ whose inner product in $L_2(\Omega, \mathscr{A}, \mathscr{P})$ is integrable on R_+, as defined in Section 1.2. The norms in H_1 and H_2 are defined by

$$\|x(t;\omega)\|_{H_i} = \left\{ \int_0^\infty \|x(t;\omega)\|_{L_2(\Omega, \mathscr{A}, \mathscr{P})}^2 \, dt \right\}^{\frac{1}{2}}, \quad i = 1, 2.$$

We consider the following stochastic integral equations for $M = 1, 2, \ldots$:

$$x(t;\omega) = h(t;\omega) + \int_0^t k(t, \tau; \omega) f(\tau, x(\tau;\omega)) \, d\tau$$

$$+ \int_0^M k_0(t, \tau; \omega) e(\tau, x(\tau;\omega)) \, d\tau, \tag{4.1.1}$$

for $t \in [0, M] \subset R_+$. Define the operators $\mathscr{K}_M$ and $\mathscr{L}_M$ from the Hilbert space H_2 into Hilbert space H_1 by

$$(\mathscr{K}_M x)(t;\omega) = \int_0^t k(t, \tau; \omega) x(\tau;\omega) \, d\tau$$

and

$$(\mathscr{L}_M x)(t;\omega) = \int_0^M k(t, \tau; \omega) x(\tau;\omega) \, d\tau$$

for $t \in [0, M]$. We now give a lemma which will be used.

Lemma 4.1.2 The integral operator $\mathscr{L}_M$ defined earlier from Hilbert space H_2 into Hilbert space H_1 is a bounded operator if the kernel $k_0(t, \tau; \omega)$ is such that

$$\int_0^\infty \int_0^M |||k_0(t, \tau; \omega)|||^2 \, d\tau \, dt$$

exists and is finite.

PROOF For each $x(t; \omega) \in H_2$ we have

$$\|(\mathscr{L}_M x)(t; \omega)\|_{H_1}^2 = \int_0^\infty \left\| \int_0^M k_0(t, \tau; \omega) x(\tau; \omega) \, d\tau \right\|_{L_2(\Omega, \mathscr{A}, \mathscr{P})}^2 dt$$

$$\leq \int_0^\infty \int_0^M |||k_0(t, \tau; \omega)|||^2 \, d\tau \cdot \int_0^M \|x(\tau; \omega)\|_{L_2(\Omega, \mathscr{A}, \mathscr{P})}^2 \, d\tau \, dt$$

by Schwarz's inequality. But the second inside integral is less than or equal to the norm squared of $x(t; \omega)$ in H_2, and thus we have

$$\|(\mathscr{L}_M x)(t; \omega)\|_{H_1}^2 \leq \int_0^\infty \|x(\tau; \omega)\|_{L_2(\Omega, \mathscr{A}, \mathscr{P})}^2 \, d\tau \cdot \int_0^\infty \int_0^M |||k_0(t, \tau; \omega)|||^2 \, d\tau \, dt$$

$$= \|x(t; \omega)\|_{H_2}^2 \cdot \int_0^\infty \int_0^M |||k_0(t, \tau; \omega)|||^2 \, d\tau \, dt,$$

which is finite by hypothesis. Hence $\mathscr{L}_M$ is bounded.

We shall now prove a theorem with respect to the existence of a random solution of Eq. (4.1.1) for $M \geq 1$. The fixed-point theorem of Krasnosel'skii given in Chapter I and the theory of admissibility are used in the proof.

Theorem 4.1.3 Consider the random integral equation (4.1.1) subject to the following conditions:

(i) H_1 and H_2 are Hilbert spaces stronger than C_c such that the pair (H_2, H_1) is admissible with respect to each of the operators

$$(\mathscr{K}_M x)(t; \omega) = \int_0^t k(t, \tau; \omega) x(\tau; \omega) \, d\tau \quad \text{and}$$

$$(\mathscr{L}_M x)(t; \omega) = \int_0^M k_0(t, \tau; \omega) x(\tau; \omega) \, d\tau,$$

where $t \in [0, M]$, $M \geq 1$, $k(t, \tau; \omega)$, and $k_0(t, \tau; \omega)$ behave as described previously and $\int_0^\infty \int_0^\infty |||k_0(t, \tau; \omega)|||^2 \, d\tau \, dt$ exists and is finite. Further assume that $\mathscr{L}_M$ is completely continuous.

(ii) $x(t; \omega) \to f(t, x(t; \omega))$ is an operator on

$$S = \{x(t; \omega) : x(t; \omega) \in H_1, \quad \|x(t; \omega)\|_{H_1} \leq \rho\}$$

4.1 EXISTENCE AND UNIQUENESS OF A RANDOM SOLUTION

for some $\rho \geq 0$ with values in H_2 satisfying the Lipschitz condition

$$\|f(t, x(t;\omega)) - f(t, y(t;\omega))\|_{H_2} \leq \lambda \|x(t;\omega) - y(t;\omega)\|_{H_1}$$

for $x(t;\omega), y(t;\omega) \in S$, and $\lambda \geq 0$ a constant.

(iii) $x(t;\omega) \to e(t, x(t;\omega))$ is a continuous operator on S with values in H_2 such that $\|e(t, x(t;\omega))\|_{H_2} \leq \gamma$, for $\gamma \geq 0$ a constant.

(iv) $h(t;\omega) \in H_1$.

Then there exists at least one random solution of Eq. (4.1.1), provided that

$$\lambda K_{1M} < 1, \qquad \|h(t;\omega)\|_{H_1} + K_{1M}\|f(t,0)\|_{H_2} + \gamma K_{2M} \leq \rho(1 - \lambda K_{1M})$$

where K_{1M} and K_{2M} are the norms of $\mathscr{K}_M$ and $\mathscr{L}_M$, respectively.

PROOF It is obvious by definition that sets of the type S are closed, bounded, convex sets in a Hilbert space. By definition, H_1 and H_2 are Banach spaces, since they are Hilbert spaces.

Let $x(t;\omega), y(t;\omega) \in S$. Define the operator U_M from S into H_1 by

$$(U_M x)(t;\omega) = h(t;\omega) + \int_0^t k(t,\tau;\omega) f(\tau, x(\tau;\omega)) \, d\tau$$

for $t \in [0, M]$. Taking the norm of both sides of this equation, we have

$$\|(U_M x)(t;\omega)\|_{H_1} = \left\| h(t;\omega) + \int_0^t k(t,\tau;\omega) f(\tau, x(\tau;\omega)) \, d\tau \right\|_{H_1}$$
$$\leq \|h(t;\omega)\|_{H_1} + K_{1M}\|f(t, x(t;\omega))\|_{H_2}, \qquad (4.1.2)$$

since K_{1M} is the norm of the integral operator $\mathscr{K}_M$, and

$$\|(\mathscr{K}_M x)(t;\omega)\|_{H_1} \leq K_{1M}\|x(t;\omega)\|_{H_2}.$$

But using the Lipschitz condition on $f(t, x)$, we find

$$\|f(t, x(t;\omega))\|_{H_2} = \|f(t, x(t;\omega)) - f(t, 0) + f(t, 0)\|_{H_2}$$
$$\leq \lambda \|x(t;\omega)\|_{H_1} + \|f(t, 0)\|_{H_2}. \qquad (4.1.3)$$

Hence Eq. (4.1.2) becomes

$$\|(U_M x)(t;\omega)\|_{H_1} \leq \|h(t;\omega)\|_{H_1} + K_{1M}[\lambda \|x(t;\omega)\|_{H_1} + \|f(t, 0)\|_{H_2}]$$
$$\leq \|h(t;\omega)\|_{H_1} + K_{1M}\|f(t, 0)\|_{H_2} + \gamma K_{2M} + \lambda K_1 \|x(t;\omega)\|_{H_1}$$
$$\leq \rho(1 - \lambda K_{1M}) + \lambda K_{1M} \rho = \rho,$$

from the fact that $\gamma K_{2M} \geq 0$ and the last condition of the theorem. Thus $U_M(S) \subset S$. Since H_1 is a Hilbert space and the difference of elements in a

Hilbert space is in the Hilbert space, $(U_M x)(t;\omega) - (U_M y)(t;\omega) \in H_1$, and

$$\|(U_M x)(t;\omega) - (U_M y)(t;\omega)\|_{H_1}$$

$$= \|h(t;\omega) + \int_0^t k(t,\tau;\omega) f(\tau, x(\tau;\omega)) \, d\tau$$

$$- h(t;\omega) - \int_0^t k(t,\tau;\omega) f(\tau, y(\tau;\omega)) \, d\tau \|_{H_1}$$

$$\leq K_{1M} \| f(t, x(t;\omega)) - f(t, y(t;\omega)) \|_{H_2}$$

$$\leq \lambda K_{1M} \| x(t;\omega) - y(t;\omega) \|_{H_1}.$$

By hypothesis, $\lambda K_{1M} < 1$, so that U_M is a contraction on S.

Define the operator V_M from S into H_1 by

$$(V_M x)(t;\omega) = \int_0^M k_0(t,\tau;\omega) e(\tau, x(\tau;\omega)) \, d\tau, \qquad t \in [0, M].$$

From Lemma 4.1.2, the operator $\mathscr{L}_M$ in Condition (i) of the theorem is a completely continuous bounded operator from H_2 into H_1 since

$$\int_0^\infty \int_0^M \|\!|k_0(t,\tau;\omega)|\!\|^2 \, d\tau \, dt \leq \int_0^\infty \int_0^\infty \|\!|k_0(t,\tau;\omega)|\!\|^2 \, d\tau \, dt < \infty.$$

The function $e(t, x(t;\omega))$ maps elements of Hilbert space H_1 into Hilbert space H_2 and is bounded in H_2 by Condition (iii). Hence e is continuous and bounded from H_1 into H_2. The operator V_M may be expressed as $V_M = \mathscr{L}_M e$. Thus V_M is a completely continuous operator from S into H_1.

We have for $x(t;\omega), y(t;\omega) \in S$

$$\|(U_M x)(t;\omega) + (V_M y)(t;\omega)\|_{H_1}$$

$$= \left\| h(t;\omega) + \int_0^t k(t,\tau;\omega) f(\tau, x(\tau;\omega)) \, d\tau \right.$$

$$\left. + \int_0^M k_0(t,\tau;\omega) e(\tau, y(\tau;\omega)) \, d\tau \right\|_{H_1}$$

$$\leq \|h(t;\omega)\|_{H_1} + K_{1M} \| f(t, x(t;\omega)) \|_{H_2} + K_{2M} \| e(t, y(t;\omega)) \|_{H_2}$$

$$\leq \|h(t;\omega)\|_{H_1} + K_{1M} \lambda \| x(t;\omega) \|_{H_1} + K_{1M} \| f(t,0) \|_{H_2} + K_{2M} \gamma$$

4.1 EXISTENCE AND UNIQUENESS OF A RANDOM SOLUTION

from (4.1.3) and Condition (iii) of the theorem. Then from the last condition of the theorem

$$\|(U_M x)(t;\omega) + (V_M y)(t;\omega)\|_{H_1} \leq \|h(t;\omega)\|_{H_1} + K_{1M}\|f(t,0)\|_{H_2}$$
$$+ K_{2M}\gamma + K_{1M}\lambda\|x(t;\omega)\|_{H_1}$$
$$\leq \rho(1 - \lambda K_{1M}) + K_{1M}\lambda\rho = \rho,$$

and we have that $(U_M x)(t;\omega) + (V_M y)(t;\omega) \in S$.

Therefore the conditions of the fixed-point theorem of Krasnosel'skii (Theorem 1.1.8) hold, and there exists at least one random solution of Eq. (4.1.1) for $M \geq 1$, which completes the proof.

In the case that $\mathscr{L}_M$ is the null operator, that is, $k_0(t, \tau; \omega) = 0$ for all t, $\tau \in R_+$, and almost all $\omega \in \Omega$, then we have for $t \in [0, M]$

$$x(t;\omega) = h(t;\omega) + \int_0^t k(t,\tau;\omega) f(\tau, x(\tau;\omega)) \, d\tau,$$

the random Volterra integral equation of Chapter II.

For the case that $\mathscr{K}_M$ is the null operator we obtain the stochastic integral equation of the Fredholm type

$$x(t;\omega) = h(t;\omega) + \int_0^M k_0(t,\tau;\omega) e(\tau, x(\tau;\omega)) \, d\tau \qquad (4.1.4)$$

for $t \in [0, M]$, $M \geq 1$. Hence we have the following corollary to Theorem 4.1.3.

Corollary 4.1.4 Consider the random integral equation (4.1.4) under the following conditions:

(i) H_1 and H_2 are Hilbert spaces stronger than C_c and the pair (H_2, H_1) is admissible with respect to the integral operator

$$(\mathscr{L}_M x)(t;\omega) = \int_0^M k_0(t,\tau;\omega) x(\tau;\omega) \, d\tau,$$

where $k_0(t, \tau; \omega)$ behaves as described and $\int_0^\infty \int_0^\infty \|\|k_0(t,\tau;\omega)\|\|^2 \, d\tau \, dt$ exists, $t \in [0, M]$, $M = 1, 2, \ldots$. Further assume $\mathscr{L}_M$ is completely continuous.

(ii) $x(t;\omega) \to e(t, x(t;\omega))$ is a continuous operator on

$$S = \{x(t;\omega) : x(t;\omega) \in H_1, \quad \|x(t;\omega)\|_{H_1} \leq \rho\}$$

for some $0 \leq \rho < \infty$ with values in H_2 such that $\|e(t, x(t;\omega))\|_{H_2} \leq \gamma$ for some $\gamma \geq 0$ a constant.

(iii) $h(t;\omega) \in H_1$.

Then there exists at least one bounded random solution of Eq. (4.1.4), provided

$$\|h(t;\omega)\|_{H_1} + \gamma K_{2M} \leq \rho.$$

PROOF   This is a special case of Theorem 4.1.3. When $\mathscr{K}_M$ is the null operator, however, the existence of at least one random solution of (4.1.1) follows from Schauder's fixed-point theorem, keeping in mind the proof of Theorem 4.1.3 (Schauder's fixed-point theorem is a special case of that of Krasnosel'skii).

Now we may note that the integral operators $\mathscr{L}_M$ on Hilbert space H_2 into H_1 converge to the operator $\mathscr{L}$ on H_2 into H_1 defined previously by

$$(\mathscr{L}x)(t;\omega) = \int_0^\infty k_0(t,\tau;\omega)x(\tau;\omega)\,d\tau, \qquad t \in R_+.$$

By a well-known theorem in functional analysis, then, $\mathscr{L}$ is a completely continuous operator from H_2 into H_1 (Bachman and Narici [1, p. 260]) under the same condition as in (i) of Corollary 4.1.4. Hence we have the following theorem.

Theorem 4.1.5   Consider the random integral equation (4.0.1) subject to the following conditions:

(i) H_1 and H_2 are Hilbert spaces stronger than C_c and the pair (H_2, H_1) is admissible with respect to the completely continuous integral operator

$$(\mathscr{L}x)(t;\omega) = \int_0^\infty k_0(t,\tau;\omega)x(\tau;\omega)\,d\tau, \qquad t \in R_+,$$

where $k_0(t,\tau;\omega)$ behaves as described previously and $\int_0^\infty \int_0^\infty \||k_0(t,\tau;\omega)\||^2\,d\tau\,dt$ exists and is finite.

(ii)   Same as Condition (ii) of Corollary 4.1.4.
(iii)   Same as Condition (iii) of Corollary 4.1.4.

Then there exists at least one bounded (by ρ) random solution of Eq. (4.0.1), provided

$$\|h(t;\omega)\|_{H_1} + \gamma L \leq \rho,$$

where L is the norm of the operator $\mathscr{L}$.

The proof of Theorem 4.1.5 follows that of Theorem 4.1.3 with the remark that $\mathscr{L}$ is a completely continuous operator.

We consider now the conditions under which the random equation (4.0.2) has a *unique* random solution. The fixed-point theorem of Banach from Chapter I is utilized in this respect. We could prove uniqueness by adding a Lipschitz condition on $e(t, x(t;\omega))$ in Theorem 4.1.3 and showing that there

4.1 EXISTENCE AND UNIQUENESS OF A RANDOM SOLUTION

is only one random solution, by a contradiction argument. However, by using Banach's theorem, we remove the condition $\|e(t, x(t;\omega))\|_{H_2} \leq \gamma$ and require only that $e(t, 0)$ be bounded in the Banach space B defined in Chapter I. Also, we use Banach spaces and the supremum norms given in Section 1.2 instead of the Hilbert spaces given previously.

Theorem 4.1.6 Suppose the random integral equation (4.0.2) satisfies the following:

(i) B and D are Banach spaces stronger than C_c such that (B, D) is admissible with respect to each of the operators

$$(\mathcal{K}x)(t;\omega) = \int_0^t k(t,\tau;\omega)x(\tau;\omega)\,d\tau,$$

$$(\mathcal{L}x)(t;\omega) = \int_0^\infty k_0(t,\tau;\omega)x(\tau;\omega)\,d\tau, \qquad t \in R_+,$$

where $k(t,\tau;\omega)$ and $k_0(t,\tau;\omega)$ behave as previously.

(ii) $x(t;\omega) \to f(t, x(t;\omega))$ is an operator on

$$S = \{x(t;\omega) : x(t;\omega) \in D, \quad \|x(t;\omega)\|_D \leq \rho\}$$

with values in B, satisfying the Lipschitz condition

$$\|f(t, x(t;\omega)) - f(t, y(t;\omega))\|_B \leq \lambda \|x(t;\omega) - y(t;\omega)\|_D$$

for $x(t;\omega), y(t;\omega) \in S$ and $\lambda \geq 0$ a constant.

(iii) $x(t;\omega) \to e(t, x(t;\omega))$ is an operator on S with values in B satisfying

$$\|e(t, x(t;\omega)) - e(t, y(t;\omega))\|_B \leq \xi \|x(t;\omega) - y(t;\omega)\|_D$$

for $x(t;\omega), y(t;\omega) \in S$ and $\xi \geq 0$ a constant.

(iv) $h(t;\omega) \in D$.

Then there exists a unique random solution of Eq. (4.0.2), provided

$$\lambda K_1 + \xi K_2 < 1,$$

$$\|h(t;\omega)\|_D + K_1 \|f(t, 0)\|_B + K_2 \|e(t, 0)\|_B \leq \rho(1 - \lambda K_1 - \xi K_2),$$

where K_1 and K_2 are the norms of $\mathcal{K}$ and $\mathcal{L}$, respectively.

PROOF The operator $\mathcal{K}$ is continuous from B into D, from the results of Chapter II. The operator $\mathcal{L}$ is continuous from $C_c(R_+, L_2(\Omega, \mathcal{A}, \mathcal{P}))$ into itself as a result of the Lemma 4.1.1 given immediately following the definition of $\mathcal{L}$. Hence $\mathcal{L}$ is bounded from Condition (i), and its norm is K_2.

Define the operators U and V from S into D by

$$(Ux)(t;\omega) = h(t;\omega) + \int_0^t k(t,\tau;\omega) f(\tau, x(\tau;\omega))\, d\tau \quad \text{and}$$

$$(Vx)(t;\omega) = \int_0^\infty k_0(t,\tau;\omega) e(\tau, x(\tau;\omega))\, d\tau, \quad t \in R_+.$$

Since D is a Banach space, $(Ux)(t;\omega) + (Vx)(t;\omega) \in D$ whenever $(Ux)(t;\omega) \in D$ and $(Vx)(t;\omega) \in D$. We must show that $(Ux)(t;\omega) + (Vx)(t;\omega) \in S$ whenever $x(t;\omega) \in S$ (inclusion property) and that $U + V$ is a contracting operator on S.

Consider another element $y(t;\omega) \in S$, and

$$(Uy)(t;\omega) + (Vy)(t;\omega) = h(t;\omega) + \int_0^t k(t,\tau;\omega) f(\tau, y(\tau;\omega))\, d\tau$$
$$+ \int_0^\infty k_0(t,\tau;\omega) e(\tau, y(\tau;\omega))\, d\tau.$$

Then we have

$$\|(Ux)(t;\omega) + (Vx)(t;\omega) - (Uy)(t;\omega) - (Vy)(t;\omega)\|_D$$

$$= \left\| h(t;\omega) + \int_0^t k(t,\tau;\omega) f(\tau, x(\tau;\omega))\, d\tau \right.$$
$$+ \int_0^\infty k_0(t,\tau;\omega) e(\tau, x(\tau;\omega))\, d\tau - h(t;\omega)$$
$$\left. - \int_0^t k(t,\tau;\omega) f(\tau, y(\tau;\omega))\, d\tau - \int_0^\infty k_0(t,\tau;\omega) e(\tau, y(\tau;\omega))\, d\tau \right\|_D$$

$$= \left\| \int_0^t k(t,\tau;\omega)[f(\tau, x(\tau;\omega)) - f(\tau, y(\tau;\omega))]\, d\tau \right.$$
$$\left. + \int_0^\infty k_0(t,\tau;\omega)[e(\tau, x(\tau;\omega)) - e(\tau, y(\tau;\omega))]\, d\tau \right\|_D$$

$$\leq K_1 \|f(t, x(t;\omega)) - f(t, y(t;\omega))\|_B + K_2 \|e(t, x(t;\omega)) - e(t, y(t;\omega))\|_B$$
$$\leq K_1 \lambda \|x(t;\omega) - y(t;\omega)\|_D + K_2 \xi \|x(t;\omega) - y(t;\omega)\|_D$$
$$= (\lambda K_1 + \xi K_2) \|x(t;\omega) - y(t;\omega)\|_D,$$

using the Lipschitz conditions on $f(t, x)$ and $e(t, x)$. Since $\lambda K_1 + \xi K_2 < 1$ by hypothesis, we have that $U + V$ is a contracting operator on S.

4.1 EXISTENCE AND UNIQUENESS OF A RANDOM SOLUTION

To show inclusion, let $x(t;\omega) \in S$. Then

$\|(Ux)(t;\omega) + (Vx)(t;\omega)\|_D$

$= \|h(t;\omega) + \int_0^t k(t,\tau;\omega)f(\tau,x(\tau;\omega))\,d\tau$

$+ \int_0^\infty k_0(t,\tau;\omega)e(\tau,x(\tau;\omega))\,d\tau\|_D$

$\leqslant \|h(t;\omega)\|_D + K_1\|f(t,x(t;\omega))\|_B + K_2\|e(t,x(t;\omega))\|_B.$

Using an inequality similar to (4.1.3) for $f(t,x(t;\omega))$ and a similar inequality for $e(t,x(t;\omega))$, we obtain

$\|(Ux)(t;\omega) + (Vx)(t;\omega)\|_D$

$\leqslant \|h(t;\omega)\|_D + K_1[\lambda\|x(t;\omega)\|_D + \|f(t,0)\|_B] + K_2[\xi\|x(t;\omega)\|_D$

$+ \|e(t,0)\|_B]$

$= \|h(t;\omega)\|_D + K_1\|f(t,0)\|_B + K_2\|e(t,0)\|_B + \|x(t;\omega)\|_D(\lambda K_1 + \xi K_2)$

$\leqslant \rho(1 - \lambda K_1 - \xi K_2) + \rho(\lambda K_1 + \xi K_2) = \rho$

from the last condition of the theorem. Thus $(Ux)(t;\omega) + (Vx)(t;\omega) \in S$.

Therefore, applying Banach's fixed-point theorem, there exists a unique random solution $x(t;\omega) \in S$ of Eq. (4.0.2), completing the proof.

For the case that $\mathscr{K}$ is the null operator we immediately obtain the following corollary to Theorem 4.1.6, which gives conditions under which the random Fredholm integral equation (4.0.1) possesses a *unique* random solution.

Corollary 4.1.7 We consider the random integral equation (4.0.1) subject to the following conditions:

(i) B and D are Banach spaces stronger than C_c such that the pair (B,D) is admissible with respect to the operator

$(\mathscr{L}x)(t;\omega) = \int_0^\infty k_0(t,\tau;\omega)x(\tau;\omega)\,d\tau, \qquad t \in R_+,$

where $k_0(t,\tau;\omega)$ behaves as described previously.

(ii) Same as Condition (iii) of Theorem 4.1.6.
(iii) Same as Condition (iv) of Theorem 4.1.6.

Then there exists a *unique* random solution of Eq. (4.0.1), provided that

$\xi K_2 < 1, \qquad \|h(t;\omega)\|_D + K_2\|e(t,0)\|_B \leqslant \rho(1 - \xi K_2),$

where K_2 is the norm of the operator $\mathscr{L}$.

PROOF A special case of Theorem 4.1.6 with $K_1 = 0$, that is, $\mathcal{K}$ is the null operator.

4.2 Some Special Cases

We now present some useful special cases of the preceding theorems and corollaries by taking C_g or C as the Banach spaces B and D.

Theorem 4.2.1 Consider the stochastic integral equation (4.0.1) subject to the following conditions:

(i) There exists a constant $Z > 0$ and a positive continuous function $g(t)$ finite on R_+ such that

$$\int_0^\infty |||k_0(t, \tau; \omega)||| g(\tau) \, d\tau \leqslant Z, \qquad t \in R_+.$$

(ii) $e(t, x)$ is continuous in $t \in R_+$ and $x \in R$ such that

$$|e(t, 0)| \leqslant \gamma g(t) \quad \text{and} \quad |e(t, x) - e(t, y)| \leqslant \xi g(t)|x - y|$$

for $\|x\|_C, \|y\|_C \leqslant \rho$ and $\gamma \geqslant 0$ and $\xi \geqslant 0$ constants.

(iii) $h(t; \omega) \in C$, the set of continuous bounded functions from R_+ into $L_2(\Omega, \mathscr{A}, \mathscr{P})$.

Then there exists a unique random solution $x(t; \omega) \in C$ of Eq. (4.0.1) such that

$$\|x(t; \omega)\|_C = \sup_{t \geqslant 0} \left\{ \int_\Omega |x(t; \omega)|^2 \, d\mathscr{P}(\omega) \right\}^{\frac{1}{2}} \leqslant \rho,$$

provided that $\|h(t; \omega)\|_C$, ξ, and γ are small enough.

PROOF We must show that under the given conditions the pair (C_g, C) is admissible with respect to the integral operator

$$(\mathscr{L}x)(t; \omega) = \int_0^\infty k_0(t, \tau; \omega) x(\tau; \omega) \, d\tau, \qquad t \in R_+.$$

Let $x(t; \omega) \in C_g$. Then we have

$$\|(\mathscr{L}x)(t; \omega)\|_{L_2(\Omega,\mathscr{A},\mathscr{P})} \leqslant \int_0^\infty \|k_0(t, \tau; \omega) x(\tau; \omega)\|_{L_2(\Omega,\mathscr{A},\mathscr{P})} \, d\tau$$

$$\leqslant \int_0^\infty |||k_0(t, \tau; \omega)||| [\|x(\tau; \omega)\|_{L_2(\Omega,\mathscr{A},\mathscr{P})} / g(\tau)] g(\tau) \, d\tau$$

where $|||k_0(t, \tau; \omega)||| = \|k_0(t, \tau; \omega)\|_{L_\infty(\Omega, \mathscr{A}, \mathscr{P})}$ is a function only of $(t, \tau) \in \Delta_1$. Using the definition of the norm in C_g, we have

$$\|(\mathscr{L}x)(t; \omega)\|_{L_2(\Omega, \mathscr{A}, \mathscr{P})}$$

$$\leq \sup_{t \geq 0} \{\|x(t; \omega)\|_{L_2(\Omega, \mathscr{A}, \mathscr{P})}/g(\tau)\} \left\{ \int_0^\infty |||k_0(t, \tau; \omega)||| g(\tau) \, d\tau \right\}$$

$$= \|x(t; \omega)\|_{C_g} \int_0^\infty |||k_0(t, \tau; \omega)||| g(\tau) \, d\tau$$

$$\leq \|x(t; \omega)\|_{C_g} Z,$$

by Condition (i) of the theorem. Thus $\|(\mathscr{L}x)(t; \omega)\|_{L_2(\Omega, \mathscr{A}, \mathscr{P})}$ is bounded, and $\mathscr{L}$ is a continuous function of x from the proof of Theorem 4.1.6. Hence $(\mathscr{L}x)(t; \omega) \in C$ and (C_g, C) is admissible with respect to $\mathscr{L}$.

From Condition (ii),

$$|e(t, x(t; \omega)) - e(t, y(t; \omega))| \leq \xi g(t) |x(t; \omega) - y(t; \omega)|$$

implies that

$$\left\{ \int_\Omega |e(t, x(t; \omega)) - e(t, y(t; \omega))|^2 \, d\mathscr{P}(\omega) \right\}^{\frac{1}{2}}$$

$$\leq \xi g(t) \left\{ \int_\Omega |x(t; \omega) - y(t; \omega)|^2 \, d\mathscr{P}(\omega) \right\}^{\frac{1}{2}}$$

and hence

$$\sup_{t \geq 0} \{\|e(t, x(t; \omega)) - e(t, y(t; \omega))\|_{L_2(\Omega, \mathscr{A}, \mathscr{P})}/g(t)\}$$

$$\leq \xi \sup_{t \geq 0} \|x(t; \omega) - y(t; \omega)\|_{L_2(\Omega, \mathscr{A}, \mathscr{P})}$$

or

$$\|e(t, x(t; \omega)) - e(t, y(t; \omega))\|_{C_g} \leq \xi \|x(t; \omega) - y(t; \omega)\|_C$$

for $\|x\|_C, \|y\|_C \leq \rho$. Likewise, $|e(t, 0)| \leq \gamma g(t)$ implies that $\|e(t, 0)\|_{C_g} \leq \gamma$.

Therefore Corollary 4.1.7 applies with $B = C_g$ and $D = C$, provided that $\|h(t; \omega)\|_C$, ξ, and γ are small enough in the sense that

$$\xi K_2 < 1, \qquad \|h(t; \omega)\|_C + K_2 \gamma \leq \rho(1 - \xi K_2).$$

Then there exists a unique random solution of Eq. (4.0.1) such that

$$\|x(t; \omega)\|_C \leq \rho,$$

completing the proof.

For $g(t) = 1$ for all $t \in R_+$ in Theorem 4.2.1 we obtain the following corollary.

Corollary 4.2.2 Consider the random integral equation (4.0.1) under the following conditions:

(i) $\int_0^\infty |||k_0(t, \tau; \omega)||| \, d\tau \leq Z, t \in R_+$, where Z is some constant greater than zero.

(ii) $e(t, x)$ is a continuous function from $R_+ \times R$ into R such that

$$|e(t, 0)| \leq \gamma \quad \text{and} \quad |e(t, x) - e(t, y)| \leq \xi |x - y|$$

for $\|x\|_C$ and $\|y\|_C$ less than or equal to $\rho \geq 0$ and $\xi \geq 0$ a constant.

(iii) $h(t; \omega) \in C$.

Then there exists a *unique* bounded (by ρ) random solution $x(t; \omega) \in C$ provided that $\|h(t; \omega)\|_C$, ξ, and γ are sufficiently small.

The following corollary is also a particular case of Theorem 4.2.1.

Corollary 4.2.3 Assume that the random integral equation (4.0.1) satisfies the following conditions:

(i) $|||k_0(t, \tau; \omega)||| \leq \Lambda$ for all $t, \tau \in R_+$ and $\int_0^\infty g(t) \, dt < \infty$.
(ii) Same as Theorem 4.2.1, Condition (ii).
(iii) Same as Theorem 4.2.1, Condition (iii).

Then there exists a unique random solution $x(t; \omega) \in C$ bounded (by ρ) on R_+ provided $\|h(t; \omega)\|_C$, ξ, and γ are small enough.

PROOF   We need only to show that the pair of Banach spaces (C_g, C) is admissible with respect to the integral operator

$$(\mathscr{L}x)(t; \omega) = \int_0^\infty k_0(t, \tau; \omega) x(\tau; \omega) \, d\tau, \qquad t \in R_+,$$

along with Condition (i) of the corollary. For $x(t; \omega) \in C_g$ we have

$$\|(\mathscr{L}x)(t; \omega)\|_{L_2(\Omega, \mathscr{A}, \mathscr{P})} \leq \int_0^\infty |||k_0(t, \tau; \omega)||| \cdot \|x(\tau; \omega)\|_{L_2(\Omega, \mathscr{A}, \mathscr{P})} \, d\tau.$$

From hypothesis (i) of the corollary we obtain

$$\|(\mathscr{L}x)(t; \omega)\|_{L_2(\Omega, \mathscr{A}, \mathscr{P})} \leq \Lambda \int_0^\infty [\|x(\tau; \omega)\|_{L_2(\Omega, \mathscr{A}, \mathscr{P})}/g(\tau)] g(\tau) \, d\tau$$

$$\leq \Lambda \|x(t; \omega)\|_{C_g} \int_0^\infty g(\tau) \, d\tau$$

$$\leq M < \infty$$

for all $t \in R_+$. Thus $(\mathscr{L}x)(t;\omega) \in C$ for $x(t;\omega) \in C_g$, and $\mathscr{L}C_g \subset C$. That is, (C_g, C) is admissible with respect to $\mathscr{L}$. Since the other hypotheses are identical to those of Theorem 4.2.1, the proof is complete.

4.3 Stochastic Asymptotic Stability of the Random Solution

In many practical situations it is of interest to know the behavior of the random solution of the random Fredholm integral equation (4.0.1) for large values of t, which may in some instances represent time. That is, if (4.0.1) describes the behavior of some physical system, it may be of interest to determine the behavior of the system after it has been operating for a long time. We shall now prove a theorem which states that under certain conditions the random solution of Eq. (4.0.1) is *stochastically asymptotically exponentially stable*, which was defined in Chapter I.

Theorem 4.3.1 Suppose that the random integral equation (4.0.1) satisfies the following conditions:

(i) $|||k_0(t, \tau; \omega)||| \leq N \exp(-\alpha t + \beta \tau)$ for $N > 0$ a constant, $\alpha > \beta > 0$, and $t, \tau \in R_+$.

(ii) $e(t, x)$ is defined on $R_+ \times R$ into R, continuous, and

$$|e(t, 0)| \leq \gamma \exp(-\alpha t)$$

on R_+, $\gamma \geq 0$ a constant, and

$$|e(t, x) - e(t, y)| \leq \xi |x - y|,$$

for $\|x\|_{C_g}, \|y\|_{C_g} \leq \rho$ and $\xi \geq 0$ a constant.

(iii) $\|h(t; \omega)\|_{L_2(\Omega, \mathscr{A}, \mathscr{P})} \leq H \exp(-\alpha t)$, $H > 0$, $t \in R_+$.

Then there exists a unique random solution $x(t; \omega)$ of Eq. (4.0.1) satisfying

$$\|x(t; \omega)\|_{L_2(\Omega, \mathscr{A}, \mathscr{P})} \leq \rho \exp(-\alpha t),$$

provided that H, ξ, and γ are sufficiently small.

PROOF We must show that the pair (C_g, C_g) is admissible with respect to the integral operator

$$(\mathscr{L}x)(t; \omega) = \int_0^\infty k_0(t, \tau; \omega) x(\tau; \omega) \, d\tau$$

IV A FREDHOLM TYPE EQUATION AND SOME APPLICATIONS

with $g(t) = \exp(-\alpha t)$, $\alpha > 0$, $t \in R_+$, and the given conditions. For $x(t;\omega) \in C_g$ we have

$$\|(\mathscr{L}x)(t;\omega)\|_{L_2(\Omega,\mathscr{A},\mathscr{P})} \leq \int_0^\infty \||k_0(t,\tau;\omega)|\| \cdot \|x(\tau;\omega)\|_{L_2(\Omega,\mathscr{A},\mathscr{P})}\, d\tau$$

$$= \int_0^\infty \||k_0(t,\tau;\omega)|\|[\|x(\tau;\omega)\|_{L_2(\Omega,\mathscr{A},\mathscr{P})}/\exp(-\alpha\tau)]$$

$$\times \exp(-\alpha\tau)\, d\tau$$

$$\leq \sup_{t\geq 0}\{\|x(t;\omega)\|_{L_2(\Omega,\mathscr{A},\mathscr{P})}/\exp(-\alpha\tau)\}$$

$$\times \int_0^\infty \||k_0(t,\tau;\omega)|\| \exp(-\alpha\tau)\, d\tau$$

$$= \|x(t;\omega)\|_{C_g} \int_0^\infty \||k_0(t,\tau;\omega)|\| \exp(-\alpha\tau)\, d\tau. \quad (4.3.1)$$

But from Condition (i) of the theorem,

$$\int_0^\infty \||k_0(t,\tau;\omega)|\| \exp(-\alpha\tau)\, d\tau \leq N \int_0^\infty \exp(-\alpha t + \beta\tau - \alpha\tau)\, d\tau$$

$$= N(\exp(-\alpha t)) \int_0^\infty \exp[(\beta - \alpha)\tau]\, d\tau$$

$$= [N/(\alpha - \beta)]\exp(-\alpha t). \quad (4.3.2)$$

Thus, combining inequalities (4.3.1) and (4.3.2), we obtain for every $t \in R_+$

$$\|(\mathscr{L}x)(t;\omega)\|_{L_2(\Omega,\mathscr{A},\mathscr{P})} \leq \|x(t;\omega)\|[N/(\alpha - \beta)]\exp(-\alpha t).$$

Therefore, by definition of C_g with $g(t) = \exp(-\alpha t)$, $(\mathscr{L}x)(t;\omega) \in C_g$, and (C_g, C_g) is admissible with respect to $\mathscr{L}$. Hence Condition (i) of Corollary 4.1.7 is satisfied with $B = D = C_g$.

Since we have (as in the proof of Theorem 4.2.1) that Condition (ii) implies

$$\|e(t, x(t;\omega)) - e(t, y(t;\omega))\|_{C_g} \leq \xi \|x(t;\omega) - y(t;\omega)\|_{C_g}$$

and $\|e(t,0)\|_{C_g} \leq \gamma$, and that Condition (iii) implies that $h(t;\omega) \in C_g = D$ for $g(t) = \exp(-\alpha t)$ by definition of C_g, all of the conditions of Corollary 4.1.7 are satisfied.

Therefore, by Corollary 4.1.7, there exists a unique random solution $x(t;\omega) \in C_g$ of Eq. (4.0.1) such that $\|x(t;\omega)\|_{C_g} \leq \rho$, provided that H, ξ, and γ are small enough in the sense that $\xi K_2 < 1$ and

$$\|h(t;\omega)\|_{C_g} + K_2\|e(t,0)\|_{C_g} \leq H + K_2\gamma \leq \rho(1 - \xi K_2).$$

From (4.3.2) the norm of $\mathscr{L}$, by definition, is $K_2 = N/(\alpha - \beta)$. Hence we must have

$$\xi N < \alpha - \beta \quad \text{and} \quad H + [\gamma N/(\alpha - \beta)] \leq \rho\{1 - [N/(\alpha - \beta)]\}.$$

Also, $\|x(t;\omega)\|_{C_g} \leq \rho$ means that

$$\|x(t;\omega)\|_{L_2(\Omega,\mathscr{A},\mathscr{P})} \leq \rho \exp(-\alpha t), \quad t \in R_+,$$

completing the proof.

We therefore have that the unique random solution of (4.0.1) is stochastically asymptotically exponentially stable. In fact, with regard to the asymptotic behavior of the random variable $x(t;\omega)$, we have

$$\|x(t;\omega)\|_{L_2(\Omega,\mathscr{A},\mathscr{P})} = \left\{\int_\Omega |x(t;\omega)|^2 \, d\mathscr{P}(\omega)\right\}^{\frac{1}{2}} \leq \rho \exp(-\alpha t)$$

$$\to 0 \quad \text{as} \quad t \to \infty,$$

that is, $\lim_{t \to \infty} E\{|x(t;\omega)|^2\} = 0$, the second absolute moment of $x(t;\omega)$ approaches zero as $t \to \infty$. Hence from Jensen's inequality we have that the expected value of the absolute solution approaches zero as $t \to \infty$, $\lim_{t \to \infty} E\{|x(t;\omega)|\} = 0$.

4.4 An Application in Stochastic Control Systems

Corduneanu [3], Desoer and Tomasian [1], and Petrovanu [1] considered the stability properties of a nonrandom linear system described by the triple $(E, F; T)$, where E is the space of inputs to the system, F is the space of outputs from the system, and T is a linear operator from E to F given by

$$(Tx)(t) = \int_0^\infty k(t,s)x(s)\,ds, \quad t \geq 0.$$

Here we have altered the function $k(t,s)$ in the paper of Corduneanu [3] to be zero whenever $s < 0$. In this section we shall study a nonlinear stochastic feedback control system for which the random output is given in terms of the random input by the nonlinear operator T defined by

$$(Tx)(t;\omega) = \int_0^\infty k(t,s;\omega)e(s,x(s;\omega))\,ds, \quad t \geq 0, \quad (4.4.1)$$

for $\omega \in \Omega$. In a feedback control system the output, or a fraction of the output, is returned as input to the system. For a stochastic system this fraction may be a random function of t in general. We shall consider the fraction of the

116   IV A FREDHOLM TYPE EQUATION AND SOME APPLICATIONS

random output to be returned as $\eta(t;\omega)(Tx)(t;\omega)$, where $0 \leqslant \eta(t;\omega) \leqslant 1$ for all $t \geqslant 0$ and $\omega \in \Omega$. See Fig. 4.4.1 for a schematic description.

The following differential system with random parameters describes the stochastic feedback control system in Fig. 4.4.1:

$$\dot{x}(t;\omega) = A(t;\omega)x(t;\omega) + \phi(t;\omega) \qquad (4.4.2)$$

with

$$\phi(t;\omega) = \eta(t;\omega)(Tx)(t;\omega), \qquad (4.4.3)$$

where the dot denotes the derivative with respect to t, T is the linear operator given by Eq. (4.4.1), $x(t;\omega)$ is an $n \times 1$ vector whose elements are random variables, $A(t;\omega)$ and $k(t,s;\omega)$ are $n \times n$ matrices whose elements are measurable functions, $\eta(t;\omega)$ is a scalar random variable for each $t \geqslant 0$, and $e(t,x)$ is an $n \times 1$ vector-valued function for each t and x. For $n = 2$ we have complex-valued random functions. By taking as the spaces E and F the space C_g, we shall study the existence of a random solution $x(t;\omega)$ and its stochastic stability properties by applying methods similar to those employed in the previous section and Theorem 4.1.6. Here we consider $n = 1$.

The random differential system (4.4.2)–(4.4.3) may be reduced to a stochastic integral equation of the mixed Volterra–Fredholm type in the form of Eq. (4.0.2). Integrating both sides of Eq. (4.4.2) and substituting the expression for $\phi(t;\omega)$ given by (4.4.3), we obtain

$$\begin{aligned}x(t;\omega) - x(0;\omega) &= \int_0^t A(\tau;\omega)x(\tau;\omega)\, d\tau + \int_0^t \phi(\tau;\omega)\, d\tau \\ &= \int_0^t A(\tau;\omega)x(\tau;\omega)\, d\tau \\ &\quad + \int_0^t \int_0^\infty \eta(\tau;\omega)k(\tau,s;\omega)e(s,x(s;\omega))\, ds\, d\tau.\end{aligned} \qquad (4.4.4)$$

In the second integral on the right-hand side of Eq. (4.4.4) the integral

$$\int_0^\infty k(\tau,s;\omega)e(s,x(s;\omega))\, ds$$

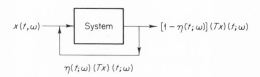

Figure 4.4.1.

4.4 AN APPLICATION IN STOCHASTIC CONTROL SYSTEMS

exists and is finite for each τ and ω; otherwise, the output of the system is infinite. Also, if for each $t \geq 0$ and $s \geq 0$, $\int_0^t \eta(\tau;\omega)k(\tau,s;\omega)\,d\tau$ exists and is finite, that is, if $\int_0^t k(\tau,s;\omega)\,d\tau$ exists and is finite, since $0 \leq \eta(t;\omega) \leq 1$, then we may interchange the order of integration by Fubini's theorem (Hewitt and Stromberg [1]) to obtain

$$\int_0^\infty \left[\int_0^t \eta(\tau;\omega)k(\tau,s;\omega)\,d\tau \right] e(s, x(s;\omega))\,ds.$$

Then Eq. (4.4.4) may be written as

$$x(t;\omega) = \int_0^t A(\tau;\omega)x(\tau;\omega)\,d\tau + \int_0^\infty k^*(t,\tau;\omega)e(\tau, x(\tau;\omega))\,d\tau, \quad (4.4.5)$$

where $x(0;\omega) = 0$ and

$$k^*(t,\tau;\omega) = \int_0^t \eta(u;\omega)k(u,\tau;\omega)\,du, \quad t,\tau \in R_+.$$

The following theorem gives conditions under which a unique random solution of Eq. (4.4.5) exists and has the property of stochastic asymptotic exponential stability.

Theorem 4.4.1 Suppose that the random equation (4.4.5) satisfies the following conditions:

(i) $|||A(\tau;\omega)||| \leq N_1 \exp(-\alpha t + \delta t)$ for $N_1 > 0$ a constant, $\alpha > \delta > 0$, and $0 \leq \tau \leq t < \infty$.

(ii) $|||k^*(t,\tau;\omega)||| \leq N_2 \exp(-\alpha t + \beta \tau)$ for $N_2 > 0$ a constant, $\alpha > \beta > 0$, and $t, \tau \in R_+$.

(iii) $e(t, x(t;\omega))$ is such that $e(t, 0) \in C_g$, is continuous in t uniformly in x, and satisfies

$$|e(t, x) - e(t, y)| \leq \xi |x - y|$$

for $\|x(t;\omega)\|_{C_g}, \|y(t;\omega)\|_{C_g} \leq \rho$, and ξ a constant.

Then there exists a unique random solution of Eq. (4.4.5) satisfying

$$\{E|x(t;\omega)|^2\}^{\frac{1}{2}} \leq \rho \exp(-\alpha t), \quad t \geq 0,$$

provided that ξ and $|e(t, 0)|$ are sufficiently small.

PROOF We must show that the pair of spaces (C_g, C_g) is admissible with respect to the integral operators

$$(\mathcal{K}x)(t;\omega) = \int_0^t A(\tau;\omega)x(\tau;\omega)\,d\tau$$

and
$$(\mathscr{L}x)(t;\omega) = \int_0^\infty k^*(t,\tau;\omega)x(\tau;\omega)\,d\tau, \qquad t \geq 0,$$

with $g(t) = \exp(-\alpha t)$ and Conditions (i) and (ii).

For $x(t;\omega)$ in C_g we have

$\|(\mathscr{L}x)(t;\omega)\|_{L_2(\Omega,\mathscr{A},\mathscr{P})}$

$$\leq \int_0^\infty \|\!|k^*(t,\tau;\omega)|\!\|[\|x(\tau;\omega)\|_{L_2(\Omega,\mathscr{A},\mathscr{P})}/\exp(-\alpha\tau)]\exp(-\alpha\tau)\,d\tau$$

$$\leq \sup_{t \geq 0}\{\|x(t;\omega)\|_{L_2(\Omega,\mathscr{A},\mathscr{P})}/\exp(-\alpha t)\}\int_0^\infty \|\!|k^*(t,\tau;\omega)|\!\|\exp(-\alpha\tau)\,d\tau.$$

But from Condition (ii) of the theorem and the definition of the norm in C_g, we have

$$\|(\mathscr{L}x)(t;\omega)\|_{L_2(\Omega,\mathscr{A},\mathscr{P})} \leq \|x(t;\omega)\|_{C_g} N_2 \int_0^\infty \exp(-\alpha t + \beta\tau - \alpha\tau)\,d\tau$$

$$= \|x(t;\omega)\|_{C_g} N_2 \exp(-\alpha t)\int_0^\infty \exp[-(\alpha-\beta)\tau]\,d\tau$$

$$= \|x(t;\omega)\|_{C_g}[N_2/(\alpha-\beta)]\exp(-\alpha t) < \infty, \qquad t \geq 0,$$

since $\alpha > \beta$. Thus, by definition of C_g, where $g(t) = \exp(-\alpha t)$, $(\mathscr{L}x)(t;\omega) \in C_g$ for all $x(t;\omega) \in C_g$, and (C_g, C_g) is admissible with respect to $\mathscr{L}$. Likewise,

$\|(\mathscr{K}x)(t;\omega)\|_{L_2(\Omega,\mathscr{A},\mathscr{P})}$

$$\leq \int_0^t \|\!|A(\tau;\omega)|\!\|[\|x(\tau;\omega)\|_{L_2(\Omega,\mathscr{A},\mathscr{P})}/\exp(-\alpha\tau)]\exp(-\alpha\tau)\,d\tau$$

$$\leq \|x(t;\omega)\|_{C_g} N_1 \int_0^t \exp(-\alpha t + \delta\tau - \alpha\tau)\,d\tau$$

$$= \|x(t;\omega)\|_{C_g} N_1 \exp(-\alpha t)\{1 - \exp[-(\alpha-\delta)t]\}/(\alpha-\delta)$$

$$\leq \|x(t;\omega)\|_{C_g}[N_1/(\alpha-\delta)]\exp(-\alpha t) < \infty, \qquad t \geq 0,$$

from Condition (i) and the definition of the norm in C_g. Hence $(\mathscr{K}x)(t;\omega) \in C_g$ whenever $x(t;\omega) \in C_g$, and the pair (C_g, C_g) is admissible with respect to $\mathscr{K}$.

Since the function $f(t,x)$ in Eq. (4.0.2) is the identity function in x in Eq. (4.4.5), the constant λ in Theorem 4.1.6 is equal to one. From Condition (iii) we have, as before, that

$$\|e(t, x(t;\omega)) - e(t, y(t;\omega))\|_{C_g} \leq \xi\|x(t;\omega) - y(t;\omega)\|_{C_g}.$$

4.4 AN APPLICATION IN STOCHASTIC CONTROL SYSTEMS

Since the stochastic free term is identically zero, all of the conditions of Theorem 4.1.6 are satisfied for $B = D = C_g$, and it follows that there exists a unique random solution of Eq. (4.4.5) in the set

$$S = \{x(t;\omega) : x(t;\omega) \in C_g, \quad \|x(t;\omega)\|_{C_g} \leq \rho\}$$

for some $\rho > 0$, provided that ξ and $|e(t,0)|$ are small enough. Hence the random solution satisfies

$$\|x(t;\omega)\|_{L_2(\Omega,\mathscr{A},\mathscr{P})} = \{E|x(t;\omega)|^2\}^{\frac{1}{2}} \leq \rho \exp(-\alpha t), \quad t \geq 0,$$

by the definition of the space C_g. The constants ξ and $|e(t,0)|$ must be small enough in the sense that

$$K_1 + \xi K_2 < 1 \quad \text{and} \quad K_2 \|e(t,0)\|_{C_g} \leq \rho(1 - K_1 - \xi K_2),$$

where K_1 and K_2 are the norms of the operators $\mathscr{H}$ and $\mathscr{L}$, respectively. From the preceding results we see that

$$K_1 = N_1^*/(\alpha - \delta) \quad \text{and} \quad K_2 = N_2^*/(\alpha - \beta),$$

where N_1^* and N_2^* are the greatest lower bounds of the constants N_1 and N_2 which satisfy Conditions (i) and (ii), respectively, and the given inequalities. Therefore we must have

$$\frac{N_1^*}{\alpha - \delta} + \frac{\xi N_2^*}{\alpha - \beta} < 1, \quad \frac{N_2^*}{\alpha - \beta}\|e(t,0)\|_{C_g} \leq \rho\left(1 - \frac{N_1^*}{\alpha - \delta} - \frac{\xi N_2^*}{\alpha - \beta}\right),$$

completing the proof.

Therefore, if the conditions of Theorem 4.4.1 hold, then the unique random solution of the system (4.4.2)–(4.4.3) satisfies

$$E[|x(t;\omega)|] \to 0$$

as $t \to \infty$.

We remark that this is a very general stochastic control system because of the generality of the stochastic kernel, the nonlinear operator T, and the functions $\eta(t;\omega)$ and $A(t;\omega)$. The operator T as given says that the system output is a function of both past and future input, which may seem a bit unrealistic at first glance. However, this operator contains all other operators of the Volterra and Fredholm types for compact or noncompact intervals in R_+, and the results obtained here have wide applicability in stochastic control systems.

4.5 A Random Perturbed Fredholm Integral Equation

We will consider in this section the existence and uniqueness of a random solution of the stochastic integral equation of the Fredholm type of the form

$$x(t;\omega) = h(t, x(t;\omega)) + \int_0^\infty k_0(t, \tau;\omega)e(\tau, x(\tau;\omega))\, d\tau \qquad (4.5.1)$$

for $t \geq 0$; $k_0(t, \tau;\omega)$ and $e(t, x(t;\omega))$ behave as described in Chapter I and in Section 4.1. Equation (4.5.1) was recently studied by Milton, Padgett, and Tsokos [1]. The function $h(t, x(t;\omega))$ is a function from R_+ into $L_2(\Omega, \mathscr{A}, \mathscr{P})$ under certain convenient conditions, and $h(t, x)$ is a scalar function of $t \geq 0$ and scalar x.

To obtain the existence and uniqueness of a random solution of (4.5.1), we shall first consider the existence and uniqueness of a random solution of the mixed Volterra–Fredholm type equation of the form

$$x(t;\omega) = h(t, x(t;\omega)) + \int_0^t k(t, \tau;\omega) f(\tau, x(\tau;\omega))\, d\tau$$

$$+ \int_0^\infty k_0(t, \tau;\omega)e(\tau, x(\tau;\omega))\, d\tau, \qquad t \geq 0, \qquad (4.5.2)$$

where in addition to the given conditions we have that the functions $k(t, \tau;\omega)$ and $f(t, x(t;\omega))$ behave as described in Chapters I and II. Note that Eq. (4.5.1) is a special case of Eq. (4.5.2); that is, when $k(t, \tau;\omega)$ is identically equal to zero for all t and τ in R_+ and almost all $\omega \in \Omega$, then Eq. (4.5.2) reduces to (4.5.1). Also note that when $k_0(t, \tau;\omega)$ is equal to zero for all $t, \tau \in R_+$ and almost all $\omega \in \Omega$ we obtain the random, nonlinear, perturbed Volterra equation studied recently by Milton and Tsokos [5].

Let $H \subset C_c$ be the space of all functions such that the inner product of elements of H are integrable on R_+ and let H_1 and H_2 be Hilbert spaces contained in H with norms defined by

$$\|x(t;\omega)\|_{H_i} = \left\{ \int_0^\infty \|x(t;\omega)\|^2_{L_2(\Omega, \mathscr{A}, \mathscr{P})}\, dt \right\}^{\frac{1}{2}}, \qquad i = 1, 2.$$

We shall consider the following stochastic integral equation for $M = 1, 2, \ldots$:

$$x(t;\omega) = h(t, x(t;\omega)) + \int_0^t k(t, \tau;\omega) f(\tau, x(\tau;\omega))\, d\tau$$

$$+ \int_0^M k_0(t, \tau;\omega)e(\tau, x(\tau;\omega))\, d\tau \qquad (4.5.3)$$

4.5 A RANDOM PERTURBED FREDHOLM INTEGRAL EQUATION

for $t \in [0, M] \subset R_+$. Define the operators T_M and W_M from the Hilbert space H_2 into Hilbert space H_1 by

$$(T_M x)(t;\omega) = \int_0^t k(t,\tau;\omega)x(\tau;\omega)\,d\tau \quad \text{and}$$

$$(W_M x)(t;\omega) = \int_0^M k_0(t,\tau;\omega)x(\tau;\omega)\,d\tau$$

for $t \in [0, M]$.

We will now prove a theorem with respect to the existence of a random solution of the stochastic integral equation (4.5.3).

Theorem 4.5.1 Consider the random integral equation (4.5.3) subject to the following conditions:

(i) H_1 and H_2 are Hilbert spaces stronger than C_c such that the pair (H_2, H_1) is admissible with respect to each of the operators

$$(T_M x)(t;\omega) = \int_0^t k(t,\tau;\omega)x(\tau;\omega)\,d\tau \quad \text{and}$$

$$(W_M x)(t;\omega) = \int_0^M k_0(t,\tau;\omega)x(\tau;\omega)\,d\tau,$$

where $t \in [0, M]$, $M = 1, 2, \ldots$; $k(t,\tau;\omega)$ and $k_0(t,\tau;\omega)$ behave as described previously; and $\int_0^\infty \int_0^\infty \|\!|k_0(t,\tau;\omega)|\!\|^2\,d\tau\,dt$ exists and is finite. Further assume that W_M is completely continuous.

(ii) $x(t;\omega) \to f(t, x(t;\omega))$ is an operator on

$$S = \{x(t;\omega) : x(t;\omega) \in H_1, \quad \|x(t;\omega)\|_{H_1} \leq \rho\}$$

for some $\rho \geq 0$, with values in H_2 satisfying

$$\|f(t, x(t;\omega)) - f(t, y(t;\omega))\|_{H_2} \leq \lambda \|x(t;\omega) - y(t;\omega)\|_{H_1}$$

for $x(t;\omega), y(t;\omega) \in S$ and $\lambda \geq 0$ a constant.

(iii) $x(t;\omega) \to e(t, x(t;\omega))$ is a continuous operator on S with values in H_2 such that $\|e(t, x(t;\omega))\|_{H_2} \leq \gamma$ for $\gamma \geq 0$ a constant.

(iv) $x(t;\omega) \to h(t, x(t;\omega))$ is an operator on S with values in H_1 such that

$$\|h(t, x(t;\omega)) - h(t, y(t;\omega))\|_{H_1} \leq \lambda_1 \|x(t;\omega) - y(t;\omega)\|_{H_1}$$

for some constant $\lambda_1 \geq 0$.

Then there exists at least one random solution of Eq. (4.5.3) provided that

$$\lambda_1 + \lambda K_{1M} < 1,$$

$$\|h(t, x(t;\omega))\|_{H_1} + K_{1M}\|f(t,0)\|_{H_2} + \lambda K_{2M} \leq \rho(1 - \lambda K_{1M}),$$

where K_{1M} and K_{2M} are the norms of T_M and W_M, respectively.

PROOF The set S is closed, bounded, and convex in H_1. By definition, H_1 and H_2 are Banach spaces since they are Hilbert spaces.

Let $x(t;\omega), y(t;\omega) \in S$. Define the operator U_M from S into H_1 by

$$(U_M x)(t;\omega) = h(t, x(t;\omega)) + \int_0^t k(t,\tau;\omega) f(\tau, x(\tau;\omega))\, d\tau$$

for $t \in [0, M]$. Taking the norm of each side of this equation we have

$$\|(U_M x)(t;\omega)\|_{H_1}$$

$$= \left\| h(t, x(t;\omega)) + \int_0^t k(t,\tau;\omega) f(\tau, x(\tau;\omega))\, d\tau \right\|_{H_1}$$

$$\leq \|h(t, x(t;\omega))\|_{H_1} + \left\| \int_0^t k(t,\tau;\omega) f(\tau, x(\tau;\omega))\, d\tau \right\|_{H_1}$$

$$\leq \|h(t, x(t;\omega))\|_{H_1} + K_{1M}\|f(t, x(t;\omega))\|_{H_2} \quad (4.5.4)$$

since K_{1M} is the norm of the operator T_M and

$$\|(T_M Z)(t;\omega)\|_{H_1} \leq K_{1M}\|Z(t;\omega)\|_{H_2}$$

for $Z(t;\omega) \in H_2$ due to Lemma 2.1.1 and the admissibility of the pair (H_2, H_1). However, using Condition (ii), we have

$$\|f(t, x(t;\omega))\|_{H_2} = \|f(t, x(t;\omega)) - f(t,0) + f(t,0)\|_{H_2}$$

$$\leq \lambda\|x(t;\omega)\|_{H_1} + \|f(t,0)\|_{H_2}. \quad (4.5.5)$$

Hence (4.5.4) becomes

$$\|(U_M x)(t;\omega)\|_{H_1} \leq \|h(t, x(t;\omega))\|_{H_1} + K_{1M}\lambda\|x(t;\omega)\|_{H_1} + K_{1M}\|f(t,0)\|_{H_2}$$

$$\leq \|h(t, x(t;\omega))\|_{H_1} + K_{1M}\|f(t,0)\|_{H_2} + \lambda K_{2M}$$

$$+ \lambda K_{1M}\|x(t;\omega)\|_{H_1}$$

$$\leq \rho(1 - \lambda K_{1M}) + \lambda K_{1M}\rho = \rho$$

by the last condition of the theorem and the fact that $x(t;\omega) \in S$. Thus $U_M(S) \subset S$. Since the difference of elements in a Hilbert space is in the Hilbert space,

$$(U_M x)(t;\omega) - (U_M y)(t;\omega) \in H_1$$

4.5 A RANDOM PERTURBED FREDHOLM INTEGRAL EQUATION

and

$$\|(U_M x)(t;\omega) - (U_M y)(t;\omega)\|_{H_1}$$

$$= \left\| h(t, x(t;\omega)) + \int_0^t k(t,\tau;\omega) f(\tau, x(\tau;\omega)) \, d\tau \right.$$

$$\left. - h(t, y(t;\omega)) - \int_0^t k(t,\tau;\omega) f(\tau, y(\tau;\omega)) \, d\tau \right\|_{H_1}$$

$$= \left\| h(t, x(t;\omega)) - h(t, y(t;\omega)) + \int_0^t k(t,\tau;\omega) [f(\tau, x(\tau;\omega)) - f(\tau, y(\tau;\omega))] \, d\tau \right\|_{H_1}$$

$$\leq \|h(t, x(t;\omega)) - h(t, y(t;\omega))\|_{H_1}$$

$$+ \left\| \int_0^t k(t,\tau;\omega)[f(\tau, x(\tau;\omega)) - f(\tau, y(\tau;\omega))] \, d\tau \right\|_{H_1}$$

$$\leq \|h(t, x(t;\omega)) - h(t, y(t;\omega))\|_{H_1} + K_{1M} \|f(t, x(t;\omega)) - f(t, y(t;\omega))\|_{H_2}.$$

Using the Lipschitz conditions given in (ii) and (iv), we have that

$$\|(U_M x)(t;\omega) - (U_M y)(t;\omega)\|_{H_1} \leq \lambda_1 \|x(t;\omega) - y(t;\omega)\|_{H_1}$$

$$+ \lambda K_{1M} \|x(t;\omega) - y(t;\omega)\|_{H_1}$$

$$= (\lambda_1 + \lambda K_{1M}) \|x(t;\omega) - y(t;\omega)\|_{H_1}.$$

Using the condition of the theorem that $\lambda_1 + \lambda K_{1M} < 1$, we can state that U_M is a contraction mapping on S.

Now define the operator V_M from S into H_1 by

$$(V_M x)(t;\omega) = \int_0^M k_0(t,\tau;\omega) e(\tau, x(\tau;\omega)) \, d\tau, \qquad t \in [0, M].$$

From Condition (i) of the theorem and Lemma 4.1.2 the operator W_M is a completely continuous bounded operator from H_2 into H_1. The function $e(t, x(t;\omega))$ is a continuous and bounded mapping from S into H_2 by Condition (iii). We may express the operator V_M as the composite $W_M e$, and therefore V_M is a completely continuous operator from S into H_1.

We have for $x(t;\omega)$, $y(t;\omega) \in S$

$$\|(U_M x)(t;\omega) + (V_M y)(t;\omega)\|_{H_1}$$
$$= \left\| h(t, x(t;\omega)) + \int_0^t k(t,\tau;\omega) f(\tau, x(\tau;\omega)) \, d\tau \right.$$
$$\left. + \int_0^M k_0(t,\tau;\omega) e(\tau, y(\tau;\omega)) \, d\tau \right\|_{H_1}$$
$$\leq \|h(t, x(t;\omega))\|_{H_1} + K_{1M}\|f(t, x(t;\omega))\|_{H_2} + K_{2M}\|e(t, y(t;\omega))\|_{H_2}$$
$$\leq \|h(t, x(t;\omega))\|_{H_1} + K_{1M}\lambda\|x(t;\omega)\|_{H_1} + K_{1M}\|f(t,0)\|_{H_2} + K_{2M}\gamma$$

from (4.5.5) and Condition (iii). Then from the last condition of the theorem

$$\|(U_M x)(t;\omega) + (V_M y)(t;\omega)\|_{H_1} \leq K_{1M}\lambda\|x(t;\omega)\|_{H_1} + \rho(1 - \lambda K_{1M})$$
$$\leq K_{1M}\lambda\rho + \rho - K_{1M}\lambda\rho = \rho$$

and we have that

$$(U_M x)(t;\omega) + (V_M y)(t;\omega) \in S.$$

Therefore, the conditions of the fixed-point theorem of Krasnosel'skii hold, and there exists at least one random solution of Eq. (4.5.3) for $M = 1, 2, \ldots$, which completes the proof.

In the case that T_M is the null operator, we obtain the perturbed stochastic integral equation of the Fredholm type

$$x(t;\omega) = h(t, x(t;\omega)) + \int_0^M k_0(t,\tau;\omega) e(\tau, x(\tau;\omega)) \, d\tau \qquad (4.5.6)$$

for $t \in [0, M]$, $M = 1, 2, \ldots$. Hence we have the following corollary to Theorem 4.5.1.

Corollary 4.5.2 Consider the random integral equation (4.5.6) under the following conditions:

(i) H_1 and H_2 are Hilbert spaces stronger than C_c and the pair (H_2, H_1) is admissible with respect to the integral operator

$$(W_M x)(t;\omega) = \int_0^M k_0(t,\tau;\omega) x(\tau;\omega) \, d\tau,$$

where $k_0(t,\tau;\omega)$ behaves as described previously and

$$\int_0^\infty \int_0^\infty \|\|k_0(t,\tau;\omega)\|\|^2 \, d\tau \, dt$$

exists and is finite, $t \in [0, M]$, $M = 1, 2, \ldots$. Also assume W_M is completely continuous.

4.5 A RANDOM PERTURBED FREDHOLM INTEGRAL EQUATION

(ii) $x(t;\omega) \to e(t, x(t;\omega))$ is a continuous operator on

$$S = \{x(t;\omega) : x(t;\omega) \in H_1, \quad \|x(t;\omega)\|_{H_1} \leq \rho\}$$

for some $\rho \geq 0$ with values in H_2 such that $\|e(t, x(t;\omega))\|_{H_2} \leq \gamma$ for some $\gamma \geq 0$ a constant.

(iii) $x(t;\omega) \to h(t, x(t;\omega))$ is a contraction on S.

Then there exists at least one bounded random solution of Eq. (4.5.6) provided that $\|h(t, x(t;\omega))\|_{H_1} + \gamma K_{2M} \leq \rho$.

PROOF All that is required is to show that under the given conditions $(U_M x)(t;\omega)$ is a contraction on S and that $(U_M x)(t;\omega) + (V_M y)(t;\omega) \in S$ whenever $x(t;\omega)$ and $y(t;\omega) \in S$. By definition

$$(U_M x)(t;\omega) = h(t, x(t;\omega)) + \int_0^t k(t, \tau;\omega) f(\tau, x(\tau;\omega)) \, d\tau.$$

Under the assumption that T_M is the null operator, $(U_M x)(t;\omega) = h(t, x(t;\omega))$. Hence by Condition (iii) U_M is a contraction on S. The proof that under the last restriction of the theorem $(U_M x)(t;\omega) + (V_M y)(t;\omega) \in S$ is analogous to that given in Theorem 4.5.1.

As $M \to \infty$, the sequence of integral operators W_M on the Hilbert space H_2 into H_1 converges to the operator W defined by

$$(Wx)(t;\omega) = \int_0^\infty k_0(t, \tau;\omega) x(\tau;\omega) \, d\tau$$

and W is a completely continuous operator from H_2 into H_1, under the same condition as in (i) of Corollary 4.5.2. Hence we have the following theorem.

Theorem 4.5.3 Consider the random integral equation (4.5.1) subject to the following conditions:

(i) H_1 and H_2 are Hilbert spaces stronger than C_c and the pair (H_2, H_1) is admissible with respect to the completely continuous integral operator

$$(Wx)(t;\omega) = \int_0^\infty k_0(t, \tau;\omega) x(\tau;\omega) \, d\tau, \qquad t \in R_+,$$

where $k_0(t, \tau;\omega)$ behaves as described previously and

$$\int_0^\infty \int_0^\infty \|\|k_0(t, \tau;\omega)\|\|^2 \, d\tau \, dt$$

exists and is finite.

(ii) Same as Condition (ii) of Corollary 4.5.2.
(iii) Same as Condition (iii) of Corollary 4.5.2. ∎

Then there exists at least one bounded (by ρ) random solution of Eq. (4.5.1) provided

$$\|h(t, x(t;\omega))\|_{H_1} + \gamma K \leqslant \rho,$$

where K is the norm of the operator W.

PROOF The proof follows the lines of that of Theorem 4.5.1.

We now consider the conditions under which the random integral equation (4.5.2) possesses a *unique* random solution. We shall use Banach's fixed-point theorem, which requires only the boundedness of $e(t, 0)$ in the space B and not that $\|e(t, x(t;\omega))\|_{H_2} \leqslant \gamma$. Also, we use the Banach spaces B and D and not the Hilbert spaces H_1 and H_2.

Theorem 4.5.4 Suppose the random integral equation (4.5.2) satisfies the following:

(i) B and D are Banach spaces stronger than C_c such that (B, D) is admissible with respect to each of the operators

$$(Tx)(t;\omega) = \int_0^t k(t, \tau;\omega) x(\tau;\omega)\, d\tau \qquad \text{and}$$

$$(Wx)(t;\omega) = \int_0^\infty k_0(t, \tau;\omega) x(\tau;\omega)\, d\tau, \qquad t \in R_+,$$

where $k(t, \tau;\omega)$ and $k_0(t, \tau;\omega)$ behave as before.

(ii) $x(t;\omega) \to f(t, x(t;\omega))$ is an operator on

$$S = \{x(t;\omega) : x(t;\omega) \in D, \quad \|x(t;\omega)\|_D \leqslant \rho\}$$

with values in B satisfying

$$\|f(t, x(t;\omega)) - f(t, y(t;\omega))\|_B \leqslant \lambda \|x(t;\omega) - y(t;\omega)\|_D$$

for $x(t;\omega), y(t;\omega) \in S$ and $\lambda \geqslant 0$ a constant.

(iii) $x(t;\omega) \to e(t, x(t;\omega))$ is an operator on S with values in B satisfying

$$\|e(t, x(t;\omega)) - e(t, y(t;\omega))\|_B \leqslant \xi \|x(t;\omega) - y(t;\omega)\|_D$$

for $x(t;\omega), y(t;\omega) \in S$ and $\xi \geqslant 0$ a constant.

(iv) $x(t;\omega) \to h(t, x(t;\omega))$ is an operator on S with values in D satisfying

$$\|h(t, x(t;\omega)) - h(t, y(t;\omega))\|_D \leqslant \gamma \|x(t;\omega) - y(t;\omega)\|_D$$

for $x(t;\omega), y(t;\omega) \in S$ and $\gamma \geqslant 0$ a constant.

Then there exists a unique random solution of Eq. (4.5.2) provided

$$K_1 \lambda + K_2 \xi + \gamma < 1$$

4.5 A RANDOM PERTURBED FREDHOLM INTEGRAL EQUATION 127

and

$$\|h(t, x(t;\omega))\|_D + K_1\|f(t, 0)\|_B + K_2\|e(t, 0)\|_B \leq \rho(1 - \lambda K_1 - \xi K_2),$$

where K_1 and K_2 are the norms of T and W, respectively.

PROOF The operators T and W are continuous operators from B into D by the continuity assumption on $k(t, \tau; \omega)$ and the continuity and integrability assumptions on $k_0(t, \tau; \omega)$ and Lemma 2.1.1. Therefore the operators T and W are bounded and their norms are finite.

Let us define the operators U and V from S into D by

$$(Ux)(t;\omega) = h(t, x(t;\omega)) + \int_0^t k(t, \tau; \omega) f(\tau, x(\tau;\omega)) \, d\tau \quad \text{and}$$

$$(Vx)(t;\omega) = \int_0^\infty k_0(t, \tau; \omega) e(\tau, x(\tau;\omega)) \, d\tau, \qquad t \geq 0.$$

Since D is a Banach space, $(Ux)(t;\omega) + (Vx)(t;\omega) \in D$ whenever $(Ux)(t;\omega)$, $(Vx)(t;\omega) \in D$. We must show that $(Ux)(t;\omega) + (Vx)(t;\omega) \in S$ whenever $x(t;\omega) \in S$ (inclusion property) and that $U + V$ is a contracting operator on S. Consider another element $y(t;\omega) \in S$. Then

$$(Uy)(t;\omega) + (Vy)(t;\omega) = h(t, y(t;\omega)) + \int_0^t k(t, \tau; \omega) f(\tau, y(\tau;\omega)) \, d\tau$$

$$+ \int_0^\infty k_0(t, \tau; \omega) e(\tau, y(\tau;\omega)) \, d\tau.$$

Then we have

$$\|(Ux)(t;\omega) + (Vx)(t;\omega) - (Uy)(t;\omega) - (Vy)(t;\omega)\|_D$$

$$= \left\| h(t, x(t;\omega)) + \int_0^t k(t, \tau; \omega) f(\tau, x(\tau;\omega)) \, d\tau \right.$$

$$+ \int_0^\infty k_0(t, \tau; \omega) e(\tau, x(\tau;\omega)) \, d\tau$$

$$- h(t, y(t;\omega)) - \int_0^t k(t, \tau; \omega) f(\tau, y(\tau;\omega)) \, d\tau$$

$$\left. - \int_0^\infty k_0(t, \tau; \omega) e(\tau, y(\tau;\omega)) \, d\tau \right\|_D.$$

128   IV   A FREDHOLM TYPE EQUATION AND SOME APPLICATIONS

Now using the fact that T and W are bounded operators from B to D, we can write

$$\|(Ux)(t;\omega) + (Vx)(t;\omega) - (Uy)(t;\omega) - (Vy)(t;\omega)\|_D$$
$$\leq K_1 \|f(t, x(t;\omega)) - f(t, y(t;\omega))\|_B + K_2 \|e(t, x(t;\omega)) - e(t, y(t;\omega))\|_B$$
$$+ \|h(t, x(t;\omega)) - h(t, y(t;\omega))\|_D.$$

Now applying the Lipschitz conditions given in (ii)–(iv), we have

$$\|(Ux)(t;\omega) + (Vx)(t;\omega) - (Uy)(t;\omega) - (Vy)(t;\omega)\|_D$$
$$\leq K_1 \lambda \|x(t;\omega) - y(t;\omega)\|_D + K_2 \xi \|x(t;\omega) - y(t;\omega)\|_D$$
$$+ \gamma \|x(t;\omega) - y(t;\omega)\|_D$$
$$= (K_1 \lambda + K_2 \xi + \gamma) \|x(t;\omega) - y(t;\omega)\|_D.$$

Since $(K_1 \lambda + K_2 \xi + \gamma) < 1$ by hypothesis, we have that $U + V$ is a contracting operator on S.

To show inclusion, let $x(t;\omega) \in S$. Then

$$\|(Ux)(t;\omega) + (Vx)(t;\omega)\|_D$$
$$= \left\| h(t, x(t;\omega)) + \int_0^t k(t, \tau; \omega) f(\tau, x(\tau;\omega)) \, d\tau \right.$$
$$\left. + \int_0^\infty k_0(t, \tau; \omega) e(\tau, x(\tau;\omega)) \, d\tau \right\|_D$$
$$\leq \|h(t, x(t;\omega))\|_D + K_1 \|f(t, x(t;\omega))\|_B + K_2 \|e(t, x(t;\omega))\|_B.$$

Using inequalities similar to inequality (4.5.5), we obtain

$$\|f(t, x(t;\omega))\|_B \leq \lambda \|x(t;\omega)\|_D + \|f(t,0)\|_B \quad \text{and}$$
$$\|e(t, x(t;\omega))\|_B \leq \xi \|x(t;\omega)\|_D + \|e(t,0)\|_B.$$

Hence we can write

$$\|(Ux)(t;\omega) + (Vx)(t;\omega)\|_D \leq \|h(t, x(t;\omega))\|_D + K_1 \lambda \|x(t;\omega)\|_D$$
$$+ K_1 \|f(t,0)\|_B + K_2 \xi \|x(t;\omega)\|_D$$
$$+ K_2 \|e(t,0)\|_B.$$

Using the condition of the theorem that

$$\|h(t, x(t;\omega))\|_D + K_1 \|f(t,0)\|_B + K_2 \|e(t,0)\|_B \leq \rho(1 - \lambda K_1 - \xi K_2),$$

4.5 A RANDOM PERTURBED FREDHOLM INTEGRAL EQUATION

we have

$$\|(Ux)(t;\omega) + (Vx)(t;\omega)\|_D \leq \rho(1 - \lambda K_1 - \xi K_2) + K_1\lambda\rho + K_2\xi\rho = \rho.$$

Thus $(Ux)(t;\omega) + (Vx)(t;\omega) \in S$.

Therefore, applying Banach's fixed-point theorem, there exists a unique random solution of (4.5.2), $x(t;\omega) \in S$, and the proof is complete.

For the case that W is the null operator, we obtain the theorem recently studied by Milton and Tsokos [5]. For the case that T is the null operator, we obtain the following corollary to Theorem 4.5.4, which gives conditions under which the perturbed random Fredholm integral equation (4.5.1) possesses a unique random solution.

Corollary 4.5.5 We consider the random integral equation (4.5.1) subject to the following conditions:

(i) B and D are Banach spaces stronger than C_c such that the pair (B, D) is admissible with respect to the operator

$$(Wx)(t;\omega) = \int_0^\infty k_0(t,\tau;\omega)x(\tau;\omega)\,d\tau, \qquad t \in R_+,$$

where $k_0(t,\tau;\omega)$ behaves as described previously.

(ii) Same as Condition (iii) of Theorem 4.5.4.
(iii) Same as Condition (iv) of Theorem 4.5.4.

Then there exists a unique random solution of the random integral equation (4.5.1) provided that $K_2\xi + \gamma < 1$, and

$$\|h(t, x(t;\omega))\|_D + K_2\|e(t,0)\|_B \leq \rho(1 - K_2\xi),$$

where K_2 is the norm of the operator W.

We now present some useful special cases of Corollary 4.5.5 by taking as the Banach spaces B and D the spaces C_g and C.

Theorem 4.5.6 Consider the stochastic integral equation (4.5.1) subject to the following conditions:

(i) There exist a constant $Z > 0$ and a positive continuous function $g(t)$ finite on R_+ such that

$$\int_0^\infty \|k_0(t,\tau;\omega)\|g(\tau)\,d\tau \leq Z, \qquad t \in R_+.$$

(ii) $e(t, x)$ is continuous in t uniformly in x from $R_+ \times R$ into R such that $|e(t, 0)| \leq \gamma g(t)$ and
$$|e(t, x) - e(t, y)| \leq \xi g(t)|x - y|$$
for $\gamma \geq 0$ and $\xi \geq 0$ constants.

(iii) $h(t, x)$ is continuous in t uniformly in x from $R_+ \times R$ into R such that $|h(t, 0)| \leq \alpha$ and
$$|h(t, x) - h(t, y)| \leq \lambda |x - y|$$
for some $\lambda \geq 0$ and $\alpha \geq 0$.

Then there exists a unique random solution $x(t; \omega) \in C$ of Eq. (4.5.1) such that $\|x(t; \omega)\|_C \leq \rho$ provided that ξ, λ, γ, and $\|h(t, x(t; \omega))\|_C$ are sufficiently small.

PROOF We must show that under the given conditions the pair (C_g, C) is admissible with respect to the integral operator
$$(Wx)(t; \omega) = \int_0^\infty k_0(t, \tau; \omega) x(\tau; \omega) \, d\tau, \qquad t \in R_+.$$

Let $x(t; \omega) \in C_g$. Then we have
$$\|(Wx)(t; \omega)\|_{L_2(\Omega, \mathscr{A}, \mathscr{P})} \leq \int_0^\infty \|k_0(t, \tau; \omega) x(\tau; \omega)\|_{L_2(\Omega, \mathscr{A}, \mathscr{P})} \, d\tau$$
$$\leq \int_0^\infty \|\!|k_0(t, \tau; \omega)|\!\|[\|x(\tau; \omega)\|_{L_2(\Omega, \mathscr{A}, \mathscr{P})}/g(\tau)] g(\tau) \, d\tau,$$
where $\|\!|k_0(t, \tau; \omega)|\!\|$ is a function only of $(t, \tau) \in \Delta_1$. Using the definition of the norm in C_g, we have
$$\|(Wx)(t; \omega)\|_{L_2(\Omega, \mathscr{A}, \mathscr{P})} \leq \|x(t; \omega)\|_{C_g} \int_0^\infty \|\!|k_0(t, \tau; \omega)|\!\| g(\tau) \, d\tau$$
$$\leq \|x(t; \omega)\|_{C_g} Z$$
by Condition (i) of the theorem. Thus W is a bounded operator and $(Wx)(t; \omega) \in C$. Hence (C_g, C) is admissible with respect to W. It can be shown that Conditions (ii) are sufficient for $e(t, x(t; \omega))$ to be in C_g for $x(t; \omega) \in S = \{x(t; \omega) \in C : \|x(t; \omega)\|_C \leq \rho\}$, and that Conditions (iii) are sufficient for $h(t, x(t; \omega))$ to be in C for $x(t; \omega) \in S$. Therefore Corollary 4.5.5 applies with $B = C_g$ and $D = C$ provided that $\|h(t, x(t; \omega))\|_C$, ξ, γ, and λ are small enough in the sense that $K_2 \xi + \lambda < 1$ and
$$\|h(t, x(t; \omega))\|_C + K_2 \gamma \leq \rho(1 - \xi K_2),$$
where K_2 is the norm of the operator W.

For the special case $g(t) = 1$ for all $t \in R_+$ in Theorem 4.5.6 we obtain the following corollary.

4.5 A RANDOM PERTURBED FREDHOLM INTEGRAL EQUATION

Corollary 4.5.7 Consider the random integral equation (4.5.1) under the following conditions:

(i) $\int_0^\infty |||k_0(t, \tau; \omega)||| \, d\tau \leq Z$, $t \in R_+$, where Z is some nonnegative constant.

(ii) $e(t, x)$ is continuous in t uniformly in x from $R_+ \times R$ into R such that $|e(t, 0)| \leq \gamma$ and
$$|e(t, x) - e(t, y)| \leq \xi |x - y|$$
for some constants $\gamma \geq 0$ and $\xi \geq 0$.

(iii) Same as Condition (iii) of Theorem 4.5.6.

Then there exists a unique random solution $x(t; \omega) \in C$ of Eq. (4.5.1) such that $\|x(t; \omega)\|_C \leq \rho$ provided that ξ, λ, γ, and $\|h(t, x(t; \omega))\|_C$ are sufficiently small.

Corollary 4.5.8 Assume that the random integral equation (4.5.1) satisfies the following conditions:

(i) $|||k_0(t, \tau; \omega)||| \leq \Lambda$ for all $t, \tau \in R_+$ and $\int_0^\infty g(t) \, dt < \infty$.
(ii) Same as Condition (ii) of Theorem 4.5.6.
(iii) Same as Condition (iii) of Theorem 4.5.6.

Then there exists a unique random solution $x(t; \omega) \in C$ of (4.5.1) such that $\|x(t; \omega)\|_C \leq \rho$ provided that ξ, λ, γ, and $\|h(t, x(t; \omega))\|_C$ are sufficiently small.

PROOF We need only to show that the pair of Banach spaces (C_g, C) is admissible with respect to the integral operator

$$(Wx)(t; \omega) = \int_0^\infty k_0(t, \tau; \omega) x(\tau; \omega) \, d\tau, \qquad t \in R_+,$$

along with Condition (i) of the corollary. For $x(t; \omega) \in C_g$ we have

$$\|(Wx)(t; \omega)\|_{L_2(\Omega, \mathscr{A}, \mathscr{P})} \leq \int_0^\infty |||k_0(t, \tau; \omega)||| \cdot \|x(\tau; \omega)\|_{L_2(\Omega, \mathscr{A}, \mathscr{P})} \, d\tau.$$

From hypothesis (i) of the corollary we obtain

$$\|(Wx)(t; \omega)\|_{L_2(\Omega, \mathscr{A}, \mathscr{P})} \leq \Lambda \int_0^\infty [\|x(\tau; \omega)\|_{L_2(\Omega, \mathscr{A}, \mathscr{P})}/g(\tau)] g(\tau) \, d\tau$$

$$\leq \Lambda \|x(t; \omega)\|_{C_g} \int_0^\infty g(\tau) \, d\tau$$

$$< \infty \qquad \text{for all} \quad t \in R_+.$$

Thus $(Wx)(t; \omega) \in C$ for $x(t; \omega) \in C_g$ and (C_g, C) is admissible with respect to W. Since the other hypotheses are identical to those of Theorem 4.5.6, the proof is complete.

CHAPTER V

Random Discrete Fredholm and Volterra Systems

5.0 Introduction

In the previous chapter we investigated the random Fredholm integral equation (4.0.1), and in Chapter II we presented the theory and some applications concerning the stochastic integral equation of the Volterra type (2.0.1). We shall now study a discrete version of the random integral equation of the Fredholm type of the form (4.0.1), which will be very useful for the application of an electronic computer in obtaining a realization of the random solution of the Fredholm equation in Chapter IV. Equation (4.0.1) may be "discretized" by replacing the integral with a sum of the functions evaluated at discrete points $t_1, t_2, \ldots, t_n, \ldots$, for example. We shall utilize again the concepts and theory of admissibility which were used in Chapters II and IV in order to show the existence and uniqueness of a random solution of the stochastic discrete Fredholm system

$$x_n(\omega) = h_n(\omega) + \sum_{j=1}^{\infty} c_{n,j}(\omega) f_j(x_j(\omega)), \qquad n = 1, 2, \ldots. \qquad (5.0.1)$$

5.1 EXISTENCE AND UNIQUENESS OF A SOLUTION OF (5.0.1)

We shall also consider some asymptotic stochastic stability properties of the random solution of Eq. (5.0.1) and investigate the approximation of the $x_n(\omega)$, $n = 1, 2, \ldots$.

The discrete version of the stochastic Volterra integral equation of the form (2.0.1) is a special case of the system (5.0.1); that is, when

$$c_{n,j}(\omega) = 0, \qquad j > n, \quad n = 1, 2, \ldots,$$

we obtain the random discrete Volterra system

$$x_n(\omega) = h_n(\omega) + \sum_{j=1}^{n} c_{n,j}(\omega) f_j(x_j(\omega)), \qquad n = 1, 2, \ldots. \tag{5.0.2}$$

The discrete version of the random Volterra integral equation that was presented in Section 3.1.3 is analogous to the system (5.0.2) whenever the numerical integration error term $\delta^{(n)}(\omega)$ is ignored.

Some of the results presented here are stochastic versions of some results of Petrovanu [1].

5.1 Existence and Uniqueness of a Random Solution of System (5.0.1)

Let the spaces X, X_g, X_1, and X_{bv} be as defined in Chapter I, Section 1.2. Let B^* and D^* be Banach spaces contained in X with the norm in B^* denoted by

$$\|\mathbf{x}\|_{B^*} = \|x_n(\omega)\|_{B^*}$$

and the norm in D^* denoted likewise. That is, $x_n(\omega) \in X$ is a function from N, the positive integers, into the space $L_2(\Omega, \mathscr{A}, \mathscr{P})$ and X_g, X_1, and X_{bv} are Banach spaces contained in X. Hence the random functions in X are discrete parameter second-order stochastic processes.

Let T be a linear operator from X into itself. With respect to T and the Banach spaces B^* and D^*, we now state and prove a lemma analogous to Lemma 2.1.1.

Lemma 5.1.1 If T is a continuous operator from X into itself, B^* and D^* are stronger than X, and the pair (B^*, D^*) is admissible with respect to T, then T is a continuous operator from B^* to D^*.

PROOF Suppose $\mathbf{x}_i \in B^*$ such that $\mathbf{x}_i \to^{B^*} \mathbf{x}$, that is, $x_{in}(\omega) \to^{B^*} x_n(\omega)$ as $i \to \infty$. Assume that $Tx_{in}(\omega) \to^{D^*} y_n(\omega)$ as $i \to \infty$. But $Tx_{in}(\omega) \to^{X} Tx_n(\omega)$, since T is continuous from X into itself, and $x_{in}(\omega) \to^{B^*} x_n(\omega)$ implies that $x_{in}(\omega) \to^{X} x_n(\omega)$. However, $Tx_{in}(\omega) \to^{D^*} y_n(\omega)$ implies that $Tx_{in}(\omega) \to^{X} y_n(\omega)$ as $i \to \infty$. Hence $Tx_n(\omega) = y_n(\omega)$ because the limit in X is unique. Therefore

T is closed and by the closed graph theorem it follows that T is continuous from B^* into D^*, completing the proof.

If T is a continuous operator from Banach space B^* into D^*, then it is bounded and, as before, there exists a constant $K > 0$ such that

$$\|Tx_n(\omega)\|_{D^*} \leq K\|x_n(\omega)\|_{B^*}.$$

We make the following assumptions concerning the functions in the random system (5.0.1): The functions $x_n(\omega)$ and $h_n(\omega)$ are functions of $n \in N$ with values in $L_2(\Omega, \mathscr{A}, \mathscr{P})$. For each value of $x_n(\omega)$, $n = 1, 2, \ldots, f_n(x_n(\omega))$ is a scalar, and for each $n = 1, 2, \ldots, f_n(x_n(\omega))$ has values in the space $L_2(\Omega, \mathscr{A}, \mathscr{P})$. For each n and j in N, $c_{n,j}(\omega)$ is assumed to be in $L_\infty(\Omega, \mathscr{A}, \mathscr{P})$ so that the product of $c_{n,j}(\omega)$ and $f_j(x_j(\omega))$ will be in $L_2(\Omega, \mathscr{A}, \mathscr{P})$. Also, for each value of n

$$\|\|c_{n,j}(\omega)\|\| = \mathscr{P}\text{-ess}\sup_\omega |c_{n,j}(\omega)| = \|c_{n,j}(\omega)\|_{L_\infty(\Omega, \mathscr{A}, \mathscr{P})}$$

and $\|\|c_{n,j}(\omega)\|\| \cdot \|x_j(\omega)\|_{L_2(\Omega, \mathscr{A}, \mathscr{P})}$ are assumed to be summable with respect to $j \in N$.

Consider the linear operator T defined by

$$Tx_n(\omega) = \sum_{j=1}^{\infty} c_{n,j}(\omega) x_j(\omega), \qquad n = 1, 2, \ldots,$$

for $x_j(\omega)$ in X. It may be shown that the operator T is continuous from the space X into itself by using an argument similar to that of Lemma 4.1.1.

The following theorem gives conditions under which there exists a unique random solution of the system (5.0.1).

Theorem 5.1.2 Consider the random discrete equation (5.0.1) subject to the following conditions:

(i) B^* and D^* are Banach spaces stronger than X such that the pair (B^*, D^*) is admissible with respect to the linear operator

$$Tx_n(\omega) = \sum_{j=1}^{\infty} c_{n,j}(\omega) x_j(\omega), \qquad n = 1, 2, \ldots,$$

where $c_{n,j}(\omega)$ has the properties given previously.

(ii) $x_n(\omega) \to f_n(x_n(\omega))$ is an operator on

$$S = \{x_n(\omega) : x_n(\omega) \in D^*, \quad \|x_n(\omega)\|_{D^*} \leq \rho\}$$

with values in B^* satisfying

$$\|f_n(x_n(\omega)) - f_n(y_n(\omega))\|_{B^*} \leq \lambda \|x_n(\omega) - y_n(\omega)\|_{D^*}$$

for $x_n(\omega), y_n(\omega) \in S$ and $\lambda \geq 0$ a constant.

(iii) $h_n(\omega) \in D^*$.

5.1 EXISTENCE AND UNIQUENESS OF A SOLUTION OF (5.0.1)

Then there exists a unique random solution $x_n(\omega) \in S$ of the random discrete equation (5.0.1), provided that

$$\lambda K < 1, \qquad \|h_n(\omega)\|_{D^*} + K\|f_n(0)\|_{B^*} \leq \rho(1 - \lambda K),$$

where K is the norm of T.

PROOF   Define the operator U from S into D^* by

$$Ux_n(\omega) = h_n(\omega) + \sum_{j=1}^{\infty} c_{n,j}(\omega) f_j(x_j(\omega)), \qquad n = 1, 2, \ldots.$$

As in Theorem 2.1.2, we show that $U(S) \subset S$ and that U is a contraction operator on S. Then Banach's fixed-point theorem applies.

Let $x_n(\omega), y_n(\omega) \in S$. Then

$$\|Ux_n(\omega)\|_{D^*} = \left\|h_n(\omega) + \sum_{j=1}^{\infty} c_{n,j}(\omega) f_j(x_j(\omega))\right\|_{D^*}$$

$$\leq \|h_n(\omega)\|_{D^*} + K\|f_n(x_n(\omega))\|_{B^*}$$

from the result following Lemma 5.1.1 that T is a bounded linear operator from S into D^*. From Condition (ii) of the theorem

$$\|f_n(x_n(\omega))\|_{B^*} \leq \|f_n(x_n(\omega)) - f_n(0)\|_{B^*} + \|f_n(0)\|_{B^*}$$

$$\leq \lambda \|x_n(\omega)\|_{D^*} + \|f_n(0)\|_{B^*}.$$

Hence

$$\|Ux_n(\omega)\|_{D^*} \leq \|h_n(\omega)\|_{D^*} + K\|f_n(0)\|_{B^*} + \lambda K \|x_n(\omega)\|_{D^*}$$

$$\leq \rho(1 - \lambda K) + \lambda K \rho = \rho$$

from the last hypothesis of the theorem. Thus $U(S) \subset S$. Since the difference of elements of a Banach space is in the Banach space,

$$Ux_n(\omega) - Uy_n(\omega) \in D^*$$

and

$$\|Ux_n(\omega) - Uy_n(\omega)\|_{D^*} = \left\|h_n(\omega) + \sum_{j=1}^{\infty} c_{n,j}(\omega) f_j(x_j(\omega))\right.$$

$$\left. - h_n(\omega) - \sum_{j=1}^{\infty} c_{n,j}(\omega) f_j(y_j(\omega))\right\|_{D^*}$$

$$= \left\|\sum_{j=1}^{\infty} c_{n,j}(\omega)[f_j(x_j(\omega)) - f_j(y_j(\omega))]\right\|_{D^*}$$

$$\leq K \|f_n(x_n(\omega)) - f_n(y_n(\omega))\|_{B^*}$$

$$\leq \lambda K \|x_n(\omega) - y_n(\omega)\|_{D^*}$$

by Condition (ii). Since $\lambda K < 1$, we have that U is a contraction mapping on S. Therefore, applying the fixed-point theorem of Banach, there exists a unique random solution $x_n(\omega) \in S$ of Eq. (5.0.1), completing the proof.

5.2 Special Cases of Theorem 5.1.2

In this section we shall present some special cases of Theorem 5.1.2 which will be very useful in practice. We take as the spaces B^* and D^* the spaces X_g, X_1, or X_{bv}.

Theorem 5.2.1 Consider the random discrete equation (5.0.1) subject to the following conditions:

(i) There exist a constant $Z > 0$ and a positive sequence $g_n, n = 1, 2, \ldots$, such that

$$\sum_{j=1}^{\infty} |||c_{n,j}(\omega)|||g_j \leq Z, \qquad n = 1, 2, \ldots.$$

(ii) $f_n(x)$ is a function defined for $n \in N$ and scalar x such that $|f_n(0)| \leq \gamma g_n$ and

$$|f_n(x_n(\omega)) - f_n(y_n(\omega))| \leq \lambda g_n |x_n(\omega) - y_n(\omega)|$$

for $\|x_n(\omega)\|_{X_1}, \|y_n(\omega)\|_{X_1} \leq \rho$ and λ and γ constants.

(iii) $h_n(\omega) \in X_1$.

Then there exists a unique random solution of (5.0.1),

$$\|x_n(\omega)\|_{X_1} \leq \rho,$$

provided that $\|h_n(\omega)\|_{X_1}$, γ, and λ are small enough.

PROOF If we show that the pair of Banach spaces (X_g, X_1) is admissible with respect to the linear operator

$$Tx_n(\omega) = \sum_{j=1}^{\infty} c_{n,j}(\omega) x_j(\omega), \qquad n = 1, 2, \ldots, \tag{5.2.1}$$

then the conclusion follows from Theorem 5.1.2 with $B^* = X_g$ and $D^* = X_1$.

5.2 SPECIAL CASES OF THEOREM 5.1.2

Let $x_n(\omega) \in X_g$. Taking the norm of both sides of Eq. (5.2.1), we have by the generalized Minkowski inequality (Beckenbach and Bellman [1, p. 22])

$$\|Tx_n(\omega)\|_{L_2(\Omega,\mathscr{A},\mathscr{P})} = \left\| \sum_{j=1}^{\infty} c_{n,j}(\omega) x_j(\omega) \right\|_{L_2(\Omega,\mathscr{A},\mathscr{P})}$$

$$\leq \sum_{j=1}^{\infty} \|c_{n,j}(\omega) x_j(\omega)\|_{L_2(\Omega,\mathscr{A},\mathscr{P})}$$

$$\leq \sum_{j=1}^{\infty} \|\|c_{n,j}(\omega)\|\|[\|x_j(\omega)\|_{L_2(\Omega,\mathscr{A},\mathscr{P})}/g_j]g_j$$

$$\leq \sup_n \{\|x_n(\omega)\|_{L_2(\Omega,\mathscr{A},\mathscr{P})}/g_n\} \sum_{j=1}^{\infty} \|\|c_{n,j}(\omega)\|\| g_j$$

$$\leq \|x_n(\omega)\|_{X_g} \sum_{j=1}^{\infty} \|\|c_{n,j}(\omega)\|\| g_j$$

by the definition of the norm in X_g. Since the last sum on the right is less than or equal to Z by Condition (i), we have

$$\|Tx_n(\omega)\|_{L_2(\Omega,\mathscr{A},\mathscr{P})} \leq Z \|x_n(\omega)\|_{X_g}, \qquad n = 1, 2, \ldots .$$

Hence $Tx_n(\omega)$ is bounded for all n, and so by definition it is in X_1. Therefore (X_g, X_1) is admissible with respect to T.

From Condition (ii) we have

$$\left\{ \int_\Omega |f_n(x_n(\omega)) - f_n(y_n(\omega))|^2 \, d\mathscr{P}(\omega) \right\}^{\frac{1}{2}} \leq \lambda g_n \left\{ \int_\Omega |x_n(\omega) - y_n(\omega)|^2 \, d\mathscr{P}(\omega) \right\}^{\frac{1}{2}}$$

or

$$\|f_n(x_n(\omega)) - f_n(y_n(\omega))\|_{L_2(\Omega,\mathscr{A},\mathscr{P})} \leq \lambda g_n \|x_n(\omega) - y_n(\omega)\|_{L_2(\Omega,\mathscr{A},\mathscr{P})} .$$

This implies that

$$\sup_n \{\|f_n(x_n(\omega)) - f_n(y_n(\omega))\|_{L_2(\Omega,\mathscr{A},\mathscr{P})}/g_n\} \leq \lambda \sup_n \|x_n(\omega) - y_n(\omega)\|_{L_2(\Omega,\mathscr{A},\mathscr{P})},$$

which means by definition that

$$\|f_n(x_n(\omega)) - f_n(y_n(\omega))\|_{X_g} \leq \lambda \|x_n(\omega) - y_n(\omega)\|_{X_1}.$$

Likewise,

$$\|f_n(0)\|_{X_g} \leq \gamma,$$

and from Theorem 5.1.2 we have that there exists a unique random solution of Eq. (5.0.1) provided that $\|h_n(\omega)\|_{X_1}$, γ, and λ are small enough in the sense that

$$\lambda Z^* < 1, \qquad \|h_n(\omega)\|_{X_1} + Z^* \gamma \leq \rho(1 - \lambda Z^*),$$

where Z^* is the infimum of all constants $Z > 0$ that satisfy Condition (i), completing the proof.

For $g_n = 1$, $n = 1, 2, \ldots$, we obtain the following corollary to Theorem 5.2.1.

Corollary 5.2.2 Consider the random equation (5.0.1) under the following conditions:

(i) There exists a constant $Z > 0$ such that

$$\sum_{j=1}^{\infty} |||c_{n,j}(\omega)||| \leq Z, \qquad n = 1, 2, \ldots.$$

(ii) $f_n(x)$ is a function of $n \in N$ and scalar x such that

$$|f_n(x) - f_n(y)| \leq \lambda |x - y|$$

and $|f_n(0)| \leq \gamma$ for λ and γ constants.

(iii) $h(t; \omega) \in X_1$.

Then there exists a unique random solution of Eq. (5.0.1) bounded for $n \in N$, provided that $\|h_n(\omega)\|_{X_1}$, γ, and λ are small enough.

We also have the following theorem as a special case of Theorem 5.1.2.

Theorem 5.2.3 Suppose that the random equation (5.0.1) satisfies the following conditions:

(i) There exists a $Z > 0$ such that

$$|||c_{n,j}(\omega)||| \leq Z, \qquad n, j \in N,$$

and a positive sequence g_n, $n = 1, 2, \ldots$, such that $\sum_{n=1}^{\infty} g_n < \infty$.

(ii) Same as Condition (ii) of Theorem 5.2.1.

(iii) Same as Condition (iii) of Theorem 5.2.1.

Then there exists a unique random solution

$$\|x_n(\omega)\|_{X_1} \leq \rho,$$

provided that $\|h_n(\omega)\|_{X_1}$, λ, and γ are small enough.

PROOF We need only to show that the pair of Banach spaces (X_g, X_1) is admissible with respect to the linear operator given by expression (5.2.1) along with Condition (i) of the theorem. Taking the norm of $Tx_n(\omega)$ for

$x_n(\omega) \in X_g$ as in the proof of Theorem 5.2.1, we obtain

$$\|Tx_n(\omega)\|_{L_2(\Omega,\mathscr{A},\mathscr{P})} \leqslant \sum_{j=1}^{\infty} \|\|c_{n,j}(\omega)\|\|[\|x_j(\omega)\|_{L_2(\Omega,\mathscr{A},\mathscr{P})}/g_j]g_j$$

$$\leqslant \sup_n \{\|x_n(\omega)\|_{L_2(\Omega,\mathscr{A},\mathscr{P})}/g_n\} \sum_{j=1}^{\infty} \|\|c_{n,j}(\omega)\|\|g_j.$$

But by the definition of the norm in X_g and Condition (i) of the theorem, we have

$$\|Tx_n(\omega)\|_{L_2(\Omega,\mathscr{A},\mathscr{P})} \leqslant \|x_n(\omega)\|_{X_g} \sum_{j=1}^{\infty} \|\|c_{n,j}(\omega)\|\|g_j$$

$$\leqslant \|x_n(\omega)\|_{X_g} Z \sum_{j=1}^{\infty} g_j < \infty.$$

Thus $Tx_n(\omega)$ is bounded from N into $L_2(\Omega, \mathscr{A}, \mathscr{P})$, and by definition is in X_1. Therefore (X_g, X_1) is admissible with respect to T. Since the other conditions are the same as those of Theorem 5.2.1, this completes the proof.

5.3 Stochastic Stability of the Random Solution

In the continuous case we examined the conditions under which the random solution $x(t; \omega)$ was stochastically asymptotically exponentially stable. We shall now consider the *stochastic geometric stability* of the random solution $x_n(\omega)$ of the stochastic discrete system (5.0.1), which is analogous to the stochastic asymptotic exponential stability in Chapters II and IV. Thus we state and prove the following discrete analog of Theorem 4.3.1.

Theorem 5.3.1 Suppose that the random equation (5.0.1) satisfies the following conditions:

(i) There exist constants $Z > 0$ and $0 < \alpha < 1$ such that, for all $n, j \in N$,

$$\|\|c_{n,j}(\omega)\|\| \leqslant Z\alpha^{n+j}.$$

(ii) $f_n(x)$ is defined for $n \in N$ and scalars x such that

$$|f_n(0)| \leqslant \gamma\alpha^n, \quad n = 1, 2, \ldots, \quad \text{and} \quad |f_n(x) - f_n(y)| \leqslant \lambda|x - y|$$

for $\|x_n(\omega)\|_{X_g}, \|y_n(\omega)\|_{X_g} \leqslant \rho$ and λ and γ constants.

(iii) $\|h_n(\omega)\|_{L_2(\Omega,\mathscr{A},\mathscr{P})} \leqslant \beta\alpha^n, \beta > 0, n = 1, 2, \ldots$.

Then there exists a unique random solution $x_n(\omega)$ of Eq. (5.0.1) satisfying

$$\{E[|x_n(\omega)|^2]\}^{1/2} \leqslant \rho\alpha^n,$$

provided that β, λ, and γ are sufficiently small.

V RANDOM DISCRETE FREDHOLM AND VOLTERRA SYSTEMS

PROOF We must show that the pair (X_g, X_g) is admissible with respect to the linear operator

$$Tx_n(\omega) = \sum_{j=1}^{\infty} c_{n,j}(\omega) x_j(\omega), \qquad n = 1, 2, \ldots,$$

with $g_n = \alpha^n$, $n = 1, 2, \ldots$, and the given conditions. For $x_n(\omega) \in X_g$, taking the norm of $Tx_n(\omega)$, we obtain as before

$$\|Tx_n(\omega)\|_{L_2(\Omega,\mathscr{A},\mathscr{P})} \leq \sum_{j=1}^{\infty} \||c_{n,j}(\omega)\|| \cdot \|x_j(\omega)\|_{L_2(\Omega,\mathscr{A},\mathscr{P})}$$

$$= \sum_{j=1}^{\infty} \||c_{n,j}(\omega)\|| [\|x_j(\omega)\|_{L_2(\Omega,\mathscr{A},\mathscr{P})}/\alpha^j] \alpha^j$$

$$\leq \sup_{n} \{\|x_n(\omega)\|_{L_2(\Omega,\mathscr{A},\mathscr{P})}/\alpha^n\} \sum_{j=1}^{\infty} \||c_{n,j}(\omega)\|| \alpha^j$$

$$= \|x_n(\omega)\|_{X_g} \sum_{j=1}^{\infty} \||c_{n,j}(\omega)\|| \alpha^j \qquad (5.3.1)$$

by the definition of the norm in X_g. But from Condition (i) of the theorem we have

$$\sum_{j=1}^{\infty} \||c_{n,j}(\omega)\|| \alpha^j \leq Z \sum_{j=1}^{\infty} \alpha^{n+j} \alpha^j = Z\alpha^n \sum_{j=1}^{\infty} (\alpha^2)^j$$

$$= Z\alpha^n \{[1/(1-\alpha^2)] - 1\} = [Z\alpha^2/(1-\alpha^2)]\alpha^n.$$

Therefore from expression (5.3.1) we get

$$\|Tx_n(\omega)\|_{L_2(\Omega,\mathscr{A},\mathscr{P})} \leq \|x_n(\omega)\|_{X_g} [Z\alpha^2/(1-\alpha^2)] \alpha^n < \infty \qquad (5.3.2)$$

since $0 < \alpha < 1$. Hence $Tx_n(\omega) \in X_g$ by definition, and the pair (X_g, X_g) is admissible with respect to T. Thus Condition (i) of Theorem 5.1.2 is satisfied with $B^* = D^* = X_g$.

From Condition (ii) we have that

$$\|f_n(x_n(\omega)) - f_n(y_n(\omega))\|_{L_2(\Omega,\mathscr{A},\mathscr{P})} \leq \lambda \|x_n(\omega) - y_n(\omega)\|_{L_2(\Omega,\mathscr{A},\mathscr{P})}$$

and hence

$$\sup_{n} [\|f_n(x_n(\omega)) - f_n(y_n(\omega))\|_{L_2(\Omega,\mathscr{A},\mathscr{P})}/\alpha^n] \leq \lambda \sup_{n} \{\|x_n(\omega) - y_n(\omega)\|_{L_2(\Omega,\mathscr{A},\mathscr{P})}/\alpha^n\}$$

which means that

$$\|f_n(x_n(\omega)) - f_n(y_n(\omega))\|_{X_g} \leq \lambda \|x_n(\omega) - y_n(\omega)\|_{X_g}.$$

In a similar manner we have $\|f_n(0)\|_{X_g} \leq \gamma$, and from Condition (iii) we get $\|h_n(\omega)\|_{X_g} \leq \beta$. Therefore all of the conditions of Theorem 5.1.2 are satisfied with $B^* = D^* = X_g$ for $g_n = \alpha^n$, $n = 1, 2, \ldots$, and the conclusion holds, provided that β, λ, and γ are small enough in the following sense:

$$\lambda K < 1, \qquad \|h_n(\omega)\|_{X_g} + K\|f_n(0)\|_{X_g} \leq \beta + K\gamma \leq \rho(1 - \lambda K),$$

where K is the norm of T. However, from inequality (5.3.2) we have that

$$\|Tx_n(\omega)\|_{X_g} \leq [Z\alpha^2/(1 - \alpha^2)]\|x_n(\omega)\|_{X_g},$$

which implies that

$$K = Z^*\alpha^2/(1 - \alpha^2),$$

where Z^* is the infimum of all constants satisfying Condition (i) of the theorem and (5.3.2). Thus we must have β, λ, and γ small in the sense that

$$\lambda Z^*\alpha^2 < 1 - \alpha^2$$

and

$$\beta + \frac{Z^*\alpha^2\gamma}{1 - \alpha^2} \leq \rho\left[\frac{1 - \alpha^2(1 + \lambda Z^*)}{1 - \alpha^2}\right],$$

completing the proof.

It follows that the random solution of the discrete system (5.0.1) is *stochastically geometrically stable* under the conditions of Theorem 5.3.1. That is,

$$\{E[|x_n(\omega)|^2]\}^{\frac{1}{2}} \leq \rho\alpha^n, \qquad n = 1, 2, \ldots.$$

Therefore, as $n \to \infty$, we have

$$E[|x_n(\omega)|^2] \to 0$$

since $0 < \alpha < 1$. Hence from Jensen's inequality we have that the expected value of the absolute random solution approaches zero as $n \to \infty$,

$$\lim_{n \to \infty} E[|x_n(\omega)|] = 0.$$

5.4 An Approximation to System (5.0.1)

In this section we shall consider the random system with only a finite number $m \in N$ of unknowns as follows:

$$x_n^{(m)}(\omega) = \begin{cases} h_n(\omega) + \sum_{j=1}^{\infty} c_{n,j}(\omega)f_j(x_j^{(m)}(\omega)), & n = 1, \ldots, m, \\ 0, & n = m+1, \ldots. \end{cases} \quad (5.4.1)$$

The following theorem gives conditions under which the system (5.4.1) possesses a unique random solution for each m that converges to the random solution of (5.0.1) in the space X_{bv} as $m \to \infty$.

Theorem 5.4.1 Consider the random system (5.4.1) subject to the following conditions:

(i) There exists a positive constant Z and a finite positive sequence g_n, $n = 1, 2, \ldots$, such that

$$\sum_{j=1}^{\infty} |||c_{1,j}(\omega)||| g_j + \sum_{i=1}^{\infty} \sum_{j=1}^{\infty} |||c_{i+1,j}(\omega) - c_{i,j}(\omega)||| g_j \leq Z.$$

(ii) $f_n(x)$ is a function of $n \in N$ and scalars x such that $|f_n(0)| \leq \gamma g_n$ and

$$|f_n(x_n(\omega)) - f_n(y_n(\omega))| \leq \lambda g_n \|\mathbf{x} - \mathbf{y}\|_{X_{bv}}, \qquad n = 1, 2, \ldots,$$

for $\|\mathbf{x}\|_{X_{bv}}, \|\mathbf{y}\|_{X_{bv}} \leq \rho$ and λ a constant.

(iii) $h_n(\omega) \in X_{bv}$.

Then there exists a unique random solution of (5.4.1) for each m, provided $\|\mathbf{h}\|_{X_{bv}}$, λ, and γ are small enough. Also, if $x_n(\omega)$ is the random solution of (5.0.1), where $c_{n,j}(\omega)$ satisfies (i) and $f_n(x)$ satisfies (ii), and $x_n^{(m)}(\omega)$ is the random solution of (5.4.1), then

$$\lim_{m \to \infty} \|\mathbf{x} - \mathbf{x}^{(m)}\|_{X_{bv}} = 0,$$

provided $\lambda Z^* < 1$, where Z^* is the infimum of all constants satisfying Condition (i).

PROOF We must show that the pair of spaces (X_g, X_{bv}) is admissible with respect to the operator

$$Tx_n(\omega) = \sum_{j=1}^{\infty} c_{n,j}(\omega) x_j(\omega), \qquad n = 1, 2, \ldots, \qquad (5.4.2)$$

along with Condition (i) of the theorem. Let $x_n(\omega) \in X_g$. Then taking the norm of expression (5.4.2), we obtain as before

$$\|Tx_n(\omega)\|_{L_2(\Omega, \mathscr{A}, \mathscr{P})} \leq \sum_{j=1}^{\infty} |||c_{n,j}(\omega)||| [\|x_j(\omega)\|_{L_2(\Omega, \mathscr{A}, \mathscr{P})}/g_j] g_j.$$

Also, for $i = 1, 2, \ldots$

$$\|Tx_{i+1}(\omega) - Tx_i(\omega)\|_{L_2(\Omega, \mathscr{A}, \mathscr{P})}$$

$$= \|\sum_{j=1}^{\infty} [c_{i+1,j}(\omega) - c_{i,j}(\omega)] x_j(\omega)\|_{L_2(\Omega, \mathscr{A}, \mathscr{P})}$$

$$\leq \sum_{j=1}^{\infty} |||c_{i+1,j}(\omega) - c_{i,j}(\omega)||| [\|x_j(\omega)\|_{L_2(\Omega, \mathscr{A}, \mathscr{P})}/g_j] g_j.$$

5.4 AN APPROXIMATION TO SYSTEM (5.0.1)

Hence

$$\|Tx_1(\omega)\|_{L_2(\Omega,\mathscr{A},\mathscr{P})} + \sum_{i=1}^{\infty} \|Tx_{i+1}(\omega) - Tx_i(\omega)\|_{L_2(\Omega,\mathscr{A},\mathscr{P})}$$

$$\leq \sum_{j=1}^{\infty} \|\|c_{1,j}(\omega)\|\|[\|x_j(\omega)\|_{L_2(\Omega,\mathscr{A},\mathscr{P})}/g_j]g_j$$

$$+ \sum_{i=1}^{\infty} \sum_{j=1}^{\infty} \|\|c_{i+1,j}(\omega) - c_{i,j}(\omega)\|\|[\|x_j(\omega)\|_{L_2(\Omega,\mathscr{A},\mathscr{P})}/g_j]/g_j$$

$$\leq \sup_n \{\|x_n(\omega)\|_{L_2(\Omega,\mathscr{A},\mathscr{P})}/g_n\} \left[\sum_{j=1}^{\infty} \|\|c_{1,j}(\omega)\|\| g_j \right.$$

$$\left. + \sum_{i=1}^{\infty} \sum_{j=1}^{\infty} \|\|c_{i+1,j}(\omega) - c_{i,j}(\omega)\|\| g_j \right]$$

$$\leq \|x_n(\omega)\|_{X_g} Z < \infty \tag{5.4.3}$$

by definition of the norm in X_g and Condition (i). Therefore by definition $Tx_n(\omega) \in X_{bv}$, and the pair (X_g, X_{bv}) is admissible with respect to T.

From Condition (ii), as in the proof of Theorem 5.2.1, we have that

$$\|f_n(0)\|_{X_g} \leq \gamma.$$

Also, from Condition (ii) we have

$$\|f_n(x_n(\omega)) - f_n(y_n(\omega))\|_{L_2(\Omega,\mathscr{A},\mathscr{P})} \leq \lambda g_n \|\mathbf{x} - \mathbf{y}\|_{X_{bv}}$$

or

$$\sup_n \{\|f_n(x_n(\omega)) - f_n(y_n(\omega))\|_{L_2(\Omega,\mathscr{A},\mathscr{P})}/g_n\} = \|f_n(x_n(\omega)) - f_n(y_n(\omega))\|_{X_g}$$

$$\leq \lambda \|\mathbf{x} - \mathbf{y}\|_{X_{bv}}.$$

Hence all of the conditions of Theorem 5.1.2 are satisfied with $B^* = X_g$ and $D^* = X_{bv}$. Since from inequality (5.4.3) and the definition of the norm in X_{bv} we have

$$\|T\mathbf{x}\|_{X_{bv}} \leq Z \|x_n(\omega)\|_{X_g},$$

there exists a unique random solution of the system (5.4.1), provided that $\|\mathbf{h}\|_{X_{bv}}$, λ, and γ are small enough in the sense that

$$\lambda Z^* < 1, \qquad \|\mathbf{h}\|_{X_{bv}} + Z^* \gamma \leq \rho(1 - \lambda Z^*),$$

where Z^* is the infimum of all constants satisfying Condition (i) and (5.4.3).

For the other part of the conclusion, we consider for fixed m

$$\|\mathbf{x} - \mathbf{x}^{(m)}\|_{X_{\text{bv}}}$$

$$= \|x_1(\omega) - x_1^{(m)}(\omega)\|_{L_2(\Omega,\mathscr{A},\mathscr{P})} + \sum_{i=1}^{m} \|x_{i+1}(\omega) - x_{i+1}^{(m)}(\omega) - x_i(\omega)$$

$$+ x_i^{(m)}(\omega)\|_{L_2(\Omega,\mathscr{A},\mathscr{P})} + \sum_{i=m+1}^{\infty} \|x_{i+1}(\omega) - x_i(\omega)\|_{L_2(\Omega,\mathscr{A},\mathscr{P})} \quad (5.4.4)$$

since $x_i^{(m)}(\omega) = 0$ for $i > m$. But for $i = 1, 2, \ldots$

$$\|x_{i+1}(\omega) - x_i(\omega) - x_{i+1}^{(m)}(\omega) + x_i^{(m)}(\omega)\|_{L_2(\Omega,\mathscr{A},\mathscr{P})}$$

$$= \left\| \sum_{j=1}^{\infty} c_{i+1,j}(\omega) f_j(x_j(\omega)) - \sum_{j=1}^{\infty} c_{i,j}(\omega) f_j(x_j(\omega)) \right.$$

$$\left. - \sum_{j=1}^{\infty} c_{i+1,j}(\omega) f_j(x_j^{(m)}(\omega)) + \sum_{j=1}^{\infty} c_{i,j}(\omega) f_j(x_j^{(m)}(\omega)) \right\|_{L_2(\Omega,\mathscr{A},\mathscr{P})}$$

$$= \left\| \sum_{j=1}^{\infty} c_{i+1,j}(\omega)[f_j(x_j(\omega)) - f_j(x_j^{(m)}(\omega))] \right.$$

$$\left. - \sum_{j=1}^{\infty} c_{i,j}(\omega)[f_j(x_j(\omega)) - f_j(x_j^{(m)}(\omega))] \right\|_{L_2(\Omega,\mathscr{A},\mathscr{P})}$$

$$= \left\| \sum_{j=1}^{\infty} [c_{i+1,j}(\omega) - c_{i,j}(\omega)][f_j(x_j(\omega)) - f_j(x_j^{(m)}))] \right\|_{L_2(\Omega,\mathscr{A},\mathscr{P})}$$

$$\leq \sum_{j=1}^{\infty} \|\|c_{i+1,j}(\omega) - c_{i,j}(\omega)\|\| \cdot \|f_j(x_j(\omega)) - f_j(x_j^{(m)}(\omega))\|_{L_2(\Omega,\mathscr{A},\mathscr{P})}$$

$$\leq \lambda \|\mathbf{x} - \mathbf{x}^{(m)}\|_{X_{\text{bv}}} \sum_{j=1}^{\infty} \|\|c_{i+1,j}(\omega) - c_{i,j}(\omega)\|\| g_j$$

from Condition (ii). Hence we obtain from (5.4.4)

$$\|\mathbf{x} - \mathbf{x}^{(m)}\|_{X_{\text{bv}}} \leq \lambda \|\mathbf{x} - \mathbf{x}^{(m)}\|_{X_{\text{bv}}} \sum_{j=1}^{\infty} \|\|c_{1,j}(\omega)\|\| g_j$$

$$+ \lambda \|\mathbf{x} - \mathbf{x}^{(m)}\|_{X_{\text{bv}}} \sum_{i=1}^{m} \sum_{j=1}^{\infty} \|\|c_{i+1,j}(\omega) - c_{i,j}(\omega)\|\| g_j$$

$$+ \sum_{i=m+1}^{\infty} \|x_{i+1}(\omega) - x_i(\omega)\|_{L_2(\Omega,\mathscr{A},\mathscr{P})},$$

5.4 AN APPROXIMATION TO SYSTEM (5.0.1)

since by a similar argument as that given previously, we have

$$\|x_1(\omega) - x_1^{(m)}(\omega)\|_{L_2(\Omega,\mathscr{A},\mathscr{P})} \leq \lambda \|\mathbf{x} - \mathbf{x}^{(m)}\|_{X_{\text{bv}}} \sum_{j=1}^{\infty} \|\|c_{1,j}(\omega)\|\| g_j.$$

Therefore

$$\|\mathbf{x} - \mathbf{x}^{(m)}\|_{X_{\text{bv}}} \leq \lambda \|\mathbf{x} - \mathbf{x}^{(m)}\|_{X_{\text{bv}}} \left[\sum_{j=1}^{\infty} \|\|c_{1,j}(\omega)\|\| g_j \right.$$

$$+ \sum_{i=1}^{m} \sum_{j=1}^{\infty} \|\|c_{i+1,j}(\omega) - c_{i,j}(\omega)\|\| g_j \right]$$

$$+ \sum_{i=m+1}^{\infty} \|x_{i+1}(\omega) - x_i(\omega)\|_{L_2(\Omega,\mathscr{A},\mathscr{P})}$$

$$\leq \lambda Z^* \|\mathbf{x} - \mathbf{x}^{(m)}\|_{X_{\text{bv}}} + \sum_{i=m+1}^{\infty} \|x_{i+1}(\omega) - x_i(\omega)\|_{L_2(\Omega,\mathscr{A},\mathscr{P})}$$

from Condition (i) of the theorem. Since $\lambda Z^* < 1$, we then have

$$\|\mathbf{x} - \mathbf{x}^{(m)}\|_{X_{\text{bv}}} \leq [1/(1 - \lambda Z^*)] \sum_{i=m+1}^{\infty} \|x_{i+1}(\omega) - x_i(\omega)\|_{L_2(\Omega,\mathscr{A},\mathscr{P})} \to 0$$

as $m \to \infty$, completing the proof.

We now consider approximating the random solution $x_n^{(m)}(\omega)$ of the system (5.4.1) for each $m = 1, 2, \ldots$. We may write the finite system (5.4.1) as

$$x_n^{(m)}(\omega) = \begin{cases} h_n(\omega) + \sum_{j=1}^{m} c_{n,j}(\omega) f_j(x_j^{(m)}(\omega)) + \sum_{j=m+1}^{\infty} c_{n,j}(\omega) f_j(0), & n = 1, \ldots, m, \\ 0, & \text{otherwise.} \end{cases}$$

From this we see that if $\sum_{j=m+1}^{\infty} c_{n,j}(\omega) f_j(0)$ is small, then we may ignore it. Suppose that at each fixed $n = 1, 2, \ldots, m$ we use the successive approximations to the solution of (5.4.1) similar to those of Chapter III as follows:

$$x_{n,0}^{(m)}(\omega) = h_n(\omega),$$

$$x_{n,i+1}^{(m)}(\omega) = h_n(\omega) + \sum_{j=1}^{m} c_{n,j}(\omega) f_j(x_{j,i}^{(m)}(\omega))$$

$$+ \sum_{j=m+1}^{\infty} c_{n,j}(\omega) f_j(0), \quad i = 0, 1, \ldots. \quad (5.4.5)$$

The rate of convergence of the sequence of random variables (5.4.5) will be investigated now under the conditions of Theorem 5.4.1. By definition

V RANDOM DISCRETE FREDHOLM AND VOLTERRA SYSTEMS

of the norm in X_{bv}, we have for $i = 1, 2, \ldots$

$$\|\mathbf{x}_{i+1}^{(m)} - \mathbf{x}_i^{(m)}\|_{X_{\text{bv}}} = \|x_{1,i+1}^{(m)}(\omega) - x_{1,i}^{(m)}(\omega)\|_{L_2(\Omega,\mathscr{A},\mathscr{P})}$$

$$+ \sum_{k=1}^{m} \|x_{k+1,i+1}^{(m)}(\omega) - x_{k+1,i}^{(m)}(\omega) - x_{k,i+1}^{(m)}(\omega)$$

$$+ x_{k,i}^{(m)}(\omega)\|_{L_2(\Omega,\mathscr{A},\mathscr{P})} \quad (5.4.6)$$

since $x_n^{(m)}(\omega) = 0$ for $n > m$. But from (5.4.5)

$$\|x_{1,i+1}^{(m)}(\omega) - x_{1,i}^{(m)}(\omega)\|_{L_2(\Omega,\mathscr{A},\mathscr{P})}$$

$$= \left\| \sum_{j=1}^{m} c_{1,j}(\omega) f_j(x_{j,i}^{(m)}(\omega)) - \sum_{j=1}^{m} c_{1,j}(\omega) f_j(x_{j,i-1}^{(m)}(\omega)) \right\|_{L_2(\Omega,\mathscr{A},\mathscr{P})}$$

$$\leq \sum_{j=1}^{m} \|\!|c_{1,j}(\omega)|\!\| \, \|f_j(x_{j,i}^{(m)}(\omega)) - f_j(x_{j,i-1}^{(m)}(\omega))\|_{L_2(\Omega,\mathscr{A},\mathscr{P})}$$

$$\leq \lambda \|\mathbf{x}_i^{(m)} - \mathbf{x}_{i-1}^{(m)}\|_{X_{\text{bv}}} \sum_{j=1}^{\infty} \|\!|c_{1,j}(\omega)|\!\| g_j$$

from Condition (ii) of Theorem 5.4.1. Also,

$$\|x_{k+1,i+1}^{(m)}(\omega) - x_{k,i+1}^{(m)}(\omega) - x_{k+1,i}^{(m)}(\omega) + x_{k,i}^{(m)}(\omega)\|_{L_2(\Omega,\mathscr{A},\mathscr{P})}$$

$$= \left\| \sum_{j=1}^{m} c_{k+1,j}(\omega) f_j(x_{j,i}^{(m)}(\omega)) - \sum_{j=1}^{m} c_{k,j}(\omega) f_j(x_{j,i}^{(m)}(\omega)) \right.$$

$$\left. - \sum_{j=1}^{m} c_{k+1,j}(\omega) f_j(x_{j,i-1}^{(m)}(\omega)) + \sum_{j=1}^{m} c_{k,j}(\omega) f_j(x_{j,i-1}^{(m)}(\omega)) \right\|_{L_2(\Omega,\mathscr{A},\mathscr{P})}$$

$$= \left\| \sum_{j=1}^{m} [c_{k+1,j}(\omega) - c_{k,j}(\omega)][f_j(x_{j,i}^{(m)}(\omega)) - f_j(x_{j,i-1}^{(m)}(\omega))] \right\|_{L_2(\Omega,\mathscr{A},\mathscr{P})}$$

$$\leq \sum_{j=1}^{m} \|\!|c_{k+1,j}(\omega) - c_{k,j}(\omega)|\!\| \cdot \|f_j(x_{j,i}^{(m)}(\omega)) - f_j(x_{j,i-1}^{(m)}(\omega))\|_{L_2(\Omega,\mathscr{A},\mathscr{P})}$$

$$\leq \lambda \|\mathbf{x}_i^{(m)} - \mathbf{x}_{i-1}^{(m)}\|_{X_{\text{bv}}} \sum_{j=1}^{\infty} \|\!|c_{k+1,j}(\omega) - c_{k,j}(\omega)|\!\| g_j$$

5.4 AN APPROXIMATION TO SYSTEM (5.0.1)

from Condition (ii) of the theorem. Hence we have from Eq. (5.4.6) and these inequalities that

$$\|\mathbf{x}_{i+1}^{(m)} - \mathbf{x}_i^{(m)}\|_{X_{bv}} \leq \lambda \|\mathbf{x}_i^{(m)} - \mathbf{x}_{i-1}^{(m)}\|_{X_{bv}} \sum_{j=1}^{\infty} \|c_{1,j}(\omega)\| g_j$$

$$+ \lambda \|\mathbf{x}_i^{(m)} - \mathbf{x}_{i-1}^{(m)}\|_{X_{bv}} \sum_{k=1}^{m} \sum_{j=1}^{\infty} \|c_{k+1,j}(\omega) - c_{k,j}(\omega)\| g_j$$

$$= \lambda \|\mathbf{x}_i^{(m)} - \mathbf{x}_{i-1}^{(m)}\|_{X_{bv}} \left[\sum_{j=1}^{\infty} \|c_{1,j}(\omega)\| g_j \right.$$

$$+ \left. \sum_{k=1}^{m} \sum_{j=1}^{\infty} \|c_{k+1,j}(\omega) - c_{k,j}(\omega)\| g_j \right].$$

The last expression in brackets is less than or equal to Z^* by Condition (i) of the theorem, and hence

$$\|\mathbf{x}_{i+1}^{(m)} - \mathbf{x}_i^{(m)}\|_{X_{bv}} \leq (\lambda Z^*) \|\mathbf{x}_i^{(m)} - \mathbf{x}_{i-1}^{(m)}\|_{X_{bv}}.$$

Repeating this argument $i - 1$ times, we obtain

$$\|\mathbf{x}_{i+1}^{(m)} - \mathbf{x}_i^{(m)}\|_{X_{bv}} \leq (\lambda Z^*)^i \|\mathbf{x}_1^{(m)} - \mathbf{x}_0^{(m)}\|_{X_{bv}}.$$

Since $\lambda Z^* < 1$ in Theorem 5.4.1, we have

$$\|\mathbf{x}_{i+1}^{(m)} - \mathbf{x}_i^{(m)}\|_{X_{bv}} \to 0$$

as $i \to \infty$, $m \in N$ fixed. Hence the sequence of successive approximations in (5.4.5) converges at a rate proportional to $(\lambda Z^*)^i$.

Now, to investigate the error of approximating $x_n^{(m)}(\omega)$ by $x_{n,i}^{(m)}(\omega)$, we consider

$$\|\mathbf{x}^{(m)} - \mathbf{x}_i^{(m)}\|_{X_{bv}} = \|x_1^{(m)}(\omega) - x_{1,i}^{(m)}(\omega)\|_{L_2(\Omega, \mathcal{A}, \mathcal{P})}$$

$$+ \sum_{k=1}^{m} \|x_{k+1}^{(m)}(\omega) - x_{k+1,i}^{(m)}(\omega) - x_k^{(m)}(\omega) + x_{k,i}^{(m)}(\omega)\|_{L_2(\Omega, \mathcal{A}, \mathcal{P})}.$$

By the same kind of argument as given previously, we have

$$\|x_{k+1}^{(m)}(\omega) - x_{k+1,i}^{(m)}(\omega) - x_k^{(m)}(\omega) + x_{k,i}^{(m)}(\omega)\|_{L_2(\Omega, \mathcal{A}, \mathcal{P})}$$

$$= \left\| \sum_{j=1}^{m} c_{k+1,j}(\omega) f_j(x_j^{(m)}(\omega)) - \sum_{j=1}^{m} c_{k,j}(\omega) f_j(x_j^{(m)}(\omega)) \right.$$

$$\left. - \sum_{j=1}^{m} c_{k+1,j}(\omega) f_j(x_{j,i-1}^{(m)}(\omega)) + \sum_{j=1}^{m} c_{k,j}(\omega) f_j(x_{j,i-1}^{(m)}(\omega)) \right\|_{L_2(\Omega, \mathcal{A}, \mathcal{P})}$$

$$= \left\| \sum_{j=1}^{m} [c_{k+1,j}(\omega) - c_{k,j}(\omega)] [f_j(x_j^{(m)}(\omega)) - f_j(x_{j,i-1}^{(m)}(\omega))] \right\|_{L_2(\Omega, \mathcal{A}, \mathcal{P})}$$

$$\leq \lambda \|\mathbf{x}^{(m)} - \mathbf{x}_{i-1}^{(m)}\|_{X_{bv}} \sum_{j=1}^{\infty} \|c_{k+1,j}(\omega) - c_{k,j}(\omega)\| g_j$$

from Condition (ii) of Theorem 5.4.1. Hence we get

$$\|\mathbf{x}^{(m)} - \mathbf{x}_i^{(m)}\|_{X_{bv}} \leq \lambda \|\mathbf{x}^{(m)} - \mathbf{x}_{i-1}^{(m)}\|_{X_{bv}} \sum_{j=1}^{\infty} \||c_{1,j}(\omega)\|\|g_j$$
$$+ \lambda \|\mathbf{x}^{(m)} - \mathbf{x}_{i-1}^{(m)}\|_{X_{bv}} \sum_{k=1}^{m} \sum_{j=1}^{\infty} \||c_{k+1,j}(\omega) - c_{k,j}(\omega)\|\|g_j$$
$$\leq (\lambda Z^*)\|\mathbf{x}^{(m)} - \mathbf{x}_{i-1}^{(m)}\|_{X_{bv}}.$$

Repeating this argument $i - 1$ times, we obtain that

$$\|\mathbf{x}^{(m)} - \mathbf{x}_i^{(m)}\|_{X_{bv}} \leq (\lambda Z^*)^i \|\mathbf{x}^{(m)} - \mathbf{x}_0^{(m)}\|_{X_{bv}} \to 0$$

as $i \to \infty$ for fixed $m = 1, 2, \ldots$. Hence the error of approximating $x_n^{(m)}(\omega)$ by the ith successive approximation is less than $(\lambda Z^*)^i$ times a positive constant.

The approximation enables one to apply the electronic digital computer to obtain realizations of the random solution of Eq. (5.4.1) for each m and therefore to obtain an approximation to a realization of the random solution of Eq. (5.0.1). Combining the results with the conclusion of Theorem 5.4.1, we have that

$$\|\mathbf{x} - \mathbf{x}_i^{(m)}\|_{X_{bv}} \leq \|\mathbf{x} - \mathbf{x}^{(m)}\|_{X_{bv}} + \|\mathbf{x}^{(m)} - \mathbf{x}_i^{(m)}\|_{X_{bv}} \geq 0$$

as $i \to \infty$ and $m \to \infty$ simultaneously.

As a final remark, the results also apply to the random discrete Volterra system (5.0.2), since it is a special case of the random discrete Fredholm system (5.0.1). Also, since the discrete system in Chapter III is a special case of system (5.0.2) when the error of approximating the integral is zero (that is, the functions are zero except on the discrete set of points $t_1 < t_2 < \cdots < t_n < \cdots$), the results obtained in this chapter apply there.

5.5 Application to Stochastic Control Systems

In this section we will present applications of the random discrete Volterra equation (5.0.2) to stochastic discrete control systems. Such systems occur when the input and output are observed or obtained only at discrete times. They are also useful in approximating the continuous processes used in order to apply digital computer methods to control theory.

5.5.1 A Discrete Stochastic System

Consider the nonlinear differential system with random parameters of the form

$$\dot{x}(t;\omega) = A(\omega)x(t;\omega) + b(\omega)\phi(\sigma(t;\omega)) \qquad (5.5.1)$$

5.5 APPLICATION TO STOCHASTIC CONTROL SYSTEMS

with

$$\sigma(t;\omega) = \langle q(t;\omega), x(t;\omega) \rangle, \qquad (5.5.2)$$

where $A(\omega)$ is an $m \times m$ matrix whose elements are measurable functions, $x(t;\omega)$ and $q(t;\omega)$ are $m \times 1$ vectors of random variables, $b(\omega)$ is an $m \times 1$ vector whose elements are measurable functions, $\sigma(t;\omega)$ is a scalar random variable for $t \in R_+$, $\phi(\sigma)$ is a scalar for each value σ, and $\langle \cdot, \cdot \rangle$ is the scalar product in Euclidean space. The system (5.5.1)–(5.5.2) has been studied by Tsokos [5] with respect to its *stochastic absolute stability* using a generalized version of Popov's frequency response method.

It is the aim of this section to investigate the discrete analog of system (5.5.1)–(5.5.2). We shall study the existence and uniqueness of a random solution of the discrete system with random parameters whose state at time $n = 1, 2, \ldots$ is the random variable $x_n(\omega)$. We will also give conditions which guarantee that the unique random solution is *stochastically geometrically stable* using Theorem 5.1.2. The discrete system to be studied is given by

$$x_{n+1}(\omega) - x_n(\omega) = A(\omega)x_n(\omega) + b(\omega)\phi(\sigma_n(\omega)) \qquad (5.5.3)$$

with

$$\sigma_n(\omega) = \langle q_n(\omega), x_n(\omega) \rangle. \qquad (5.5.4)$$

We may write Eq. (5.5.3) as

$$x_{n+1}(\omega) = [A(\omega) + I]x_n(\omega) + b(\omega)\phi(\sigma_n(\omega)) = B(\omega)x_n(\omega) + b(\omega)\phi(\sigma_n(\omega)),$$

where I is the $m \times m$ identity matrix, $A(\omega)$ and $b(\omega)$ are as defined previously, $q_n(\omega)$ and $x_n(\omega)$ are $m \times 1$ vectors of random variables, and $\sigma_n(\omega)$ is a scalar random variable for $n = 1, 2, \ldots$.

We may reduce the system (5.5.3)–(5.5.4) to a stochastic discrete Volterra equation in the form of Eq. (5.0.2). For $n = 1$ we have

$$x_2(\omega) = B(\omega)x_1(\omega) + b(\omega)\phi(\sigma_1(\omega)).$$

It may be shown by induction (Hildebrand [1, pp. 73–74]) that

$$x_n(\omega) = B^{n-1}(\omega)x_1(\omega) + \sum_{k=1}^{n-1} B^{k-1}(\omega)b(\omega)\phi(\sigma_{n-k}(\omega)), \qquad n > 1.$$

Substituting this expression for $x_n(\omega)$ into Eq. (5.5.4), we obtain

$$\sigma_n(\omega) = q_n^T(\omega)B^{n-1}(\omega)x_1(\omega) + \sum_{k=1}^{n-1} q_n^T(\omega)B^{k-1}(\omega)b(\omega)\phi(\sigma_{n-k}(\omega)),$$

where T denotes the transpose. Letting $n - k = j$, this expression becomes

$$\sigma_n(\omega) = q_n^T(\omega)B^{n-1}(\omega)x_1(\omega) + \sum_{j=1}^{n-1} q_n^T(\omega)B^{n-j-1}(\omega)b(\omega)\phi(\sigma_j(\omega)), \qquad (5.5.5)$$

which is in the form of Eq. (5.0.2) with

$$h_n(\omega) = q_n^T(\omega)B^{n-1}(\omega)x_1(\omega),$$

$$c_{n,j}(\omega) = \begin{cases} q_n^T(\omega)B^{n-j-1}(\omega)b(\omega), & j = 1, 2, \ldots, n-1, \\ 0, & j \geq n, \end{cases}$$

$$B(\omega) = A(\omega) + I,$$

and

$$f_j = \phi, \quad j = 1, 2, \ldots, n.$$

We will now show that Eq. (5.5.5) has a unique random solution which is stochastically geometrically stable under certain conditions.

Theorem 5.5.1 Suppose that the random equation (5.5.5) satisfies the following conditions:

(i) There exist constants $Z > 0$ and $0 < \alpha < 1$ such that

$$\|q_n^T(\omega)B^{n-j-1}(\omega)b(\omega)\| \leq Z\alpha^{n+j}, \quad j \leq n - 1.$$

(ii) $\phi(\sigma)$ satisfies $|\phi(0)| = 0$, and

$$|\phi(x_n(\omega)) - \phi(y_n(\omega))| \leq \lambda |x_n(\omega) - y_n(\omega)|$$

for $\|x_n(\omega)\|_{X_g}$ and $\|y_n(\omega)\|_{X_g} \leq \rho$ and λ a constant.

(iii)

$$\|q_n^T(\omega)B^{n-1}(\omega)x_1(\omega)\|_{L_2(\Omega, \mathscr{A}, \mathscr{P})} \leq \beta\alpha^n, \quad \beta > 0, \quad n = 1, 2, \ldots.$$

Then there exists a unique random solution $\sigma_n(\omega)$ of Eq. (5.5.5) satisfying

$$\{E|\sigma_n(\omega)|^2\}^{\frac{1}{2}} \leq \rho\alpha^n,$$

provided that β and λ are sufficiently small.

PROOF We must show that the pair (X_g, X_g) is admissible with respect to the linear operator

$$W\sigma_n(\omega) = \sum_{j=1}^{n-1} q_n^T(\omega)B^{n-j-1}(\omega)b(\omega)\sigma_j(\omega) \tag{5.5.6}$$

5.5 APPLICATION TO STOCHASTIC CONTROL SYSTEMS

for $n = 1, 2, \ldots$, with $g_n = \alpha^n$, $n = 1, 2, \ldots$, and Condition (i). For $\sigma_n(\omega) \in X_g$, taking the norm of both sides of (5.5.6), we have

$$\|W\sigma_n(\omega)\| \leq \sum_{j=1}^{n-1} \|q_n^T(\omega) B^{n-j-1}(\omega) b(\omega)\| [\|\sigma_j(\omega)\|/\alpha^j] \alpha^j$$

$$\leq \sup_n \{\|\sigma_n(\omega)\|/\alpha^n\} \sum_{j=1}^{n-1} \|q_n^T(\omega) B^{n-j-1}(\omega) b(\omega)\| \alpha^j$$

$$\leq \|\sigma_n(\omega)\|_{X_g} Z \sum_{j=1}^{n-1} \alpha^{n+2j}$$

$$\leq \|\sigma_n(\omega)\|_{X_g} Z \alpha^n \{[1/(1-\alpha^2)] - 1\}$$

$$= \|\sigma_n(\omega)\|_{X_g} [Z\alpha^2/(1-\alpha^2)] \alpha^n < \infty$$

from the definition of the norm in X_g and Condition (i) of the theorem. Hence $W\sigma_n(\omega) \in X_g$ for $\sigma_n(\omega) \in X_g$; that is, (X_g, X_g) is admissible with respect to W.

From Condition (ii) we have that

$$\sup_n \{\|\phi(\sigma_n(\omega)) - \phi(y_n(\omega))\|/\alpha^n\} \leq \lambda \sup_n \{\|\sigma_n(\omega) - y_n(\omega)\|/\alpha^n\}$$

or

$$\|\phi(\sigma_n(\omega)) - \phi(y_n(\omega))\|_{X_g} \leq \lambda \|\sigma_n(\omega) - y_n(\omega)\|_{X_g}.$$

Likewise, from Condition (iii) we have that the stochastic free term is in the space X_g.

Therefore all conditions of Theorem 5.1.2 are satisfied with $B^* = D^* = X_g$, and there exists a unique random solution of Eq. (5.5.5) which satisfies

$$\{E|\sigma_n(\omega)|^2\}^{\frac{1}{2}} \leq \rho \alpha^n, \quad n = 1, 2, \ldots,$$

provided that β and λ are small enough, completing the proof.

Thus from Theorem 5.5.1

$$E\{|\sigma_n(\omega)|^2\} \to 0 \quad \text{as} \quad n \to \infty,$$

and hence from Eq. (5.5.4) and Jensen's inequality

$$E^2\{|\langle q_n(\omega), x_n(\omega)\rangle|\} \leq E\{|\langle q_n(\omega), x_n(\omega)\rangle|^2\} \to 0 \quad \text{as} \quad n \to \infty.$$

That is,

$$E\{|q_n^T(\omega) x_n(\omega)|\} = E\{|\sigma_n(\omega)|\} \to 0$$

as $n \to \infty$, and $E\{|\phi(\sigma_n(\omega))|\} \to 0$ as $n \to \infty$, since $\phi(0) = 0$ and Condition (ii) of the theorem holds.

5.5.2 Another Discrete Stochastic System[†]

Consider the nonlinear differential system with random parameters of of the form
$$\dot{x}(t;\omega) = A(\omega)x(t;\omega) + b(\omega)\phi(\sigma(t;\omega)) \tag{5.5.7}$$
with
$$\sigma(t;\omega) = e(t;\omega) + \int_0^t \langle q(t - \tau;\omega), x(\tau;\omega)\rangle \, d\tau, \tag{5.5.8}$$
where $A(\omega)$ is an $m \times m$ matrix whose elements are measurable functions, $x(t;\omega)$ and $q(t;\omega)$ are $m \times 1$ vectors whose elements are random variables, $b(\omega)$ is an $m \times 1$ vector whose elements are measurable functions, $\sigma(t;\omega)$ and $e(t;\omega)$ are scalar random variables for $t \in R_+$, $\phi(\sigma)$ is a scalar for each value σ, and $\langle \cdot, \cdot \rangle$ denotes the scalar product in Euclidean space. The system (5.5.7)–(5.5.8) has been studied by Tsokos [2] with respect to its *stochastic absolute stability* using a generalized version of Popov's frequency response method.

It is the aim of this section to study a discrete version of the system (5.5.7)–(5.5.8). We shall investigate the existence of a unique random solution of the discrete system with random parameters whose state at time $n = 1, 2, \ldots$ is the random variable $x_n(\omega)$. We will also give conditions which guarantee that the random solution is *stochastically geometrically stable* using Theorem 5.1.2.

The discrete version of system (5.5.7)–(5.5.8) to be studied is given by
$$x_{n+1}(\omega) - x_n(\omega) = A(\omega)x_n(\omega) + b(\omega)\phi(\sigma_n(\omega)) \tag{5.5.9}$$
with
$$\sigma_n(\omega) = e_n(\omega) + \sum_{j=1}^n \langle q_{n-j+1}(\omega), x_j(\omega)\rangle \tag{5.5.10}$$
for $n = 1, 2, \ldots$. We may write Eq. (5.5.9) as
$$x_{n+1}(\omega) = [A(\omega) + I]x_n(\omega) + b(\omega)\phi(\sigma_n(\omega)) = B(\omega)x_n(\omega) + b(\omega)\phi(\sigma_n(\omega)),$$
where I is the $m \times m$ identity matrix, $B(\omega) = A(\omega) + I$, $A(\omega)$ and $b(\omega)$ are as defined previously, $q_n(\omega)$ and $x_n(\omega)$ are $m \times 1$ vectors of random variables, and $\sigma_n(\omega)$ and $e_n(\omega)$ are scalar random variables for $n = 1, 2, \ldots$.

We may reduce the system (5.5.9)–(5.5.10) to a stochastic discrete Volterra equation in the form of Eq. (5.0.2). For $n = 1$ we have
$$x_2(\omega) = B(\omega)x_1(\omega) + b(\omega)\phi(\sigma_1(\omega)).$$

[†] Section 5.5.2 adapted from Padgett and Tsokos [16] with permission of Taylor and Francis, Ltd.

5.5 APPLICATION TO STOCHASTIC CONTROL SYSTEMS

It may be shown by induction (Hildebrand, [1, pp. 73–74]) that

$$x_n(\omega) = B^{n-1}(\omega)x_1(\omega) + \sum_{k=1}^{n-1} B^{k-1}(\omega)b(\omega)\phi(\sigma_{n-k}(\omega))$$

for $n > 1$. Letting $n - k = j$, we obtain

$$x_n(\omega) = B^{n-1}(\omega)x_1(\omega) + \sum_{j=1}^{n-1} B^{n-j-1}(\omega)b(\omega)\phi(\sigma_j(\omega)).$$

Substituting into (5.5.10), we have that

$$\sigma_n(\omega) = e_n(\omega) + \sum_{j=1}^{n} q_{n-j-1}^{T}(\omega)B^{j-1}(\omega)x_1(\omega)$$
$$+ \sum_{j=1}^{n} q_{n-j+1}^{T}(\omega) \sum_{k=1}^{j-1} B^{j-k-1}(\omega)b(\omega)\phi(\sigma_k(\omega)), \quad (5.5.11)$$

where T denotes the transpose and the sums are zero whenever the upper limit is less than one. In the second sum we may interchange the order of summation and change the variable of summation to get

$$\sum_{j=1}^{n} \sum_{k=1}^{j-1} q_{n-j+1}^{T}(\omega)B^{j-k-1}(\omega)b(\omega)\phi(\sigma_k(\omega))$$
$$= \sum_{k=1}^{n-1} \left[\sum_{j=k+1}^{n} q_{n-j+1}^{T}(\omega)B^{j-k-1}(\omega)b(\omega) \right] \phi(\sigma_k(\omega)).$$

Let $j - k = i$ in the inside sum. Then

$$\sum_{j=1}^{n} \sum_{k=1}^{j-1} q_{n-j+1}^{T}(\omega)B^{j-k-1}(\omega)b(\omega)\phi(\sigma_k(\omega))$$
$$= \sum_{k=1}^{n-1} \left[\sum_{i=1}^{n-k} q_{n-k-i+1}^{T}(\omega)B^{i-1}(\omega)b(\omega) \right] \phi(\sigma_k(\omega)).$$

Thus Eq. (5.5.11) may be written as

$$\sigma_n(\omega) = h_n(\omega) + \sum_{j=1}^{n} c_{n,j}(\omega)\phi(\sigma_j(\omega)),$$

where

$$h_n(\omega) = e_n(\omega) + \sum_{j=1}^{n} q_{n-j+1}^{T}(\omega)B^{j-1}(\omega)x_1(\omega),$$

$$c_{n,j}(\omega) = \begin{cases} \sum_{i=1}^{n-j} q_{n-j-i+1}^{T}(\omega)B^{i-1}(\omega)b(\omega), & 1 \leq j \leq n-1, \\ 0, & \text{otherwise}, \end{cases} \quad (5.5.12)$$

which is of the form (5.0.2) with $f_j = \phi$, $j = 1, 2, \ldots$.

We will now present a theorem that gives conditions under which the stochastic equation (5.5.11) possesses a unique random solution that is stochastically geometrically stable.

Theorem 5.5.2 Suppose that the random equation (5.5.11) satisfies the following conditions:

(i) There exist constants $Z > 0$ and $0 < \alpha < 1$ such that
$$\left\| \sum_{i=1}^{n-j} q_{n-j-i+1}^{T}(\omega) B^{i-1}(\omega) b(\omega) \right\| \leqslant Z\alpha^{n+j}, \qquad 1 \leqslant j \leqslant n-1.$$

(ii) $\phi(\sigma)$ satisfies $\phi(0) = 0$ and
$$|\phi(x_n(\omega)) - \phi(y_n(\omega))| \leqslant \lambda |x_n(\omega) - y_n(\omega)|$$
for $\|x_n(\omega)\|_{X_g}$ and $\|y_n(\omega)\|_{X_g} \leqslant \rho$ and λ a constant.

(iii)
$$\left\| e_n(\omega) + \sum_{j=1}^{n} q_{n-j+1}^{T}(\omega) B^{j-1}(\omega) x_1(\omega) \right\|_{L_2(\Omega, \mathscr{A}, \mathscr{P})} \leqslant \beta \alpha^n,$$
$$\beta > 0, \quad n = 1, 2, \ldots .$$

Then there exists a unique random solution $\sigma_n(\omega)$ of Eq. (5.5.11) satisfying
$$\{E|\sigma_n(\omega)|^2\}^{\frac{1}{2}} \leqslant \rho \alpha^n, \qquad n = 1, 2, \ldots,$$
provided that β and λ are sufficiently small.

PROOF We must show that the pair (X_g, X_g) is admissible with respect to the linear operator
$$W\sigma_n(\omega) = \sum_{j=1}^{n-1} c_{n,j}(\omega) \sigma_j(\omega), \tag{5.5.13}$$
where $c_{n,j}(\omega)$ is given by expression (5.5.12), with $g_n = \alpha^n$, $n = 1, 2, \ldots$, and Condition (i). For $\sigma_n(\omega) \in X_g$, taking the norm of both sides of (5.5.13), we have

$$\|W\sigma_n(\omega)\| \leqslant \sum_{j=1}^{n-1} \|\|c_{n,j}(\omega)\|\| [\|\sigma_j(\omega)\|/\alpha^j] \alpha^j$$

$$\leqslant \sup\{\|\sigma_n(\omega)\|/\alpha^n\} \sum_{j=1}^{n-1} \|\|c_{n,j}(\omega)\|\| \alpha^j$$

$$= \|\sigma_n(\omega)\|_{X_g} \sum_{j=1}^{n-1} \left\| \sum_{i=1}^{n-j} q_{n-j-i+1}^{T}(\omega) B^{i-1}(\omega) b(\omega) \right\| \alpha^j$$

$$\leqslant \|\sigma_n(\omega)\|_{X_g} Z\alpha^n \sum_{j=1}^{\infty} \alpha^{2j}$$

$$= \|\sigma_n(\omega)\|_{X_g} Z\alpha^n \{[1/(1-\alpha^2)] - 1\}$$

$$= \|\sigma_n(\omega)\|_{X_g} [Z\alpha^2/(1-\alpha^2)] \alpha^n < \infty, \qquad n = 1, 2, \ldots,$$

5.5 APPLICATION TO STOCHASTIC CONTROL SYSTEMS

by definition of the norm in X_g and Condition (i) of the theorem. Hence $W\sigma_n(\omega) \in X_g$ for all $\sigma_n(\omega) \in X_g$; that is, (X_g, X_g) is admissible with respect to W.

From Condition (ii) we have that

$$\sup_n \{\|\phi(x_n(\omega)) - \phi(y_n(\omega))\|/\alpha^n\} \leq \lambda \sup_n \{\|x_n(\omega) - y_n(\omega)\|/\alpha^n\}$$

or

$$\|\phi(x_n(\omega)) - \phi(y_n(\omega))\|_{X_g} \leq \lambda \|x_n(\omega) - y_n(\omega)\|_{X_g}$$

for $x_n(\omega), y_n(\omega)$ in the set S of Theorem 5.1.2 with $D^* = X_g$. Likewise, Condition (iii) implies that the stochastic free term is in X_g.

Therefore all conditions of Theorem 5.1.2 are satisfied with $D^* = B^* = X_g$, and there exists a unique random solution of (5.5.11) which satisfies

$$\{E|\sigma_n(\omega)|^2\}^{1/2} \leq \rho\alpha^n, \quad n = 1, 2, \ldots,$$

provided that λ and β are small enough, completing the proof.

The foregoing result implies that as $n \to \infty$

$$E[|\sigma_n(\omega)|^2] \to 0,$$

and hence that $E[|\sigma_n(\omega)|]$ and $E[|\phi(\sigma_n(\omega))|]$ approach zero as $n \to \infty$, from Condition (ii).

CHAPTER VI

Nonlinear Perturbed Random Integral Equations and Application to Biological Systems

6.0 Introduction

In this chapter we shall study a nonlinear perturbed random integral equation of the Volterra type of the form

$$x(t;\omega) = h(t, x(t;\omega)) + \int_0^t k(t, \tau;\omega) f(\tau, x(\tau;\omega))\, d\tau, \qquad t \geq 0, \quad (6.0.1)$$

which has recently been studied by Milton and Tsokos [5].

From a deterministic point of view such integral equations are important in many physical problems, especially in the field of theoretical physics. Chandrasekhar [1], among others, has applied the deterministic version of (6.0.1) in determining the relative changes of radiative intensity in a radiation field due to the presence of the atmosphere. Equations of this general form also arise in neutron transport theory. Furthermore, due to the difficulty of handling random equations such as (6.0.1) mathematically, in many instances a simple deterministic model is used instead of a more realistic stochastic model which results in a random equation in the form (6.0.1).

We shall first be concerned with conditions which will ensure the existence and uniqueness of a random solution of (6.0.1). In Sections 6.2.1–6.2.3 we shall illustrate some recent applications (Milton and Tsokos [4, 6, 7]) of the theoretical results to biological problems, namely a *metabolizing system*, a *physiological model*, and a *stochastic model for communicable diseases*.

6.1 The Random Integral Equation

$$x(t;\omega) = h(t, x(t;\omega)) + \int_0^t k(t, \tau;\omega)f(\tau, x(\tau;\omega))\,d\tau$$

We shall be using the topological spaces defined in Chapter I, and the random functions which constitute Eq. (6.0.1) are defined as for the non-perturbed random integral equation. The random perturbed term $h(t, x(t;\omega))$ is a map from R_+ into $L_2(\Omega, \mathscr{A}, \mathscr{P})$. We shall denote the norm of the operator T defined in Lemma 2.1.1 as follows:

$$\|T\|^* = \sup\{\|(Tx)(t;\omega)\|_D/\|x(t;\omega)\|_B : x(t;\omega) \in B, \quad \|x(t;\omega)\|_B \neq 0\}.$$

The norm guarantees that

$$\|(Tx)(t;\omega)\|_D \leq \|T\|^*\|x(t;\omega)\|_B.$$

6.1.1 Existence and Uniqueness of a Random Solution

With respect to the aim of this chapter, the following theorems give conditions under which a unique random solution—a second-order stochastic process—exists.

Theorem 6.1.1 Consider the random integral equation (6.0.1) subject to the following conditions:

(i) B and D are Banach spaces stronger than $C_c(R_+, L_2(\Omega, \mathscr{A}, \mathscr{P}))$ such that (B, D) is admissible with respect to the operator T given by

$$(Tx)(t;\omega) = \int_0^t k(t, \tau;\omega)x(\tau;\omega)\,d\tau, \qquad t \geq 0.$$

(ii) $x(t;\omega) \to f(t, x(t;\omega))$ is an operator on $S = \{x(t;\omega): x(t;\omega) \in D, \|x(t;\omega)\|_D \leq \rho\}$ with values in B satisfying

$$\|f(t, x(t;\omega)) - f(t, y(t;\omega))\|_B \leq \lambda \|x(t;\omega) - y(t;\omega)\|_D$$

for $x(t;\omega)$, $y(t;\omega)$ elements of S, and λ and ρ positive constants;

(iii) $x(t;\omega) \to h(t, x(t;\omega))$ is an operator on S with values in D satisfying

$$\|h(t, x(t;\omega)) - h(t, y(t;\omega))\|_D \leq \gamma \|x(t;\omega) - y(t;\omega)\|_D$$

for $x(t;\omega)$, $y(t;\omega)$ elements of S and γ a positive constant.

Then there exists a unique random solution of Eq. (6.0.1) in S provided

$$\gamma + \lambda K < 1, \qquad \|h(t, x(t;\omega))\|_D + K\|f(t, 0)\|_B \leq \rho(1 - \lambda K),$$

where $K = \|T\|^*$.

PROOF Let U be an operator from S into D defined as follows:

$$(Ux)(t;\omega) = h(t, x(t;\omega)) + \int_0^t k(t, \tau;\omega) f(\tau, x(\tau;\omega))\, d\tau. \qquad (6.1.1)$$

We must show that U is a contracting operator on the set S. Consider $y(t;\omega) \in S$. We can write

$$(Uy)(t;\omega) = h(t, y(t;\omega)) + \int_0^t k(t, \tau;\omega) f(\tau, y(\tau;\omega))\, d\tau. \qquad (6.1.2)$$

Subtracting Eq. (6.1.2) from Eq. (6.1.1), we obtain

$$(Ux)(t;\omega) - (Uy)(t;\omega) = h(t, x(t;\omega)) - h(t, y(t;\omega))$$
$$+ \int_0^t k(t, \tau;\omega)[f(\tau, x(\tau;\omega)) - f(\tau, y(\tau;\omega))]\, d\tau.$$

Since B is a Banach space, $f(\tau, x(\tau;\omega)) - f(\tau, y(\tau;\omega))$ is an element of B. The admissibility of (B, D) with respect to T implies that $(Ux)(t;\omega) - (Uy)(t;\omega)$ is an element of D. From Lemma 2.1.1, T is continuous from B to D, implying that

$$\|(Ux)(t;\omega) - (Uy)(t;\omega)\|_D \leq \|h(t, x(t;\omega)) - h(t, y(t;\omega))\|_D$$
$$+ K\|f(t, x(t;\omega)) - f(t, y(t;\omega))\|_B.$$

Applying the Lipschitz conditions given in (ii) and (iii), we have that

$$\|(Ux)(t;\omega) - (Uy)(t;\omega)\|_D \leq \gamma \|x(t;\omega) - y(t;\omega)\|_D$$
$$+ \lambda K \|x(t;\omega) - y(t;\omega)\|_D$$
$$= (\gamma + \lambda K)\|x(t;\omega) - y(t;\omega)\|_D.$$

6.1 THE RANDOM INTEGRAL EQUATION

Since $\gamma + \lambda K < 1$, the first condition of the definition of contraction mapping is satisfied.

It now remains to show that $U(S) \subset S$. Let $x(t;\omega) \in S$. Then

$$(Ux)(t;\omega) = h(t, x(t;\omega)) + \int_0^t k(t,\tau;\omega) f(\tau, x(\tau;\omega))\, d\tau.$$

Taking norms, we obtain

$$\|(Ux)(t;\omega)\|_D = \left\| h(t, x(t;\omega)) + \int_0^t k(t,\tau;\omega) f(\tau, x(\tau;\omega))\, d\tau \right\|_D$$

$$\leq \|h(t, x(t;\omega))\|_D + \left\| \int_0^t k(t,\tau;\omega) f(\tau, x(\tau;\omega))\, d\tau \right\|_D$$

$$\leq \|h(t, x(t;\omega))\|_D + K\|f(t, x(t;\omega))\|_B.$$

Note that $\|f(t, x(t;\omega))\|_B$ can be written as

$$\|f(t, x(t;\omega))\|_B = \|f(t, x(t;\omega)) - f(t,0) + f(t,0)\|_B$$

$$\leq \|f(t, x(t;\omega)) - f(t,0)\|_B + \|f(t,0)\|_B.$$

Again applying the Lipschitz condition, we have

$$\|f(t, x(t;\omega))\|_B \leq \lambda \|x(t;\omega)\|_D + \|f(t,0)\|_B.$$

Thus

$$\|(Ux)(t;\omega)\|_D \leq \|h(t, x(t;\omega))\|_D + K\lambda \|x(t;\omega)\|_D + K\|f(t,0)\|_B.$$

By definition of S, $\|x(t;\omega)\|_D \leq \rho$. Hence we have

$$\|(Ux)(t;\omega)\|_D \leq \|h(t, x(t;\omega))\|_D + K\lambda\rho + K\|f(t,0)\|_B.$$

Using the condition that

$$\|h(t, x(t;\omega))\|_D + K\|f(t,0)\|_B \leq \rho(1 - \lambda K),$$

we have that

$$\|(Ux)(t;\omega)\|_D \leq \rho(1 - \lambda K) + K\lambda\rho = \rho.$$

Hence $(Ux)(t;\omega) \in S$ for $x(t;\omega)$ an element of S or $U(S) \subset S$. Therefore the conditions of Banach's fixed-point theorem have been satisfied and we can conclude that there exists a unique element $x(t;\omega) \in S$ such that $(Ux)(t;\omega) = x(t;\omega)$.

6.1.2 Some Special Cases

We shall now consider some special cases of the random integral equation (6.0.1) which are useful in various applications.

160   VI   NONLINEAR PERTURBED RANDOM INTEGRAL EQUATIONS

Theorem 6.1.2   Suppose that the random integral equation (6.0.1) satisfies the following conditions:

(i) There exists a number $A > 0$ and a positive-valued continuous function $g(t)$ on R_+ such that

$$\int_0^t \|k(t, \tau; \omega)\| g(\tau)\, d\tau \leq A.$$

(ii) $f(t, x)$ is continuous in t uniformly in x from $R_+ \times R$ into R; there exists a constant Λ such that $|f(t, 0)| \leq \Lambda g(t)$, $t \in R_+$, and $|f(t, x) - f(t, y)| \leq \lambda g(t)|x - y|$ for some $\lambda > 0$.

(iii) $h(t, x)$ is continuous in t uniformly in x from $R_+ \times R$ into R such that $|h(t, 0)| \leq Q$, $t \in R_+$, and $Q > 0$; $|h(t, x) - h(t, y)| \leq \gamma |x - y|$ for some $\gamma > 0$. Then there exists a unique random solution $x(t; \omega) \in C$ of the random integral equation (6.0.1) such that

$$\|x(t; \omega)\|_C \leq \rho, \text{ provided } \|h(t, x(t; \omega))\|_C, \lambda, \gamma, \text{ and}$$

$\|f(t, 0)\|_{C_g}$ are sufficiently small.

REMARK   By sufficiently small, we mean that $\gamma + \lambda K < 1$ and

$$\|h(t, x(t; \omega))\|_C + K\|f(t, 0)\|_{C_g} \leq \rho(1 - \lambda K).$$

PROOF   The proof consists basically in showing three things. First, that under the conditions assumed in (i) the pair (C_g, C) is admissible with respect to the operator

$$(Tx)(t; \omega) = \int_0^t k(t, \tau; \omega) x(\tau; \omega)\, d\tau;$$

second, that Conditions (ii) are sufficient for Conditions (ii) of Theorem 6.1.1 to hold; and third, that Conditions (iii) are sufficient for Conditions (iii) of Theorem 6.1.1 to hold.

Let $x(t; \omega) \in C_g$. Then

$$\|(Tx)(t; \omega)\| = \left\| \int_0^t k(t, \tau; \omega) x(\tau; \omega)\, d\tau \right\|$$

$$\leq \int_0^t \|k(t, \tau; \omega)\| [\|x(\tau; \omega)\|/g(\tau)] g(\tau)\, d\tau$$

$$\leq \|x(t; \omega)\|_{C_g} \int_0^t \|k(t, \tau; \omega)\| g(\tau)\, d\tau$$

$$\leq A \|x(t; \omega)\|_{C_g}.$$

6.1   THE RANDOM INTEGRAL EQUATION

Thus $\|(Tx)(t;\omega)\|$ is bounded, which implies that $(Tx)(t;\omega) \in C$ for $x(t;\omega) \in C_g$. Thus (C_g, C) is admissible with respect to T. Now let $t_n \to t$ in R_+. We must show that $f(t_n, x(t_n;\omega)) \to f(t, x(t;\omega))$ in $L_2(\Omega, \mathscr{A}, \mathscr{P})$. That is, we must show that given $\varepsilon > 0$, there exists an N_ε such that $n > N_\varepsilon$ implies

$$\|f(t_n, x(t_n;\omega)) - f(t, x(t;\omega))\| < \varepsilon$$

or equivalently that

$$\lim_{n \to \infty} \|f(t_n, x(t_n;\omega)) - f(t, x(t;\omega))\| = 0,$$

where $\|\cdot\| = \|\cdot\|_{L_2(\Omega, \mathscr{A}, \mathscr{P})}$ throughout this section. Consider

$$\|f(t_n, x(t_n;\omega)) - f(t, x(t;\omega))\|$$
$$\leq \|f(t_n, x(t_n;\omega)) - f(t_n, x(t;\omega)) + f(t_n, x(t;\omega)) - f(t, x(t;\omega))\|$$
$$\leq \|f(t_n, x(t_n;\omega)) - f(t_n, x(t;\omega))\| + \|f(t_n, x(t;\omega)) - f(t, x(t;\omega))\|.$$

(6.1.3)

We only need to show that each term on the right side of this inequality can be made arbitrarily small. Consider $\|f(t_n, x(t_n;\omega)) - f(t_n, x(t;\omega))\|$. For each n

$$|f(t_n, x(t_n;\omega)) - f(t_n, x(t;\omega))| \leq \lambda g(t_n)|x(t_n;\omega) - x(t;\omega)|$$

by the Lipschitz condition given in (ii). Squaring and integrating over Ω, we obtain that

$$\int_\Omega |f(t_n, x(t_n;\omega)) - f(t_n, x(t;\omega))|^2 \, d\mathscr{P}(\omega)$$
$$\leq \int_\Omega \lambda^2 g^2(t_n)|x(t_n;\omega) - x(t;\omega)|^2 \, d\mathscr{P}(\omega).$$

Thus

$$\|f(t_n, x(t_n;\omega)) - f(t_n, x(t;\omega))\| \leq \lambda g(t_n)\|x(t_n;\omega) - x(t;\omega)\|.$$

Taking limits, we obtain that

$$\lim_{n \to \infty} \|f(t_n, x(t_n;\omega)) - f(t_n, x(t;\omega))\| \leq \lim_{n \to \infty} \lambda g(t_n)\|x(t_n;\omega) - x(t;\omega)\|$$
$$= \lambda g(t) \cdot 0 = 0,$$

where the limit on the right is due to the fact that g is continuous at t and $x(t;\omega) \in C$. Hence there exists an N_1 such that $n > N_1$ implies that

$$\|f(t_n, x(t_n;\omega)) - f(t_n, x(t;\omega))\| < \varepsilon/2.$$

Since $f(t, x)$ is continuous in t uniformly in x, there exists an N_2 such that $n > N_2$ implies that

$$|f(t_n, x(t; \omega)) - f(t, x(t; \omega))| < \varepsilon/2.$$

Squaring and integrating over Ω, we obtain that

$$\int_\Omega |f(t_n, x(t; \omega)) - f(t, x(t; \omega))|^2 \, d\mathcal{P}(\omega) < (\varepsilon/2)^2$$

for $n > N_2$. Hence for $n > N_2$

$$\|f(t_n, x(t; \omega)) - f(t, x(t; \omega))\| < \varepsilon/2.$$

Let $N_\varepsilon = \max(N_1, N_2)$; then for $n > N_\varepsilon$ inequality (6.1.3) becomes

$$\|f(t_n, x(t_n; \omega)) - f(t, x(t; \omega))\| < \varepsilon/2 + \varepsilon/2 = \varepsilon.$$

Thus under Conditions (ii), $t \to f(t, x(t; \omega))$ is continuous from R_+ into $L_2(\Omega, \mathcal{A}, \mathcal{P})$. Now fix $\omega \in \Omega$. For each $t \in R_+$

$$|f(t, x(t; \omega))| = |f(t, x(t; \omega)) - f(t, 0) + f(t, 0)|$$
$$\leq |f(t, x(t; \omega)) - f(t, 0)| + |f(t, 0)|$$
$$\leq \lambda g(t)|x(t; \omega)| + |f(t, 0)| \leq \lambda g(t)|x(t; \omega)| + \Lambda g(t).$$

Again squaring and integrating over Ω, we obtain that

$$\int_\Omega |f(t, x(t; \omega))|^2 \, d\mathcal{P}(\omega) \leq \lambda^2 g^2(t) \int_\Omega |x(t; \omega)|^2 \, d\mathcal{P}(\omega)$$
$$+ 2\lambda \Lambda g^2(t) \int_\Omega |x(t; \omega)| \, d\mathcal{P}(\omega) + \Lambda^2 g^2(t).$$

Using the Cauchy–Bunyakovskii–Schwarz inequality, we obtain

$$\|f(t, x(t; \omega))\|^2 \leq \lambda^2 g^2(t) \|x(t; \omega)\|^2 + 2\lambda \Lambda g^2(t) \|x(t; \omega)\| + \Lambda^2 g^2(t).$$

Since $x(t; \omega) \in S$, where $S = \{x(t; \omega) \in C : \|x(t; \omega)\|_C \leq \rho\}$, $\|x(t; \omega)\| \leq \|x(t; \omega)\|_C \leq \rho$. Thus

$$\|f(t, x(t; \omega))\|^2 \leq \lambda^2 g^2(t) \rho^2 + 2\lambda \Lambda g^2(t) \rho + \Lambda^2 g^2(t)$$
$$= z^2 g^2(t) \quad \text{where} \quad z^2 = \lambda^2 \rho^2 + 2\lambda \Lambda \rho + \Lambda^2.$$

Therefore

$$\|f(t, x(t; \omega))\| \leq z g(t).$$

This implies that $f(t, x(t; \omega)) \in C_g$ for $x(t; \omega) \in S$. Let $x(t; \omega), y(t; \omega) \in S$. Then

$$|f(t, x(t; \omega)) - f(t, y(t; \omega))| \leq \lambda |x(t; \omega) - y(t; \omega)|,$$

6.1 THE RANDOM INTEGRAL EQUATION

implying that
$$\|f(t, x(t;\omega)) - f(t, y(t;\omega))\| \leq Ag(t)\|x(t;\omega) - y(t;\omega)\|$$
or that
$$\|f(t, x(t;\omega)) - f(t, y(t;\omega))\|/g(t) \leq \lambda\|x(t;\omega) - y(t;\omega)\|.$$

Hence by definition of the norms in C_g and C we have that
$$\|f(t, x(t;\omega)) - f(t, y(t;\omega))\|_{C_g} \leq \lambda\|x(t;\omega) - y(t;\omega)\|_C.$$

Thus Conditions (ii) are sufficient for Conditions (ii) of Theorem 6.1.1 to hold. The proof that Conditions (iii) are sufficient for Conditions (iii) of Theorem 6.1.1 to hold is analogous to that just given and will be omitted. The remainder of the proof is identical to that of Theorem 6.1.1.

When $g(t) \equiv 1$ we have the following corollary.

Corollary 6.1.3 Suppose that the random integral equation (6.0.1) satisfies the following conditions:

(i) $\int_0^t \||k(t, \tau;\omega)\|| \, d\tau \leq A$ for $t \in R_+$, $A > 0$.
(ii) $f(t, x)$ is continuous in t uniformly in x from $R_+ \times R$ into R; there exists a constant Λ such that $|f(t, 0)| \leq \Lambda$, $t \in R_+$; and $|f(t, x) - f(t, y)| \leq \lambda|x - y|$ for some $\lambda > 0$.
(iii) Same as Theorem 6.1.2, Condition (iii).

Then there exists a unique random solution $x(t;\omega) \in C$ of the random integral equation (6.0.1) such that $\|x(t;\omega)\|_C \leq \rho$, provided $\|h(t, x(t;\omega))\|_C$, λ, γ, and $\|f(t, 0)\|_C$ are sufficiently small.

Corollary 6.1.4 Assume that the random integral equation (6.0.1) satisfies the following conditions:

(i) $\||k(t, \tau;\omega)\|| \leq \Lambda_1$ for $(t, \tau) \in \Delta$ and $\int_0^\infty g(t) \, dt = M < \infty$.
(ii) Same as Theorem 6.1.2, Condition (ii).
(iii) Same as Theorem 6.1.2, Condition (iii).

Then there exists a unique random solution $x(t;\omega) \in C$ of the random integral equation (6.0.1) such that $\|x(t;\omega)\|_C \leq \rho$ provided that $\|h(t, x(t;\omega))\|_C$, λ, γ, and $\|f(t, 0)\|_{C_g}$ are sufficiently small.

PROOF It is necessary only to show that (C_g, C) is admissible with respect to the integral operator
$$(Tx)(t;\omega) = \int_0^t k(t, \tau;\omega)x(\tau;\omega) \, d\tau.$$

Corollary 6.1.5 Consider Eq. (6.0.1) under the following conditions:

(i) $\|\|k(t,\tau;\omega)\|\| \leq \Lambda_2 e^{-\alpha(t-\tau)}$, $0 \leq \tau \leq t < \infty$, and $\sup_{t \in R_+}\{\int_t^{t+1} g(\tau)\,d\tau\}$ $< \infty$, where Λ_2 and α are positive constants.

(ii) Same as Theorem 6.1.2, Condition (ii).

(iii) Same as Theorem 6.1.2, Condition (iii).

Then there exists a unique random solution $x(t;\omega) \in C$ of Eq. (6.0.1) such that $\|x(t;\omega)\|_C \leq \rho$, provided $\|h(t,x(t;\omega))\|_C$, λ, γ, and $\|f(t,0)\|_{C_g}$ are sufficiently small.

PROOF Since Conditions (ii) and (iii) are identical to those of Theorem 6.1.2, it is sufficient to show that (C_g, C) is admissible with respect to the integral operator

$$(Tx)(t;\omega) = \int_0^t k(t,\tau;\omega) x(\tau;\omega)\,d\tau.$$

Let $x(t;\omega) \in C_g$. Then

$$\|(Tx)(t;\omega)\| \leq \int_0^t \|\|k(t,\tau;\omega)\|\|\,\|x(\tau;\omega)\|\,d\tau$$

$$\leq \Lambda_2 \int_0^t e^{-\alpha(t-\tau)} \|x(\tau;\omega)\|\,d\tau$$

$$\leq \Lambda_2 \int_0^t e^{-\alpha(t-\tau)} [\|x(\tau;\omega)\|/g(\tau)] g(\tau)\,d\tau.$$

Using the definition of the norm in C_g, we have

$$\|(Tx)(t;\omega)\| \leq \Lambda_2 \|x(t;\omega)\|_{C_g} \int_0^t e^{-\alpha(t-\tau)} g(\tau)\,d\tau.$$

However,

$$\sup_{0 \leq t} \left\{ \int_t^{t+1} g(\tau)\,d\tau \right\} < \infty$$

implies that

$$\int_0^t e^{-\alpha(t-\tau)} g(\tau)\,d\tau = N < \infty$$

and thus that $\|(Tx)(t;\omega)\| \leq \|x(t;\omega)\|_{C_g} N \Lambda_2$, $t \in R_+$. Therefore $(Tx)(t;\omega) \in C$, which implies that (C_g, C) is admissible with respect to T, and the proof is complete.

6.2 Applications to Biological Systems

In this section we shall present some biological systems which are characterized by random integral equations of the form studied in Section 6.1. Specifically, we shall present a stochastic formulation of mathematical models for the study of blood flow in a circulatory system which was investigated in the deterministic setting by Stephenson [1]; a stochastic version of a deterministic equation arising in the mathematical description of a biochemical metabolizing system which was originally studied by Branson [1, 2], Wijsman [1], and Hearon [1]; and finally a random formulation of a model which arises in the study of the spread of a communicable disease through a finite population, which was treated deterministically by Landau and Rapoport [1].

In each case we present the manner in which the deterministic model arises and why such models should more realistically be characterized from a stochastic point of view. The studies given in this section are due to Milton and Tsokos [4, 6, 7].

6.2.1 A Random Integral Equation in a Metabolizing System

Biochemists are concerned with the study of metabolizing systems and have made repeated attempts to describe such systems mathematically. Generally speaking, a metabolizing system can be thought of as an irregularly shaped region of complex structure where a substance called the metabolite is being produced, consumed, transported, modified, or stored. The multitude and complexity of the reactions which take place simultaneously in any biological system make a deterministic mathematical description of the metabolizing process virtually impossible and at best highly speculative. Biochemists have, however, made various attempts to describe these reaction systems and have in many instances used as their mathematical models deterministic integral equations (Branson [1, 2], Wijsman [1], and Hearon [1]). The integral equation description seems to be especially suited to biological models in that they are well able to handle situations in which the state of the system depends not only on the immediately preceding state but on all previous states.

In a typical experiment on metabolism the experimenter is interested in the evolution in time of the amount of some substance present in the system. The function of time which describes this evaluation shall be denoted by M. Also associated with any metabolizing system will be two functions F and R which we shall call the metabolizing function and the rate function, respectively. These functions physically have the following interpretation:

$M(t)$ = amount of metabolite present in the system at time t;

$R(t)$ = rate at which the metabolite is entering the system from the outside at time t;

$F(t - \tau, M(\tau))$ = the fraction left at time t of any amount of metabolite which entered the system at time τ, $0 \leq \tau \leq t$.

The essential idea in the integral equation description is that the amount of metabolite present in the system at time t is attributable to two sources: the amount remaining from the initial amount present and the amount remaining from that which has entered the system from outside sources at any time $\tau \leq t$. Under the assumption that this is a good description of the metabolizing system under study, Branson [1] proposed that the system be characterized by the following deterministic integral equation:

$$M(t) = M(0)F(t, M(0)) + \int_0^t R(\tau)F(t - \tau, M(\tau))\, d\tau, \qquad t \geq 0, \quad (6.2.1)$$

where the unknown function is M and F and R are considered as being known. There has been considerable discussion of the general validity of this equation as a description of an arbitrary metabolizing system, for example, see Hearon [1] and Wijsman [1]. However, there seems to be general agreement that Eq. (6.2.1) is a valid model in the case of a first-order reaction. We shall discuss this case in depth.

In many metabolizing systems, especially those which occur in nature as opposed to carefully controlled laboratory experiments, it is virtually impossible to know exactly the amount of metabolite present at time $t = 0$, the beginning of our observation of the system. This is due in part to the fact that this amount will be influenced to some extent by conditions existing in the system prior to our observation and also to the fact that we must estimate this amount using experimental techniques. A usual procedure is to obtain several experimental values for $M(0)$ and solve the deterministic equation using as the "true" value of $M(0)$ the mean of the values so obtained. However, if this procedure were repeated many times, the mean values so obtained would vary and the variation could be quite large. Thus the mean value actually used could be quite unrepresentative of the true state of the system and its use could lead to incorrect results. Thus it is indeed realistic to assume that the amount of metabolite present at time $t = 0$ is not a fixed constant but rather a random variable whose behavior is governed by some probability distribution function. We shall denote this random variable by $M(0; \omega)$.

Consider the function R. By definition $R(\tau)$ is the rate at which metabolite is entering the system from outside sources at time τ. In carefully controlled laboratory experiments it could perhaps be argued that this is a deterministic

6.2 APPLICATIONS TO BIOLOGICAL SYSTEMS

function; however, in a metabolizing system occurring spontaneously in nature this is certainly not the case. Thus it could be more realistic in general to assume that at each time τ, $0 \leqslant \tau \leqslant t < \infty$, $R(\tau)$ is not a fixed constant but in reality a random variable. That is, $R(\tau)$ is not a deterministic function but rather a random function which we shall denote by $R(\tau; \omega)$.

With these remarks in mind we can formulate the following random equation analogous to Eq. (6.2.1):

$$M(t; \omega) = M(0; \omega)F(t, M(0; \omega)) + \int_0^t R(\tau; \omega)F(t - \tau, M(\tau; \omega))\,d\tau, \qquad t \geqslant 0.$$

(6.2.2)

This equation is of the general form given by Eq. (6.0.1).

Wijsman [1] showed that in the case of a first-order reaction the metabolizing function $F(t - \tau, M(\tau))$ takes the form of an exponential function, namely

$$F(t - \tau, M(\tau)) = e^{-c(t-\tau)}, \qquad c > 0.$$

Thus in this case Eq. (6.2.2) reduces to the following form:

$$M(t; \omega) = M(0; \omega)e^{-ct} + \int_0^t R(\tau; \omega)e^{-c(t-\tau)}\,d\tau, \qquad t \geqslant 0. \quad (6.2.3)$$

In order to facilitate our theoretical presentation of Eq. (6.2.3), we make the following identifications:

$$x(t; \omega) = M(t; \omega), \qquad h(t, x(t; \omega)) = H(t; \omega) \equiv M(0; \omega)e^{-ct},$$

$$k(t, \tau; \omega) \equiv R(\tau; \omega)e^{-c(t-\tau)}, \qquad f(t, x(t; \omega)) \equiv 1.$$

Thus Eq. (6.2.3) can be written as (6.0.1), that is,

$$x(t; \omega) = h(t, x(t; \omega)) + \int_0^t k(t, \tau; \omega)f(\tau, x(\tau; \omega))\,d\tau, \qquad t \geqslant 0. \quad (6.2.4)$$

With respect to the functions which constitute Eq. (6.2.3) we shall make the following assumptions: For each $t \geqslant 0$ the random variable $M(t; \omega)$ has finite variance and there exists a constant Q independent of τ such that $|R(\tau; \omega)| \leqslant Q$ for almost all ω. These restrictions are necessary for our theoretical presentations, and their validity in the physical sense will be commented on later.

To show that the given stochastic integral equation possesses a unique random solution, we must show that the basic assumptions concerning the functions which constitute the formulation of Eq. (6.0.1) and the conditions of Corollary 2.1.4 are satisfied. We shall assume that $\{e^{-c(t-\tau)}R(\tau; \omega) : \omega \in \Omega\}$ is an equicontinuous family of functions from Δ into R and show that this is sufficient for our purpose. To show that the functions of the model meet the given conditions, we proceed as follows: For $t \in R_+$, $M(t;\omega)$ by assumption has finite variance. This implies that for each t, $M(t;\omega) \in L_2(\Omega, \mathscr{A}, \mathscr{P})$. Since $f(t, x(t;\omega)) \equiv 1$ and $(\Omega, \mathscr{A}, \mathscr{P})$ is a probability space and hence a finite measure space, $f(t, x(t;\omega))$ is in $L_2(\Omega, \mathscr{A}, \mathscr{P})$ for each $t \in R_+$. For fixed t, e^{ct} is a constant and since $M(0;\omega)$ has finite variance, we can conclude that $M(0;\omega) e^{-ct} \in L_2(\Omega, \mathscr{A}, \mathscr{P})$. Fix $(t, \tau) \in \Delta$. Recall that $k(t, \tau; \omega) = R(\tau; \omega) \times e^{-c(t-\tau)}$. By assumption

$$|R(\tau;\omega)| \leqslant Q, \quad \mathscr{P}\text{-a.e.},$$

$$|k(t,\tau;\omega)| = |R(\tau;\omega) e^{-c(t-\tau)}| \leqslant Q e^{-c(t-\tau)}, \quad \mathscr{P}\text{-a.e.}$$

This implies that $(t, \tau) \to k(t, \tau; \omega)$ is a map from Δ into $L_\infty(\Omega, \mathscr{A}, \mathscr{P})$. Now let $(t_n, \tau_n) \to (t, \tau)$. Choose $\varepsilon > 0$. By the equicontinuity condition, there exists an N_ε such that $n > N_\varepsilon$ implies

$$|e^{-c(t_n - \tau_n)} R(\tau_n; \omega) - e^{-c(t-\tau)} R(\tau; \omega)| < \varepsilon$$

for each $\omega \in \Omega$. Thus by definition the infinity norm, for $n > N_\varepsilon$, becomes

$$|||e^{-c(t_n - \tau_n)} R(\tau_n; \omega) - e^{-c(t-\tau)} R(\tau; \omega)||| < \varepsilon.$$

However, this implies that for $n > N_\varepsilon$ we have

$$|||k(t_n, \tau_n; \omega) - k(t, \tau; \omega)||| < \varepsilon,$$

as was to be shown.

To show that the hypotheses of Corollary 2.1.4 are satisfied, we proceed as follows: For Conditions (i) we have

$$\int_0^t |||k(t,\tau;\omega)||| \, d\tau = \int_0^t |||e^{-c(t-\tau)} R(\tau;\omega)||| \, d\tau$$

$$= \int_0^t e^{-ct} e^{c\tau} |||R(\tau;\omega)||| \, d\tau \leqslant Q e^{-ct} \int_0^t e^{c\tau} \, d\tau$$

$$= (Q e^{-ct}/c)(e^{ct} - 1) = (Q/c)(1 - e^{-ct})$$

$$\leqslant Q/c.$$

It is easy to see that Condition (ii) is satisfied. To see that $M(0;\omega) e^{-ct} \in C$, let $t_n \to t$ in R_+ and choose $\varepsilon > 0$. If $\|M(0;\omega)\| \neq 0$, choose N such that $n > N$ implies

$$|e^{-ct_n} - e^{ct}| < \varepsilon/\|M(0;\omega)\|.$$

Then for $n > N$

$$\|M(0;\omega) e^{-ct_n} - M(0;\omega) e^{-ct}\| = |e^{-ct_n} - e^{-ct}| \|M(0;\omega)\|$$

$$< [\varepsilon/\|M(0;\omega)\|] \|M(0;\omega)\| = \varepsilon.$$

6.2 APPLICATIONS TO BIOLOGICAL SYSTEMS

If $\|M(0;\omega)\| = 0$, then

$$\|M(0;\omega)e^{-ct_n} - M(0;\omega)e^{-ct}\| = 0 < \varepsilon.$$

Thus $t \to M(0;\omega)e^{-ct}$ is continuous from R_+ into $L_2(\Omega, \mathscr{A}, \mathscr{P})$. To see that the map is bounded, consider

$$\|M(0;\omega)e^{-ct}\| = e^{-ct}\|M(0;\omega)\| \leqslant \|M(0;\omega)\|.$$

Thus the conditions of Corollary 2.1.4 are satisfied, and we can conclude that there exists a unique random solution of Eq. (6.2.3) provided $\|M(0;\omega) \times e^{-ct}\|_C$, λ, and $\|f(t,0)\|_C$ are sufficiently small.

REMARK When we say that the quantities are sufficiently small we mean that

$$\lambda K < 1, \qquad \|M(0;\omega)e^{-ct}\|_C + K\|f(t,0)\|_C \leqslant \rho(1 - \lambda K),$$

where $K = \|T\|^*$, the norm of the operator T defined in Section 6.1. Note also that

$$\|M(0;\omega)e^{-ct}\|_C = \sup_{0 \leqslant t} e^{-ct}\|M(0;\omega)\| = \|M(0;\omega)\|$$

and that $\|f(t,0)\|_C = 1$. Hence we are actually requiring that

$$\lambda K < 1, \qquad \|M(0;\omega)\| + K \leqslant \rho(1 - \lambda K).$$

Since in this case λ can be any positive number, the first condition can easily be satisfied. Hence we will have a unique random solution $M(t;\omega)$ such that $E|M(t;\omega)|^2 \leqslant \rho$ for each t provided $E|M(0;\omega)|^2$ is sufficiently small.

In formulating the stochastic model, we have been forced to make certain assumptions on $M(t;\omega)$ and $R(t;\omega)$. Namely, we assume that $\{M(t;\omega) : t \in R_+\}$ is a second-order stochastic process and that for each t, $R(t;\omega)$ is $\mathscr{P}$-essentially bounded and furthermore that the bound is uniform over R_+. These restrictions make our particular approach to the problem possible and may or may not be satisfied by a particular given metabolizing system under study by a biologist or biochemist. The feasibility of these assumptions must be determined in *each instance by the experimenter*. Although these requirements appear on the surface to be quite strong, in practice they are in many cases quite easily satisfied due to the physical or chemical characteristics of the system under study. For example, if the amount of metabolite present at any time were limited due to space considerations, we would automatically satisfy the condition that $\{M(t;\omega) : t \in R_+\}$ be a second-order stochastic process. As a simple illustration, visualize the "metabolizing" system as being a reservoir and the "metabolite" as being the amount of water present at time t. If, on the other hand, the amount of metabolite present at any given time were limited due to some chemical characteristic of the system, we could come to the same conclusion. That is, visualize the metabolizing system as being perhaps a lake or stream and the metabolite of interest the amount of dissolved KCl present per gallon. This amount will be limited by chemical considerations due to the fact that there is a maximum amount of the salt which can be dissolved in a given amount of water at a given temperature. In any situation similar to this the assumption of finite variance on $M(t;\omega)$ for each t will be quite naturally satisfied. Similarly, in many systems the rate at which metabolite enters the system from outside sources at any time will be restricted due to physical limitations, especially in systems where metabolite is simply being transported, or to chemical limitations in systems where metabolite is being consumed or produced. Hence the assumption that there exists a Q such that $|R(\tau;\omega)| \leqslant Q$ $\mathscr{P}$-a.e. is an assumption which can often be realistically met. The point to be made here is twofold. First, the proposed random model is a more realistic description of a general metabolizing process than is the deterministic formulation and should be used whenever possible. Second, in many cases the restrictions placed on the functions in the random model are not extremely difficult to meet in practice, but whether or not they are met is a question which must be considered carefully by the experimenter in each case. Note that there is a certain degree of flexibility in the random formulation in that for each t we require no knowledge of the particular form of the distribution functions for the random variables involved but only that they are elements of certain spaces, namely either $L_2(\Omega, \mathscr{A}, \mathscr{P})$ or $L_\infty(\Omega, \mathscr{A}, \mathscr{P})$.

6.2.2 A Stochastic Physiological Model

Physiologists quite often are faced with the problem of constructing mathematical models which attempt to describe the complex processes that take place within the human body. Such models are very difficult to obtain and must necessarily be oversimplified due to the inability of scientists to understand fully all of the factors which can influence even the simplest process taking place in a living organism. Thus mathematical models in use are constantly subject to refinement as more insight is gained into the true nature of the process taking place. Along this line, Milton and Tsokos [4] proposed a stochastic formulation for a model used to study blood flow in a simplified circulatory system which has been studied from the deterministic point of view by Stephenson [1]. The stochastic approach yields a more realistic characterization of the system than the nonrandom formulation and should be used whenever possible.

A simplified circulatory system is visualized as consisting of the heart, a capillary bed, and connecting vessels. Schematically it is pictured in Fig.

6.2 APPLICATIONS TO BIOLOGICAL SYSTEMS

6.2.1. The points X and Y represent an inflow and an outflow point, respectively. It is assumed that a given amount M of indicator substance I is suddenly injected in the system at point X, the inflow to the capillary bed, thus producing a fixed concentration of indicator $C_1(0) > 0$ at time $t = 0$. The fraction of the amount M flowing out of the capillary bed at time t is considered to be a deterministic function of time and is denoted by $p(t)$. This function is determined by taking instantaneous measurements on the concentration of indicator at point Y. The concentration at time t is also considered to be a deterministic function of time which is denoted by $C_2(t)$. The model allows for recirculation and hence the concentration of indicator at point X will also depend on time and will be denoted $C_1(t)$. Stephenson [1] makes the following assumptions concerning these functions based upon experimental results and working experience with such models: (a) $C_1(t)$ and $C_2(t)$ are twice differentiable; (b) there exist constants $G > 0$ and $H > 0$ such that

$$|C_1'(t)| \leqslant G \quad \text{and} \quad |C_2'(t)| \leqslant H$$

for all t, where $C_1'(t) = dC_1(t)/dt$ and $C_2'(t) = dC_2(t)/dt$. Under the assumption that the model provides a reasonable description of the physical situation at hand, the following deterministic integral equation of the Volterra type is obtained:

$$p(t) = [C_2'(t)/C_1(0)] - [1/C_1(0)] \int_0^t C_1'(t - \tau)p(\tau)\, d\tau, \quad t \geqslant 0. \quad (6.2.5)$$

Let us first consider the function $C_2(t)$. The usual technique for determining the value of this function at a given time t_1 is to obtain experimentally several observations on $C_2(t_1)$. These experimental values are then averaged and the mean value of these experimental observations is used as the "true" value for $C_2(t_1)$. Due to the possibility of some diffusion in the capillary bed together with the inherent difficulties of accurately measuring instantaneous concentrations at a given point, as well as the natural variability of the circulatory system in general, if the same experiment were repeated, the mean value obtained would most likely differ from the first determination.

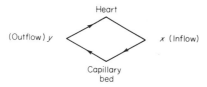

Figure 6.2.1.

If this variability is large, the actual mean value used could be quite unsatisfactory. Therefore it is more realistic to assume that the concentration of indicator at Y is for each t a random variable and that $C_2(t)$ is in fact a random function which we shall denote by $C_2(t;\omega)$. This tacitly implies that the derivative of $C_2(t;\omega)$ is also random and it will be denoted by $C_2'(t;\omega)$. The function $C_1(t)$ can be considered random because of the fact that it is obtained experimentally in a manner identical to that described for $C_2(t)$ and also due to the fact that we are allowing for recirculation. Thus $C_1(t)$ will be influenced after a certain point by the same factors which influenced $C_2(t)$, namely diffusion and natural variability. The function $p(t)$ is being expressed in terms of the random functions $C_1'(t;\omega)$ and $C_2'(t;\omega)$, and thus it should also be considered as random and will be denoted by $p(t;\omega)$.

In view of these remarks we have the following random version of Eq. (6.2.5):

$$p(t;\omega) = [C_2'(t;\omega)/C_1(0)] - [1/C_1(0)]\int_0^t C_1'(t-\tau;\omega)$$
$$\times p(\tau;\omega)\,d\tau, \qquad t \geq 0. \qquad (6.2.6)$$

We shall make the following identifications in order to simplify our theoretical investigation:

$$x(t;\omega) \equiv p(t;\omega);$$
$$h(t, x(t;\omega)) \equiv C_2'(t;\omega)/C_1(0);$$
$$k(t,\tau;\omega) \equiv C_1'(t-\tau;\omega); \qquad \text{and}$$
$$f(t, x(t;\omega)) \equiv -x(t;\omega)/C_1(0) \equiv -[1/C_1(0)]p(t;\omega).$$

Using the notational changes, Eq. (6.2.6) takes the familiar form given by Eq. (6.0.1).

With respect to the functions appearing in Eq. (6.2.6) we make the assumption that there exist constants $G > 0$ and $H > 0$ such that

$$|C_1'(t;\omega)| \leq G, \quad \mathscr{P}\text{-a.e.} \quad \text{and} \quad |C_2'(t;\omega)| \leq H, \quad \mathscr{P}\text{-a.e.}$$

for each $t \geq 0$.

To show that the model possesses a random solution, we must verify that the random functions which constitute Eq. (6.2.6) meet the required assumptions of Eq. (6.0.1). We begin by assuming that the families

$$\{C_1'(t-\tau;\omega):\omega\in\Omega\} \quad \text{and} \quad \{C_2'(t;\omega):\omega\in\Omega\}$$

are equicontinuous families from Δ into R. Fix $t \in R_+$. Then $|p(t;\omega)| \leq 1$ $\mathscr{P}$-a.e. for each t due to the fact that $p(t;\omega)$ represents the *fraction* of the original amount of indicator in the outflow at time t. Hence

$$E|p(t;\omega)|^2 = \int_\Omega |p(t;\omega)|^2 \, d\mathscr{P}(\omega) \leq 1,$$

implying that for each $t \geq 0$, $p(t;\omega) \in L_2(\Omega, \mathscr{A}, \mathscr{P})$. Furthermore, $f(t, x(t;\omega)) = -p(t;\omega)/C_1(0)$. This implies that

$$|f(t, x(t;\omega))| = |-p(t;\omega)/C_1(0)| = [1/C_1(0)]|p(t;\omega)|$$

or

$$\int_\Omega |f(t, x(t;\omega))|^2 \, d\mathscr{P}(\omega) = [1/C_1(0)]^2 \int_\Omega |p(t;\omega)|^2 \, d\mathscr{P}(\omega) \leq [1/C_1(0)]^2 < \infty.$$

Thus for $t \in R_+$, $f(t, x(t;\omega)) \in L_2(\Omega, \mathscr{A}, \mathscr{P})$. Also, for $t \in R_+$

$$h(t, x(t;\omega)) = C_2'(t;\omega)/C_1(0),$$

implying that

$$\int_\Omega |h(t, x(t;\omega))|^2 \, d\mathscr{P}(\omega) = \int_\Omega |C_2'(t;\omega)/C_1(0)|^2 \, d\mathscr{P}(\omega)$$

$$= [1/C_1(0)]^2 \int_\Omega |C_2'(t;\omega)|^2 \, d\mathscr{P}(\omega)$$

$$\leq [1/C_1(0)]^2 \int_\Omega H^2 \, d\mathscr{P}(\omega) = [1/C_1(0)]^2 H^2 < \infty,$$

that is, for fixed $t \in R_+$, $h(t, x(t;\omega)) \in L_2(\Omega, \mathscr{A}, \mathscr{P})$. Finally, fix $(t, \tau) \in \Delta$ and consider $k(t, \tau; \omega) = C_1'(t - \tau; \omega) = C_1'(u; \omega)$, where $u = t - \tau$. By our previous assumption, $|C_1'(u; \omega)| \leq G$ $\mathscr{P}$-a.e. This implies that $\mathscr{P}\{\omega : |C_1'(u; \omega)| > G\} = 0$ and hence that $C_1'(u; \omega) \in L_\infty(\Omega, \mathscr{A}, \mathscr{P})$. Thus $(t, \tau) \to k(t, \tau; \omega)$ is a map from $\Delta \to L_\infty(\Omega, \mathscr{A}, \mathscr{P})$. It remains only to show that this map is continuous. To this end, let $(t_n, \tau_n) \to (t, \tau)$. By the equicontinuity condition, given $\varepsilon > 0$, there exists an N such that $n > N$ implies

$$|C_1'(t_n - \tau_n; \omega) - C_1'(t - \tau; \omega)| < \varepsilon$$

for each $\omega \in \Omega$. Hence for $n > N$,

$$\mathscr{P}\{\omega : |C_1'(t_n - \tau_n; \omega) - C_1'(t - \tau; \omega)| > \varepsilon\} = 0,$$

implying that for $n > N$,

$$\|C_1'(t_n - \tau_n; \omega) - C_1'(t - \tau; \omega)\| < \varepsilon.$$

VI NONLINEAR PERTURBED RANDOM INTEGRAL EQUATIONS

Therefore the basic assumptions of our theoretical development are met. We shall now show that for an appropriate choice of the function g the pair (C_g, C_g) is admissible with respect to T as required in Theorem 2.1.2, Condition (i) and furthermore that Conditions (ii) and (iii) are also satisfied by the functions of the stochastic model. This allows us to apply the theorem to obtain the existence of a unique random solution.

Choose $g(t) = \alpha\, e^{\beta t}$ for $\beta > 0$ and $\alpha \geqslant 1$. Note that g is positive valued and continuous on R_+ and hence the space $C_g(R_+, L_2(\Omega, \mathscr{A}, \mathscr{P}))$ is well defined. Let $x(t;\omega) \in C_g(R_+, L_2(\Omega, \mathscr{A}, \mathscr{P}))$. Consider

$$\|(Tx)(t;\omega)\| = \left\|\int_0^t k(t,\tau;\omega)x(\tau;\omega)\,d\tau\right\|$$

$$\leqslant \int_0^t \|\!|k(t,\tau;\omega)|\!\|\,\|x(\tau;\omega)\|\,d\tau$$

$$= \int_0^t \|\!|k(t,\tau;\omega)|\!\|[\|x(\tau;\omega)\|/\alpha\, e^{\beta t}]\alpha\, e^{\beta \tau}\,d\tau$$

$$\leqslant \|x(t;\omega)\|_{C_g}\int_0^t \|\!|k(t,\tau;\omega)|\!\|\alpha\, e^{\beta \tau}\,d\tau$$

$$\leqslant [G\|x(t;\omega)\|_{C_g}/\beta]\alpha(e^{\beta t} - 1)$$

$$\leqslant [G\|x(t;\omega)\|_{C_g}/\beta]\alpha\, e^{\beta t} = [G\|x(t;\omega)\|_{C_g}/\beta]g(t).$$

Thus for $x(t;\omega) \in C_g(R_+, L_2(\Omega, \mathscr{A}, \mathscr{P}))$, $(Tx)(t;\omega) \in C_g(R_+, L_2(\Omega, \mathscr{A}, \mathscr{P}))$, implying that the pair (C_g, C_g) is admissible with respect to T.

Now, let $x(t;\omega) \in S$, where S is as defined in Theorem 2.1.2, Condition (ii) with $C_g \equiv D$. Then

$$f(t, x(t;\omega)) = [1/C_1(0)]x(t;\omega),$$

which implies that

$$\|f(t, x(t;\omega))\| = [1/C_1(0)]\|x(t;\omega)\| \leqslant [1/C_1(0)]A\alpha\, e^{\beta t} = Zg(t),$$

where Z is a constant, $Z \equiv [1/C_1(0)]A$. Thus by definition of $C_g(R_+, L_2(\Omega, \mathscr{A}, \mathscr{P}))$, $f(t, x(t;\omega)) \in C_g(R_+, L_2(\Omega, \mathscr{A}, \mathscr{P}))$. To show that the Lipschitz condition is satisfied, consider

$$\|f(t, x(t;\omega)) - f(t, y(t;\omega))\|_{C_g}$$

$$= \|-[1/C_1(0)]x(t;\omega) + [1/C_1(0)]y(t;\omega)\|_{C_g}$$

$$= [1/C_1(0)]\|x(t;\omega) - y(t;\omega)\|_{C_g}.$$

6.2 APPLICATIONS TO BIOLOGICAL SYSTEMS

Hence we can choose $\lambda = 1/C_1(0)$ and Condition (ii) is satisfied.

Let $x(t;\omega) \in S$. Then

$$\|h(t, x(t;\omega))\| = \|[1/C_1(0)]C_2'(t;\omega)\| = [1/C_1(0)]\|C_2'(t;\omega)\|$$
$$\leqslant [1/C_1(0)]H \leqslant [H/C_1(0)](\alpha/\alpha)e^{\beta t}$$
$$= Zg(t).$$

Since the equicontinuity condition implies the continuity of the map $t \to h(t, x(t;\omega))$, we have that $h(t, x(t;\omega)) \in C_g(R_+, L_2(\Omega, \mathscr{A}, \mathscr{P}))$ as was desired.

Therefore we have shown that Theorem 2.1.2 is applicable with $B \equiv D \equiv C_g$ where $g(t) = \alpha e^{\beta t}$. We can thus conclude that there exists a unique random solution $x(t;\omega)$ of Eq. (6.2.6) such that $\|x(t;\omega)\|_{C_g} \leqslant \rho$ provided $\lambda K < 1$ and $\|h(t, x(t;\omega))\|_{C_g} + K\|f(t, 0)\|_{C_g} \leqslant \rho(1 - \lambda K)$.

Note that $\|f(t, 0)\|_{C_g} = 0$ and that

$$\|h(t, x(t;\omega))\|_{C_g} = \sup_{0 \leqslant t}\{[1/C_1(0)]\|C_2'(t;\omega)\|/\alpha e^{\beta t}\}$$
$$\leqslant [1/C_1(0)]\sup_{0 \leqslant t}\{\|C_2'(t;\omega)\|\} \leqslant [1/C_1(0)]H.$$

Furthermore,

$$\|(Tx)(t;\omega)\|_{C_g} = \sup_{t \geqslant 0}[\|(Tx)(t;\omega)\|/g(t)]$$
$$\leqslant \sup_{0 \leqslant t}[G\|x(t;\omega)\|_{C_g}g(t)/\beta g(t)] = G\|x(t;\omega)\|_{C_g}/\beta.$$

Here

$$K = \sup[\|(Tx)(t;\omega)\|_{C_g}/\|x(t;\omega)\|_{C_g} : \|x(t;\omega)\|_{C_g} \neq 0]$$

and hence

$$K \leqslant G\|x(t;\omega)\|_{C_g}/\beta\|x(t;\omega)\|_{C_g} = G/\beta.$$

Thus we can conclude that there exists a unique random solution $x(t;\omega) \in S$ of Eq. (6.2.6) provided ρ is such that

$$[1/C_1(0)](G/\beta) < 1 \quad \text{and} \quad [1/C_1(0)]H \leqslant \rho\{1 - [1/C_1(0)](G/\beta)\}.$$

The mathematical restrictions placed on the functions which constitute the stochastic model do not appear to be overly restrictive in that they closely parallel the assumptions which Stephenson [1] made. That is, there exist constants G and H such that $|C_1'(t;\omega)| \leqslant G$ and $|C_2'(t;\omega)| \leqslant H$ $\mathscr{P}$-a.e. for all $t \geqslant 0$, and the assumption that $p(t;\omega) \in L_2(\Omega, \mathscr{A}, \mathscr{P})$ for each $t \geqslant 0$ is quite naturally satisfied due to the fact that the physical meaning of $p(t;\omega)$ implies $|p(t;\omega)| \leqslant 1$ $\mathscr{P}$-a.e.

The choice of the function $g(t)$ is arbitrary and could be replaced at the discretion of the investigator.

6.2.3 A Stochastic Model for Communicable Disease

Early efforts to describe the course of the spread of a communicable disease throughout a population of given size were concerned with deterministic models. Recently statisticians have begun to reconsider those models in an attempt to take into consideration the essentially random nature of many of the processes which are involved in the spread of diseases. Along this line the deterministic models are being replaced by newer, stochastic models which more realistically characterize the true nature of such a system.

Landau and Rapoport [1] have examined a mathematical model for the spread of communicable disease through a finite population, from a deterministic point of view. Their work was an extension of the study done by Kermack and McKendrick [1]. Recently Milton and Tsokos [7] have formulated a stochastic version of the communicable disease model which more realistically describes the physical situation. We shall formulate such a model and show, using the technique of admissibility theory, that a random solution, a second-order stochastic process, exists for the system and that it is also unique.

Consider a population whose members are thoroughly mixed, that is, each individual has the ability to make contact with every other individual in the population. Assume that the probability of contact between any pair of individuals is the same for each pair. A "state" will signify any quality that can be transmitted from one individual to another by contact, and an individual acquires the state by coming into contact with another individual who has already acquired the quality and has not yet lost it. The word "population" is simply used to mean any collection of objects which can be described as being thoroughly mixed, and an individual is any member of the population under study.

There are of course many variations of the general situation which can be considered. For example, there may or may not be "recovery"; there may or may not be "immunity"; there may be only a definite fixed period during which it is possible to transmit the state even though the "carrier" still possesses it himself; there may be a varying number of individuals in the population; the frequency of contact may or may not be constant; and so forth. Each of these considerations influences the rate of spread of the state and should be accounted for in the formulation of the mathematical model.

Landau and Rapoport [1] considered a model of this problem under the following conditions:

(i) The number of individuals who become affected up to and including time $t = 0$, the start of the process, is a fixed constant denoted $R(0)$.

(ii) The total number of individuals affected up to and including time t is a deterministic function of time denoted by $R(t)$, $t \geq 0$.

(iii) There is no recovery.

(iv) Each individual can acquire the state at most once.

(v) The number of individuals in the total population is a known constant N throughout the spread of the state.

(vi) The frequency of contact γ is assumed to be the same for each pair and constant throughout the spread of the state.

(vii) The probability of transmission depends on t (the time of the whole process, called the *clock time*) and s (the time elapsed since the person in the state acquired the state, called the *private time*) and is denoted $p(t, s)$.

A simple example of the feasibility of allowing the transmission probability to depend on both t and s is provided in epidemiology, when the infectiousness of a diseased individual may decrease with time, so that if contact is made shortly after the carrier acquires the disease, the probability of transmission is relatively high.

With these ideas in mind, Landau and Rapoport arrived at the following deterministic integral equation of the Volterra type in $R'(t)$, the instantaneous rate at which the number of affected individuals is changing at time $t \geq 0$:

$$R'(t) = \gamma[N - R(t)]\{R(0)p(t, t) + \int_0^t R'(\tau)p(t, t - \tau)\, d\tau\}, \qquad t \geq 0. \quad (6.2.7)$$

With respect to our aim, consider the constant function $\gamma(t) = \gamma$, which represents the number of contacts per unit time per pair and which is assumed to be the same for each pair of individuals in the population. This is a constant which in a sense characterizes the population. In order to apply Eq. (6.2.7) to a given situation, the value of the constant γ must be determined based on the experimenter's prior knowledge of the population or on direct observation of the population. The usual procedure is to obtain several estimates for γ based on observations of samples drawn from the population and then solve the deterministic equation using as the "true" value of γ the mean of the experimental values. If this procedure were repeated many times, the mean values obtained will usually vary considerably and hence the value actually used could be quite unrepresentative of the true state of nature γ. Thus it would appear to be more realistic to assume that this "constant" γ is in reality a stochastic variable whose behavior is governed by some probability distribution function. We shall denote this stochastic variable by $\gamma(\omega)$. It is clear that the number of individuals affected up to time t, $R(t)$, will be influenced by the frequency of contacts between pairs, which we are now considering to be random. Thus we shall assume that for each $t \geq 0$ the quantity $R(t)$ is no longer a fixed constant but is in reality a random variable which we shall denote by $R(t;\omega)$. That is, $R(t)$ should not be considered as a deterministic function of time but rather as a random function of time. This tacitly implies that the derivative of $R(t;\omega)$ giving the instantaneous rate of change of the number of affected individuals in the population is also a random function, which we shall denote by $R'(t;\omega)$. That is, for each $t \geq 0$, $R'(t;\omega)$ is a random variable. Therefore we can now formulate the following stochastic model for communicable diseases:

$$R'(t;\omega) = \gamma(\omega)[N - R(t;\omega)][R(0;\omega)p(t,t)]$$
$$+ \int_0^t R'(\tau;\omega)p(t, t - \tau)\,d\tau], \qquad t \geq 0. \qquad (6.2.8)$$

As in the previous models, in order to simplify our theoretical investigation, we shall employ the following identifications:

$$x(t;\omega) \equiv R'(t;\omega),$$
$$h(t, x(t;\omega)) \equiv H(t;\omega) \equiv \gamma(\omega)[N - R(t;\omega)][R(0;\omega)p(t,t)],$$
$$k(t, \tau;\omega) \equiv \gamma(\omega)[N - R(t;\omega)]p(t, t - \tau),$$
$$f(\tau, x(\tau;\omega)) \equiv x(\tau;\omega) \equiv R'(\tau;\omega).$$

Using these identifications Eq. (6.2.8) is essentially Eq. (6.0.1) without random perturbation.

We shall make certain assumptions with regard to the behavior of the random functions which constitute the communicable diseases model: We shall assume that there exists a constant Q_1 such that $|\gamma(\omega)| \leq Q_1$ $\mathscr{P}$-a.e. and that for each t, $R(t;\omega)$ and $R'(t;\omega)$ have finite variances. The assumption that the number of individuals in the population during the spread of the state is a known constant N and the interpretation of $p(t, s)$ are retained as stated earlier.

In order to show that the random integral equation (6.2.8) possesses a unique random solution, we must show that the model satisfies the conditions of Corollary 2.1.6. To show that Condition (i) holds, consider

$$|||k(t, \tau;\omega)||| = |||\gamma(\omega)[N - R(t;\omega)]p(t, t - \tau)|||$$
$$= p(t, t - \tau)|||\gamma(\omega)[N - R(t;\omega)]|||.$$

However, due to the physical meaning of the random function $R(t;\omega)$, we have that $|N - R(t;\omega)| \leq N$ $\mathscr{P}$-a.e. for each t. Hence

$$|\gamma(\omega)[N - R(t;\omega)]| = |\gamma(\omega)||N - R(t;\omega)|$$
$$\leq |||\gamma(\omega)|||N \quad \mathscr{P}\text{-a.e.} \qquad \text{for each} \quad t.$$

6.2 APPLICATIONS TO BIOLOGICAL SYSTEMS

This implies that

$$|||\gamma(\omega)[N - R(t;\omega)]||| \leq |||\gamma(\omega)|||N$$

and that

$$|||k(t,\tau;\omega)||| \leq p(t, t-\tau)|||\gamma(\omega)|||N.$$

Taking $\Lambda_2 = |||\gamma(\omega)|||N$, Condition (i) can be satisfied if $p(t, t-\tau)$ is such that

$$p(t, t-\tau) \leq e^{-\alpha(t-\tau)} \quad \text{for some} \quad \alpha > 0$$

and $0 \leq \tau \leq t < \infty$. Take the function $g \equiv g(t) \equiv 1$. Then

$$\sup_{t \in R_+}\left\{\int_t^{t+1} g(\tau)\,d\tau\right\} = 1 < \infty.$$

The function $f(t, x)$ is defined by $f(t, x) = x$. It is clear that this function is continuous in t uniformly in x from $R_+ \times R \to R$ and $|f(t, 0)| = 0$. The Lipschitz condition is obviously satisfied, and thus Condition (ii) has been met.

By assumption $t \to H(t;\omega)$ is continuous from $R_+ \to L_2(\Omega, \mathscr{A}, \mathscr{P})$. It remains only to show that this map is bounded. Observe that

$$\left[\int_\Omega |H(t;\omega)|^2\,d\mathscr{P}(\omega)\right]^{\frac{1}{2}} = \left[\int_\Omega |\gamma(\omega)[N - R(t;\omega)]R(0;\omega)p(t,t)|^2\,d\mathscr{P}(\omega)\right]^{\frac{1}{2}}$$

$$\leq \left[\int_\Omega N^2|\gamma(\omega)R(0;\omega)|^2\,d\mathscr{P}(\omega)\right]^{\frac{1}{2}}$$

$$= N\|\gamma(\omega)R(0;\omega)\|$$

$$\leq N|||\gamma(\omega)|||\,\|R(0;\omega)\|.$$

Since this argument is independent of $t \geq 0$, we have that the map is bounded. Thus we can conclude that the random integral equation (6.2.8) possesses a unique random solution provided that $\|H(t;\omega)\|_C$, λ, and $\|f(t, 0)\|_{C_g}$ are sufficiently small. That is, small in the sense that

$$\|H(t;\omega)\|_C + K\|f(t, 0)\|_{C_g} \leq \rho(1 - \lambda K) \quad \text{and} \quad \lambda K < 1,$$

where $K = \|T\|^*$. Since $f(t, 0) = 0$, we have $\|f(t, 0)\|_{C_g} \equiv \|f(t, 0)\|_C = 0$ and the model requires only that $\lambda K < 1$ and $\|H(t;\omega)\|_C \leq \rho(1 - \lambda K)$, $\rho > 0$.

Although the model has been formulated for communicable diseases, a similar model can be developed in the case of the spread of a rumor throughout a community, when the "news value" of the rumor decreases with time so that a person would be less likely to communicate the rumor, say, two weeks after hearing it than he would only a few hours after hearing it.

CHAPTER VII

On a Nonlinear Random Integral Equation with Application to Stochastic Chemical Kinetics

7.0 Introduction

The aim of this chapter is to study a random vector integral equation of the form given by

$$x(t;\omega) = h(t, x(t;\omega)) + \int_0^t k(\tau, x(\tau;\omega);\omega)\,d\tau, \qquad t \geq 0, \qquad (7.0.1)$$

where $\omega \in \Omega$, the supporting set of a probability measure space $(\Omega, \mathscr{A}, \mathscr{P})$, $x(t;\omega)$ is the unknown m-dimensional vector-valued random function defined on R_+, the stochastic kernel $k(t, x(t;\omega);\omega)$ is an m-dimensional vector-valued function on R_+, and for each $t \in R_+$ and each m-dimensional vector-valued random function $x(t;\omega)$, $h(t, x(t;\omega))$ is an m-dimensional vector-valued random variable. More specifically, we are interested in the existence and uniqueness of a solution, a random vector-valued function

$x(t;\omega)$, and some special cases which are important for studying certain physical problems.

In addition, we shall give a stochastic formulation of a classical chemical kinetics problem. The formulation of such a model results in a random integral equation of the form given by (7.0.1).

In Section 7.1 we shall introduce some topological spaces and definitions and state and prove certain lemmas which are essential in our study. An existence theorem and some special cases are given in Section 7.2. In Section 7.3 we shall give a complete and precise formulation of a chemical kinetics problem which is quite realistic for describing the physical phenomenon. The results given in this chapter are due to Milton and Tsokos [1, 5].

7.1 Mathematical Preliminaries

In this section we shall give some definitions and concepts which are basic to our study.

Definition 7.1.1 Two random vectors $x(\omega) = \{x_1(\omega), x_2(\omega), \ldots, x_m(\omega)\}$ and $y(\omega) = \{y_1(\omega), y_2(\omega), \ldots, y_m(\omega)\}$ are said to be *equal* if and only if $x_i(\omega) = y_i(\omega)$ $\mathscr{P}$-a.e. for each $i = 1, 2, \ldots, m$.

Definition 7.1.2 We shall denote by $\psi(\Omega, \mathscr{A}, \mathscr{P})$ the set of all random vectors of the form $z(\omega) = \{z_1(\omega), z_2(\omega), \ldots, z_m(\omega)\}$ where, for each $i = 1, 2, \ldots, m$, $z_i(\omega)$ is an element of $L_\infty(\Omega, \mathscr{A}, \mathscr{P})$.

Lemma 7.1.1 The space $\psi(\Omega, \mathscr{A}, \mathscr{P})$ is a normed linear space over the real numbers with the usual definition of componentwise addition and scalar multiplication, where the norm in $\psi(\Omega, \mathscr{A}, \mathscr{P})$ is defined by

$$\|z(\omega)\|_{\psi(\Omega, \mathscr{A}, \mathscr{P})} = \max_i \|\|z_i\|\|.$$

PROOF The fact that $\psi(\Omega, \mathscr{A}, \mathscr{P})$ is a linear space follows from the fact that $L_\infty(\Omega, \mathscr{A}, \mathscr{P})$ is a linear space. That $\|\cdot\|_{\psi(\Omega, \mathscr{A}, \mathscr{P})}$ is a norm follows from the fact that $\|\|\cdot\|\|$ is a norm.

Definition 7.1.3 $C_\psi(R_+, \psi(\Omega, \mathscr{A}, \mathscr{P}))$ will denote the set of all continuous functions from R_+ into $\psi(\Omega, \mathscr{A}, \mathscr{P})$.

We remark that Definition 7.1.3 simply states that $t \to x(t;\omega) = \{x_1(t;\omega), x_2(t;\omega), \ldots, x_m(t;\omega)\}$ is continuous and that for each $t \in R_+$ and each $i = 1, 2, \ldots, m$, $x_i(t;\omega) \in L_\infty(\Omega, \mathscr{A}, \mathscr{P})$. Therefore for fixed $t \in R_+$

$$\|x(t;\omega)\|_{\psi(\Omega, \mathscr{A}, \mathscr{P})} = \max_i \|\|x_i(t;\omega)\|\|.$$

Also an element of the space $C_\psi(R_+, \psi(\Omega, \mathcal{A}, \mathcal{P}))$ is a random function. We shall be assuming that for each i the sample function $x_i(t; \omega)$ is continuous in t for each $\omega \in \Omega$. Thus, since we are working with a finite measure space, for each t and i, $E|x_i(t; \omega)| < \infty$. Defining the norm of $\psi(\Omega, \mathcal{A}, \mathcal{P})$ as in Lemma 7.1.1 will enable us to obtain a relatively simple norm defined in terms of the components of the random vector.

Lemma 7.1.2 The space $C_\psi(R_+, \psi(\Omega, \mathcal{A}, \mathcal{P}))$ is a linear space over the real numbers with the usual definitions of addition and scalar multiplication for continuous functions.

Lemma 7.1.3 The collection

$$F = \{\|x(t; \omega)\|_n : \|x(t; \omega)\|_n = \sup_{0 \leq t \leq n} \|x(t; \omega)\|_{\psi(\Omega, \mathcal{A}, \mathcal{P})}\}$$

for $n = 1, 2, \ldots$, is a family of semi-norms defined on $C_\psi(R_+, \psi(\Omega, \mathcal{A}, \mathcal{P}))$.

PROOF By definition, $x(t; \omega)$ is continuous from $R_+ \to \psi(\Omega, \mathcal{A}, \mathcal{P})$. Thus, given the compact set $[0, n] \subseteq R_+$ there exists some constant M_n such that $t \in [0, n]$ implies $\|x(t; \omega)\|_{\psi(\Omega, \mathcal{A}, \mathcal{P})} \leq M_n$. Hence $\{\|x(t; \omega)\|_{\psi(\Omega, \mathcal{A}, \mathcal{P})} : t \in [0, n]\}$ is bounded above by a constant which implies that the supremum exists, and thus that $\|x(t; \omega)\|_n$ is uniquely determined and also nonnegative. Furthermore,

$$\|\lambda x(t; \omega)\|_n = \sup_{0 \leq t \leq n} \{\|\lambda x(t; \omega)\|_{\psi(\Omega, \mathcal{A}, \mathcal{P})}\} = \sup_{0 \leq t \leq n} \{|\lambda| \|x(t; \omega)\|_{\psi(\Omega, \mathcal{A}, \mathcal{P})}\}$$

$$= |\lambda| \sup_{0 \leq t \leq n} \{\|x(t; \omega)\|_{\psi(\Omega, \mathcal{A}, \mathcal{P})}\} = |\lambda| \|x(t; \omega)\|_n.$$

Also,

$$\|x(t; \omega) + y(t; \omega)\|_n = \sup_{0 \leq t \leq n} \{\|x(t; \omega) + y(t; \omega)\|_{\psi(\Omega, \mathcal{A}, \mathcal{P})}\}$$

$$\leq \sup_{0 \leq t \leq n} \{\|x(t; \omega)\|_{\psi(\Omega, \mathcal{A}, \mathcal{P})} + \|y(t; \omega)\|_{\psi(\Omega, \mathcal{A}, \mathcal{P})}\}$$

$$\leq \sup_{0 \leq t \leq n} \{\|x(t; \omega)\|_{\psi(\Omega, \mathcal{A}, \mathcal{P})}\}$$

$$+ \sup_{0 \leq t \leq n} \{\|y(t; \omega)\|_{\psi(\Omega, \mathcal{A}, \mathcal{P})}\}$$

$$= \|x(t; \omega)\|_n + \|y(t; \omega)\|_n.$$

Therefore since the argument is independent of the choice of n, $\|\cdot\|_n$ is a semi-norm on $C_\psi(R_+, \psi(\Omega, \mathcal{A}, \mathcal{P}))$ for $n = 1, 2, 3, \ldots$.

Lemma 7.1.4 The family of semi-norms defined on $C_\psi(R_+, \psi(\Omega, \mathcal{A}, \mathcal{P}))$ satisfies the axiom of separation.

7.1 MATHEMATICAL PRELIMINARIES

PROOF Choose $x(t;\omega) \in C_\psi(R_+, \psi(\Omega, \mathscr{A}, \mathscr{P}))$, $x(t;\omega) \neq 0$. Thus there exists a $t_0 \in R_+$ such that $x(t_0;\omega) \neq 0$. Now choose a natural number such that $0 \leq t_0 \leq N$, and consider

$$\|x(t;\omega)\|_N = \sup_{0 \leq t \leq N} \{\|x(t;\omega)\|_{\psi(\Omega,\mathscr{A},\mathscr{P})}\} \geq \|x(t_0;\omega)\|_{\psi(\Omega,\mathscr{A},\mathscr{P})} > 0.$$

Therefore we have $\|x(t;\omega)\|_N \neq 0$, which implies that the family F of semi-norms satisfies the axiom of separation.

Lemma 7.1.5 The space $C_\psi(R_+, \psi(\Omega, \mathscr{A}, \mathscr{P}))$ can be topologized by the family of semi-norms F and the topology obtained is locally convex and Hausdorff.

The proof follows from Theorems 1.1.11 and 1.1.12.

Lemma 7.1.6 The topology τ on $C_\psi(R_+, \psi(\Omega, \mathscr{A}, \mathscr{P}))$ induced by the family of semi-norms F is metrizable and the metric is defined by

$$\rho[x(t;\omega), y(t;\omega)] = \sum_{n=1}^{\infty} \frac{1}{2^n} \frac{\|x(t;\omega) - y(t;\omega)\|_n}{1 + \|x(t;\omega) - y(t;\omega)\|_n}.$$

The proof follows from the fact that $(\|\cdot\|_n)$ is an increasing sequence of semi-norms on $C_\psi(R_+, \psi(\Omega, \mathscr{A}, \mathscr{P}))$ and Theorem 1.1.13.

Lemma 7.1.7 The topology τ on $C_\psi(R_+, \psi(\Omega, \mathscr{A}, \mathscr{P}))$ induced by F is the topology of uniform convergence.

PROOF Assume that

$$\lim_{m \to \infty} \|x^m(t;\omega) - x(t;\omega)\|_{\psi(\Omega,\mathscr{A},\mathscr{P})} = 0$$

uniformly on every compact interval. We shall show that given an $\varepsilon > 0$, there exists a natural number N_ε such that $m > N_\varepsilon$ implies that

$$\rho(x^m(t;\omega), x(t;\omega)) < \varepsilon.$$

That is, we need to show that for $m > N_\varepsilon$

$$\sum_{n=1}^{\infty} \frac{1}{2^n} \frac{\|x^m(t;\omega) - x(t;\omega)\|_n}{1 + \|x^m(t;\omega) - x(t;\omega)\|_n} < \varepsilon.$$

Since $\sum_{n=1}^{\infty} 1/2^n$ is finite, there exists a natural number M such that $\sum_{n=M+1}^{\infty} 1/2^n < \varepsilon/2$. Also for each n and m we have

$$\|x^m(t;\omega) - x(t;\omega)\|_n / [1 + \|x^m(t;\omega) - x(t;\omega)\|_n] < 1.$$

Thus we have

$$\sum_{n=M+1}^{\infty} \frac{1}{2^n} \frac{\|x^m(t;\omega) - x(t;\omega)\|_n}{1 + \|x^m(t;\omega) - x(t;\omega)\|_n} < \sum_{n=M+1}^{\infty} \frac{1}{2^n} < \frac{\varepsilon}{2}.$$

On an interval $[0, M]$ there exists a natural number $N_{\varepsilon,M}$ such that for $m > N_{\varepsilon,M}$ we have

$$\|x^m(t;\omega) - x(t;\omega)\|_{\psi(\Omega,\mathscr{A},\mathscr{P})} < q/(1-q)$$

for every $t \in [0, M]$ where $q = (\varepsilon/2)(1/\sum_{n=1}^{M} 1/2^n)$. Since

$$\|x^m(t;\omega) - x(t;\omega)\|_M = \sup_{0 \leq t \leq M} \{\|x^m(t;\omega) - x(t;\omega)\|_{\psi(\Omega,\mathscr{A},\mathscr{P})}\},$$

for $m > N_{\varepsilon,M}$ we have

$$\|x^m(t;\omega) - x(t;\omega)\|_M \leq q/(1-q).$$

Because the sequence $(\|\cdot\|_n)$ of semi-norms is increasing, for $n = 1, 2, \ldots, M$ and $m > N_{\varepsilon,M}$, $\|x^m(t;\omega) - x(t;\omega)\|_n \leq q/(1-q)$. Choose $N_\varepsilon = \max(M, N_{\varepsilon,M})$. Then for $m > N_\varepsilon$ we have

$$\rho(x^m(t;\omega), x(t;\omega)) = \sum_{n=1}^{\infty} \frac{1}{2^n} \frac{\|x^m(t;\omega) - x(t;\omega)\|_n}{1 + \|x^m(t;\omega) - x(t;\omega)\|_n}$$

$$= \sum_{n=1}^{M} \frac{1}{2^n} \frac{\|x^m(t;\omega) - x(t;\omega)\|_n}{1 + \|x^m(t;\omega) - x(t;\omega)\|_n}$$

$$+ \sum_{n=M+1}^{\infty} \frac{1}{2^n} \frac{\|x^m(t;\omega) - x(t;\omega)\|_n}{1 + \|x^m(t;\omega) - x(t;\omega)\|_n}$$

$$< q \sum_{n=1}^{M} \frac{1}{2^n} + \frac{\varepsilon}{2}$$

$$= \left(\sum_{n=1}^{M} \frac{1}{2^n}\right)\left(\frac{\varepsilon}{2}\right)\left(\sum_{n=1}^{M} \frac{1}{2^n}\right)^{-1} + \frac{\varepsilon}{2}$$

$$= \varepsilon.$$

Therefore

$$\lim_{m \to \infty} \|x^m(t;\omega) - x(t;\omega)\|_{\psi(\Omega,\mathscr{A},\mathscr{P})} = 0$$

uniformly on closed intervals implies that $x^m(t;\omega) \to x(t;\omega)$ in the metric topology on $C_\psi(R_+, \psi(\Omega, \mathscr{A}, \mathscr{P}))$. Now assume that $x^m(t;\omega) \to x(t;\omega)$ in the metric topology on $C_\psi(R_+, \psi(\Omega, \mathscr{A}, \mathscr{P}))$, but $\|x^m(t;\omega) - x(t;\omega)\|_{\psi(\Omega,\mathscr{A},\mathscr{P})}$ does not converge uniformly to zero on some interval $[0, Q]$. Without loss of generality assume that Q is a natural number. Thus by definition of uniform convergence there exists an $\varepsilon > 0$ such that for each natural number Z there exists a point $t_Z \in [0, Q]$ and a $k > Z$ such that

$$\|x^k(t_Z;\omega) - x(t_Z;\omega)\|_{\psi(\Omega,\mathscr{A},\mathscr{P})} > \varepsilon.$$

Since $x^m(t;\omega) \to x(t;\omega)$ in the metric topology, there exists an M such that, for $m > M$, we have

$$\sum_{n=1}^{\infty} \frac{1}{2^n} \frac{\|x^m(t;\omega) - x(t;\omega)\|_n}{1 + \|x^m(t;\omega) - x(t;\omega)\|_n} < \frac{1}{2^Q} \frac{\varepsilon}{1+\varepsilon}.$$

In particular, $m > M$ implies

$$\frac{1}{2^Q} \frac{\|x^m(t;\omega) - x(t;\omega)\|_Q}{1 + \|x^m(t;\omega) - x(t;\omega)\|_Q} < \frac{1}{2^Q} \frac{\varepsilon}{1+\varepsilon}.$$

However, since

$$\|x^m(t;\omega) - x(t;\omega)\|_Q \geq \|x^m(t;\omega) - x(t;\omega)\|_{\psi(\Omega,\mathscr{A},\mathscr{P})}$$

for $t \in [0, Q]$ we have for $m > M$ and $t \in [0, Q]$ the following inequality:

$$\frac{1}{2^Q} \frac{\|x^m(t;\omega) - x(t;\omega)\|_{\psi(\Omega,\mathscr{A},\mathscr{P})}}{1 + \|x^m(t;\omega) - x(t;\omega)\|_{\psi(\Omega,\mathscr{A},\mathscr{P})}}$$
$$\leq \frac{1}{2^Q} \frac{\|x^m(t;\omega) - x(t;\omega)\|_Q}{1 + \|x^m(t;\omega) - x(t;\omega)\|_Q} < \frac{1}{2^Q} \frac{\varepsilon}{1+\varepsilon}. \qquad (7.1.1)$$

Inequality (7.1.1) holds because of the fact that the function $g(t) = t/(1+t)$ is increasing. But since uniform convergence does not exist on the interval $[0, Q]$, there exist a $t_0 \in [0, Q]$ and $r > M$ such that

$$\|x^r(t_0;\omega) - x(t_0;\omega)\|_{\psi(\Omega,\mathscr{A},\mathscr{P})} > \varepsilon,$$

implying that

$$\frac{1}{2^Q} \frac{\|x^r(t_0;\omega) - x(t_0;\omega)\|_{\psi(\Omega,\mathscr{A},\mathscr{P})}}{1 + \|x^r(t_0;\omega) - x(t_0;\omega)\|_{\psi(\Omega,\mathscr{A},\mathscr{P})}} \geq \frac{1}{2^Q} \frac{\varepsilon}{1+\varepsilon}.$$

This is a contradiction to inequality (7.1.1) and hence the proof is complete.

The following lemma is analogous to Lemma 2.1.1.

Lemma 7.1.8 Let T be a continuous linear operator from

$$C_\psi(R_+, \psi(\Omega, \mathscr{A}, \mathscr{P})) \to C_\psi(R_+, \psi(\Omega, \mathscr{A}, \mathscr{P})).$$

If B and D are Banach spaces stronger than $C_\psi(R_+, \psi(\Omega, \mathscr{A}, \mathscr{P}))$ and if (B, D) is admissible with respect to T, then T is continuous from B to D. Thus, as stated in Chapters II and IV, $T: B \to D$ is bounded and there exists a constant K such that

$$\|(Tx)(t;\omega)\|_D \leq K \|x(t;\omega)\|_B.$$

Thus we use the usual method to define the norm of T by

$$M = \|T\|^* = \sup\{\|(Tx)(t;\omega)\|_D / \|x(t;\omega)\|_B : x(t;\omega) \in B, \quad \|x(t;\omega)\|_B \neq 0\}.$$

Definition 7.1.4 The random vector-valued function $x(t;\omega)$ on R_+ is a *random solution* of Eq. (7.0.1) if for each $t \in R_+$, $x(t;\omega)$ is a vector random variable and satisfies Eq. (7.0.1) $\mathscr{P}$-a.e.

The following lemma is essential for obtaining the main results of the chapter.

Lemma 7.1.9 The operator T,

$$(Tx)(t;\omega) = \int_0^t x(\tau;\omega)\,d\tau,$$

defined on $C_\psi(R_+, \psi(\Omega, \mathscr{A}, \mathscr{P}))$ is a continuous linear operator from $C_\psi(R_+, \psi(\Omega, \mathscr{A}, \mathscr{P}))$ into $C_\psi(R_+, \psi(\Omega, \mathscr{A}, \mathscr{P}))$.

PROOF For fixed t we shall show that $(Tx)(t;\omega) \in \psi(\Omega, \mathscr{A}, \mathscr{P})$. It is sufficient to show that for fixed t and each i, the function of ω

$$(Tx_i)(t;\omega) = \int_0^t x_i(\tau;\omega)\,d\tau$$

is $\mathscr{P}$-essentially bounded. Consider

$$|(Tx_i)(t;\omega)| = \left|\int_0^t x_i(\tau;\omega)\,d\tau\right| \leq \int_0^t |x_i(\tau;\omega)|\,d\tau \leq \int_0^t |||x_i(\tau;\omega)|||\,d\tau$$

$\mathscr{P}$-a.e. Now, $x(\tau;\omega)$ is continuous from R_+ into $\psi(\Omega, \mathscr{A}, \mathscr{P})$, since $x(\tau;\omega) \in C_\psi(R_+, \psi(\Omega, \mathscr{A}, \mathscr{P}))$ by assumption. This implies that $x(\tau;\omega)$ is continuous on $[0, t]$ and hence there exists an M such that $\tau \in [0, t]$ implies $\|x(\tau;\omega)\|_{\psi(\Omega,\mathscr{A},\mathscr{P})} \leq M$. Thus, for $\tau \in [0, t]$, $\max_i |||x_i(\tau;\omega)||| \leq M$, implying that, for $\tau \in [0, t]$, $|||x_i(\tau;\omega)||| \leq M$ for each i. Therefore

$$|(Tx_i)(t;\omega)| \leq \int_0^t |||x_i(\tau;\omega)|||\,d\tau, \qquad \mathscr{P}\text{-a.e.}$$

$$\leq \int_0^t M\,d\tau = tM.$$

Thus for fixed t, $|(Tx_i)(t;\omega)| \leq tM$ $\mathscr{P}$-a.e., implying that $(Tx_i)(t;\omega)$ is $\mathscr{P}$-essentially bounded. Hence $(Tx)(t;\omega)$ is a function from R_+ into $\psi(\Omega, \mathscr{A}, \mathscr{P})$.

To show that $(Tx)(t;\omega) \in C_\psi(R_+, \psi(\Omega, \mathscr{A}, \mathscr{P}))$, it is also necessary to show that $(Tx)(t;\omega)$ is continuous on R_+. Therefore we must show that $t_n \to t$ in R_+ implies that $(Tx)(t_n;\omega) \to (Tx)(t;\omega)$ in $\psi(\Omega, \mathscr{A}, \mathscr{P})$ as $n \to \infty$. This means that we must show that $\|(Tx)(t_n;\omega) - (Tx)(t;\omega)\|_{\psi(\Omega,\mathscr{A},\mathscr{P})}$ can be made arbitrarily small for large enough n. Since

$$\|(Tx)(t_n;\omega) - (Tx)(t;\omega)\|_{\psi(\Omega,\mathscr{A},\mathscr{P})} = \max_i |||(Tx_i)(t_n;\omega) - (Tx_i)(t;\omega)|||,$$

7.1 MATHEMATICAL PRELIMINARIES

it is sufficient to say that for each i, $|||(Tx_i)(t_n;\omega) - (Tx_i)(t;\omega)|||$ can be made arbitrarily small. Take $\varepsilon > 0$ and consider the set $[0, t + \varepsilon]$. Since $x(\tau;\omega) \in C_\psi(R_+, \psi(\Omega, \mathscr{A}, \mathscr{P}))$, it is continuous from R_+ into $\psi(\Omega, \mathscr{A}, \mathscr{P})$. Hence it is also continuous on the compact set $[0, t + \varepsilon]$ and there exists a constant M_ε such that $\tau \in [0, t + \varepsilon]$ implies $\|x(\tau;\omega)\|_{\psi(\Omega,\mathscr{A},\mathscr{P})} < M_\varepsilon$. Without loss of generality we may assume $M_\varepsilon > 1$. Thus for $\tau \in [0, t + \varepsilon]$, $\max_i |||x_i(\tau;\omega)||| < M_\varepsilon$. This, in turn, implies that for $\tau \in [0, t + \varepsilon]$ and each i, $|||x_i(\tau;\omega)||| < M_\varepsilon$. Hence

$$|x_i(\tau;\omega)| \leq |||x_i(\tau;\omega)||| \leq M_\varepsilon, \qquad \mathscr{P}\text{-a.e.} \qquad (7.1.2)$$

for $\tau \in [0, t + \varepsilon]$. Since $t_n \to t$ in R_+, there exists an N_ε such that for $n > N_\varepsilon$, $|t_n - t| < \varepsilon/M_\varepsilon < \varepsilon$. Consider $|(Tx_i)(t_n;\omega) - (Tx_i)(t;\omega)|$ for $n > N_\varepsilon$ where i is arbitrary but fixed. By definition

$$|(Tx_i)(t_n;\omega) - (Tx_i)(t;\omega)| = \left| \int_0^{t_n} x_i(\tau;\omega)\,d\tau - \int_0^t x_i(\tau;\omega)\,d\tau \right|.$$

Then for $t_n > t$ we have

$$|(Tx_i)(t_n;\omega) - (Tx_i)(t;\omega)| = \left| \int_t^{t_n} x_i(\tau;\omega)\,d\tau \right| \leq \int_t^{t_n} |x_i(\tau;\omega)|\,d\tau.$$

Since $[t, t_n] \subseteq [0, t + \varepsilon]$, we have by inequality (7.1.2) that

$$|(Tx_i)(t_n;\omega) - (Tx_i)(t;\omega)| \leq \int_t^{t_n} M_\varepsilon\,d\tau, \qquad \mathscr{P}\text{-a.e.}$$

implying that

$$|(Tx_i)(t_n;\omega) - (Tx_i)(t;\omega)| \leq M_\varepsilon t_n - M_\varepsilon t = |t_n - t|M_\varepsilon < (\varepsilon/M_\varepsilon)M_\varepsilon = \varepsilon.$$

That is, for $n > N_\varepsilon$, $|(Tx_i)(t_n;\omega) - (Tx_i)(t;\omega)|$ is arbitrarily small for almost all ω. By definition we have

$$|||(Tx_i)(t_n;\omega) - (Tx_i)(t;\omega)||| = \inf S,$$

where

$$S = \{Z : \mathscr{P}\{\omega : |(Tx_i)(t_n;\omega) - (Tx_i)(t;\omega)| > Z\} = 0, \; Z > 0\}.$$

Thus we have shown that $\mathscr{P}\{\omega : |(Tx_i)(t_n;\omega) - (Tx_i)(t;\omega)| > \varepsilon\} = 0$ for $n > N_\varepsilon$. Hence $\varepsilon \in S$ and

$$|||(Tx_i)(t_n;\omega) - (Tx_i)(t;\omega)||| < \varepsilon$$

for $n > N_\varepsilon$, as was desired. Thus $(Tx)(t;\omega)$ is continuous from R_+ to $\psi(\Omega, \mathscr{A}, \mathscr{P})$, implying that T does map $C_\psi(R_+, \psi(\Omega, \mathscr{A}, \mathscr{P}))$ into $C_\psi(R_+, \psi(\Omega, \mathscr{A}, \mathscr{P}))$.

We now need to show that the mapping T itself is continuous. Let $x^n(t;\omega) \to^\tau x(t;\omega)$ as $n \to \infty$. We need to show that $(Tx^n)(t;\omega) \to^\tau (Tx)(t;\omega)$ as $n \to \infty$. The topology on the space $C_\psi(R_+, \psi(\Omega, \mathscr{A}, \mathscr{P}))$ is the topology of uniform convergence. Select $\varepsilon > 0$ and pick an interval $[0, Q] \subset R_+$ where Q is arbitrary but fixed. Since $x^n(t;\omega) \to^\tau x(t;\omega)$, there exists an $N_{\varepsilon,Q}$ such that $n > N_{\varepsilon,Q}$ implies that

$$\|x^n(t;\omega) - x(t;\omega)\|_{\psi(\Omega,\mathscr{A},\mathscr{P})} < \varepsilon/Q$$

for all $t \in [0, Q]$. By definition of the norm in $\psi(\Omega, \mathscr{A}, \mathscr{P})$,

$$\max_i \||x_i^n(t;\omega) - x_i(t;\omega)\|| < \varepsilon/Q$$

for $t \in [0, Q]$ and $n > N_{\varepsilon,Q}$. Thus

$$\||x_i^n(t;\omega) - x_i(t;\omega)\|| < \varepsilon/Q$$

for $t \in [0, Q]$ and $n > N_{\varepsilon,Q}$. This implies that

$$|x_i^n(t;\omega) - x_i(t;\omega)| < \varepsilon/Q$$

for $n > N_{\varepsilon,Q}$, and $t \in [0, Q]$ $\mathscr{P}$-a.e., Consider $|(Tx_i^n)(t;\omega) - (Tx_i)(t;\omega)|$ for $n > N_{\varepsilon,Q}$ and $t \in [0, Q]$. We can write

$$|(Tx_i^n)(t;\omega) - (Tx_i)(t;\omega)| = \left|\int_0^t x_i^n(\tau;\omega)\,d\tau - \int_0^t x_i(\tau;\omega)\,d\tau\right|$$

$$= \left|\int_0^t [x_i^n(\tau;\omega) - x_i(\tau;\omega)]\,d\tau\right|$$

$$\leq \int_0^t |x_i^n(\tau;\omega) - x_i(\tau;\omega)|\,d\tau$$

$$\leq \int_0^Q |x_i^n(\tau;\omega) - x_i(\tau;\omega)|\,d\tau$$

$$\leq \int_0^Q (\varepsilon/Q)\,d\tau = \varepsilon, \qquad \mathscr{P}\text{-a.e.}$$

That is, for $n > N_{\varepsilon,Q}$ and $t \in [0, Q]$

$$|(Tx_i^n)(t;\omega) - (Tx)(t;\omega)| < \varepsilon, \qquad \mathscr{P}\text{-a.e.}$$

Thus for $n > N_{\varepsilon,Q}$ and $t \in [0, Q]$

$$\mathscr{P}\{\omega : |(Tx_i^n)(t;\omega) - (Tx)(t;\omega)| > \varepsilon\} = 0.$$

This implies that for $n > N_{\varepsilon,Q}$ and $t \in [0, Q]$

$$\||(Tx_i^n)(t;\omega) - (Tx_i)(t;\omega)\|| < \varepsilon.$$

7.1 MATHEMATICAL PRELIMINARIES

Since the argument is independent of the choice of i, it holds for each i. Therefore we can write that for $n > N_{\varepsilon,Q}$ and $t \in [0, Q]$

$$\|(Tx^n)(t;\omega) - (Tx)(t;\omega)\|_{\psi(\Omega,\mathscr{A},\mathscr{P})} = \max_i \|\|(Tx_i^n)(t;\omega) - (Tx)(t;\omega)\|\| < \varepsilon.$$

This implies that $\|(Tx^n)(t;\omega) - (Tx)(t;\omega)\|_{\psi(\Omega,\mathscr{A},\mathscr{P})}$ converges to zero uniformly on $[0, Q]$, which is equivalent to saying that $(Tx^n)(t;\omega) \to^\tau (Tx)(t;\omega)$ and the proof is complete.

Definition 7.1.5 $C'_g = C'_g(R_+, \psi(\Omega, \mathscr{A}, \mathscr{P}))$ will denote the set of all continuous functions $x(t;\omega)$ from R_+ into $\psi(\Omega, \mathscr{A}, \mathscr{P})$ such that for g a positive-valued continuous function on R_+ we have

$$\|x(t;\omega)\|_{\psi(\Omega,\mathscr{A},\mathscr{P})} \leq Zg(t),$$

where Z is some positive constant which depends on $x(t;\omega)$.

Definition 7.1.6 $C' = C'(R_+, \psi(\Omega, \mathscr{A}, \mathscr{P}))$ will denote the set of all continuous and bounded functions $x(t;\omega)$ from R_+ into $\psi(\Omega, \mathscr{A}, \mathscr{P})$.

Lemma 7.1.10 The space $C'_g(R_+, \psi(\Omega, \mathscr{A}, \mathscr{P}))$ is a normed linear subspace of $C_\psi(R_+, \psi(\Omega, \mathscr{A}, \mathscr{P}))$ where the norm in C'_g is given by

$$\|x(t;\omega)\|_{C'_g} = \sup_{0 \leq t}\{\|x(t;\omega)\|_{\psi(\Omega,\mathscr{A},\mathscr{P})}/g(t)\}.$$

PROOF Let $x(t;\omega), y(t;\omega) \in C'_g(R_+, \psi(\Omega, \mathscr{A}, \mathscr{P}))$. Then we can write

$$\|x(t;\omega) + y(t;\omega)\|_{\psi(\Omega,\mathscr{A},\mathscr{P})} \leq \|x(t;\omega)\|_{\psi(\Omega,\mathscr{A},\mathscr{P})} + \|y(t;\omega)\|_{\psi(\Omega,\mathscr{A},\mathscr{P})}$$

$$\leq Z_1 g(t) + Z_2 g(t) \leq (Z_1 + Z_2)g(t),$$

$$\|\alpha x(t;\omega)\| = |\alpha|\,\|x(t;\omega)\| \leq |\alpha|Zg(t) = Bg(t).$$

Thus the space $C'_g(R_+, \psi(\Omega, \mathscr{A}, \mathscr{P}))$ is a linear subspace of $C_\psi(R_+, \psi(\Omega, \mathscr{A}, \mathscr{P}))$. Also from the definition we know that $\|\cdot\|_{C'_g} \geq 0$. The fact that $\|\cdot\|_{C'_g} = 0$ if and only if $x(t;\omega) = 0$ follows from the fact that $\|\cdot\|_{\psi(\Omega,\mathscr{A},\mathscr{P})}$ is a norm by Lemma 7.1.1. The fact that

$$\|\alpha x(t;\omega)\|_{C'_g} = |\alpha|\,\|x(t;\omega)\|_{C'_g}$$

also follows directly from the fact that $\|\cdot\|_{\psi(\Omega,\mathscr{A},\mathscr{P})}$ is a norm. To show that the triangle inequality holds, consider

$$\|x(t;\omega) + y(t;\omega)\|_{C'_g} = \sup_{0 \leq t}\{\|x(t;\omega) + y(t;\omega\|_{\psi(\Omega,\mathscr{A},\mathscr{P})}/g(t)\}$$

$$\leq \sup_{0 \leq t}\{[\|x(t;\omega)\|_{\psi(\Omega,\mathscr{A},\mathscr{P})}/g(t)] + [\|y(t;\omega)\|_{\psi(\Omega,\mathscr{A},\mathscr{P})}/g(t)]\}$$

$$\leq \sup_{0 \leq t}\{\|x(t;\omega)\|_{\psi(\Omega,\mathscr{A},\mathscr{P})}/g(t)\}$$

$$+ \sup_{0 \leq t}\{\|y(t;\omega)\|_{\psi(\Omega,\mathscr{A},\mathscr{P})}/g(t)\}$$

$$= \|x(t;\omega)\|_{C'_g} + \|y(t;\omega)\|_{C'_g}.$$

VII AN APPLICATION TO STOCHASTIC CHEMICAL KINETICS

Lemma 7.1.11 The normed linear space $\psi(\Omega, \mathscr{A}, \mathscr{P})$ is complete.

PROOF Let $Z^n(\omega)$ be a Cauchy sequence in $\psi(\Omega, \mathscr{A}, \mathscr{P})$. We can write the sequence in the form

$$(Z^n(\omega)) = (\{Z_1^n(\omega), Z_2^n(\omega), \ldots, Z_m^n(\omega)\}).$$

We shall show first that for each $i = 1, 2, \ldots, m$ the sequence $(Z_i^n(\omega))$ is Cauchy in $L_\infty(\Omega, \mathscr{A}, \mathscr{P})$. Since $Z^n(\omega)$ is Cauchy in $\psi(\Omega, \mathscr{A}, \mathscr{P})$, given $\varepsilon > 0$, there exists an N_ε such that $s, n > N_\varepsilon$ implies that

$$\|Z^s(\omega) - Z^n(\omega)\|_{\psi(\Omega, \mathscr{A}, \mathscr{P})} < \varepsilon.$$

Hence, by definition, for $s, n > N_\varepsilon$ we have that

$$\max_i \|\!|Z_i^s(\omega) - Z_i^n(\omega)|\!\| < \varepsilon.$$

This implies that for $s, n > N_\varepsilon$ and each i

$$\|\!|Z_i^s(\omega) - Z_i^n(\omega)|\!\| < \varepsilon.$$

Therefore for each i the sequence $Z_i^n(\omega)$ is Cauchy in $L_\infty(\Omega, \mathscr{A}, \mathscr{P})$. Since $L_\infty(\Omega, \mathscr{A}, \mathscr{P})$ is a Banach space, $Z_i^n(\omega)$ converges in $L_\infty(\Omega, \mathscr{A}, \mathscr{P})$ to an element $Z_i(\omega)$. We shall now show that the sequence $Z^n(\omega)$ converges in $\psi(\Omega, \mathscr{A}, \mathscr{P})$ to the element $Z(\omega)$, where $Z(\omega) = (Z_1(\omega), Z_2(\omega), \ldots, Z_m(\omega))$. Since for each i, $Z_i^n(\omega) \to Z_i(\omega)$ given $\varepsilon > 0$, there exists an $N_{i,\varepsilon}$ such that $n > N_{i,\varepsilon}$ implies that

$$\|\!|Z_i^n(\omega) - Z_i(\omega)|\!\| < \varepsilon.$$

Choose $N_\varepsilon = \max_i\{N_{i,\varepsilon}\}$; then for $n > N_\varepsilon$

$$\max_i \|\!|Z_i^n(\omega) - Z_i(\omega)|\!\| < \varepsilon.$$

Hence, by definition, for $n > N_\varepsilon$

$$\|Z^n(\omega) - Z(\omega)\|_{\psi(\Omega, \mathscr{A}, \mathscr{P})} < \varepsilon,$$

implying that $Z^n(\omega)$ converges to $Z(\omega)$, and the proof is complete.

Lemma 7.1.12 The space $C'_g(R_+, \psi(\Omega, \mathscr{A}, \mathscr{P}))$ is complete.

PROOF Let $x^n(t; \omega)$ be a Cauchy sequence in $C'_g(R_+, \psi(\Omega, \mathscr{A}, \mathscr{P}))$. Fix $t_0 \in R_+$ and consider $x^n(t_0; \omega)$. We shall show that the sequence $x^n(t_0; \omega)$ is Cauchy in $\psi(\Omega, \mathscr{A}, \mathscr{P})$. Choose $\varepsilon > 0$. Since $x^n(t; \omega)$ is Cauchy in $C'_g(R_+, \psi(\Omega, \mathscr{A}, \mathscr{P}))$, there exists an N_{t_0} such that $s, n > N_{t_0}$ implies

$$\|x^n(t; \omega) - x^s(t; \omega)\|_{C'_g} < \varepsilon/g(t_0).$$

Thus by definition

$$\sup_{0 \leq t}\{\|x^n(t; \omega) - x^s(t; \omega)\|_{\psi(\Omega, \mathscr{A}, \mathscr{P})}/g(t)\} < \varepsilon/g(t_0),$$

7.1 MATHEMATICAL PRELIMINARIES

implying that in particular for $n, s > N_{t_0}$

$$\|x^n(t_0;\omega) - x^s(t_0;\omega)\|_{\psi(\Omega,\mathscr{A},\mathscr{P})}/g(t_0) < \varepsilon/g(t_0).$$

That is, for $n, s > N_{t_0}$, $\|x^n(t_0;\omega) - x^s(t_0;\omega)\|_{\psi(\Omega,\mathscr{A},\mathscr{P})} < \varepsilon$, implying that $x^n(t_0;\omega)$ is a Cauchy sequence in $\psi(\Omega, \mathscr{A}, \mathscr{P})$. Since $\psi(\Omega, \mathscr{A}, \mathscr{P})$ is a Banach space, there exists an $x(t_0;\omega)$ such that $x^n(t_0;\omega) \to x(t_0;\omega)$. Since the argument is independent of the choice of t_0, $x^n(t;\omega) \to x(t;\omega)$ in $\psi(\Omega, \mathscr{A}, \mathscr{P})$ for each $t \in R_+$. We claim that the function $t \to x(t;\omega)$ is an element of $C'_g(R_+, \psi(\Omega, \mathscr{A}, \mathscr{P}))$. Choose $\varepsilon > 0$. Since $x^n(t;\omega)$ is Cauchy in $C'_g(R_+, \psi(\Omega, \mathscr{A}, \mathscr{P}))$, there exists an N_ε such that $s, n > N_\varepsilon$ implies

$$\|x^s(t;\omega) - x^n(t;\omega)\|_{C'_g} < \varepsilon/3.$$

Choose $t_0 \in R_+$ arbitrary but fixed and $s(t_0, \varepsilon)$ such that $s(t_0, \varepsilon) > N_\varepsilon$ and also such that

$$\|x^{s(t_0,\varepsilon)}(t_0;\omega) - x(t_0;\omega)\|_{\psi(\Omega,\mathscr{A},\mathscr{P})} < (\varepsilon/3)g(t_0).$$

Now we can write

$$\|x^n(t_0;\omega) - x(t_0;\omega)\|_{\psi(\Omega,\mathscr{A},\mathscr{P})}/g(t_0)$$
$$= \|x^n(t_0;\omega) - x^{s(t_0,\varepsilon)}(t_0;\omega) + x^{s(t_0,\varepsilon)}(t_0;\omega) - x(t_0;\omega)\|_{\psi(\Omega,\mathscr{A},\mathscr{P})}/g(t_0)$$
$$\leq [\|x^n(t_0;\omega) - x^{s(t_0;\varepsilon)}(t_0;\omega)\|_{\psi(\Omega,\mathscr{A},\mathscr{P})}/g(t_0)]$$
$$+ [\|x^{s(t_0,\varepsilon)}(t_0;\omega) - x(t_0;\omega)\|_{\psi(\Omega,\mathscr{A},\mathscr{P})}/g(t_0)] \leq \tfrac{1}{3}\varepsilon + \tfrac{1}{3}\varepsilon < \varepsilon$$

for all $n > N_\varepsilon$. Note that the integer N_ε is independent of the choice of t_0 and since t_0 was arbitrary, we may conclude that for $n > N_\varepsilon$

$$\sup_{0 \leq t}\{\|x^n(t;\omega) - x(t;\omega)\|_{\psi(\Omega,\mathscr{A},\mathscr{P})}/g(t)\} < \varepsilon.$$

To show that there exists a constant Z such that $\|x(t;\omega)\|_{\psi(\Omega,\mathscr{A},\mathscr{P})} \leq Zg(t)$, consider the following. By the preceding argument we can choose an n such that

$$\sup_{0 \leq t}\{\|x^n(t;\omega) - x(t;\omega)\|_{\psi(\Omega,\mathscr{A},\mathscr{P})}/g(t)\} < 1.$$

Now

$$\|x(t;\omega)\|_{\psi(\Omega,\mathscr{A},\mathscr{P})}/g(t) = \|x(t;\omega) - x^n(t;\omega) + x^n(t;\omega)\|_{\psi(\Omega,\mathscr{A},\mathscr{P})}/g(t)$$
$$\leq [\|x(t;\omega) - x^n(t;\omega)\|_{\psi(\Omega,\mathscr{A},\mathscr{P})}/g(t)]$$
$$+ [\|x^n(t;\omega)\|_{\psi(\Omega,\mathscr{A}\mathscr{P})}/g(t)]$$
$$\leq 1 + [\|x^n(t;\omega)\|_{\psi(\Omega,\mathscr{A},\mathscr{P})}/g(t)]$$
$$\leq 1 + \|x^n(t;\omega)\|_{C'_g} = Z.$$

Therefore
$$\|x(t;\omega)\|_{\psi(\Omega,\mathscr{A},\mathscr{P})} \leqslant Zg(t).$$

To show that the function $t \to x(t;\omega)$ is continuous, we must show that $t_m \to t$ in R_+ implies that $x(t_m;\omega) \to x(t;\omega)$ in $\psi(\Omega,\mathscr{A},\mathscr{P})$ as $m \to \infty$. Fix $t_0 \in R_+$. Let $t_m \to t_0$. Choose $\varepsilon > 0$. There exists an N_1 such that $m > N_1$ implies $|g(t_m) - g(t_0)| < g(t_0)$. By the first argument we can choose an n large enough so that
$$\sup_{0 \leqslant t} \{\|x^n(t;\omega) - x(t;\omega)\|_{\psi(\Omega,\mathscr{A},\mathscr{P})}/g(t)\} < \tfrac{1}{4}\varepsilon/g(t_0).$$

Since $x^n(t;\omega) \in C'_g(R_+, \psi(\Omega,\mathscr{A},\mathscr{P}))$, there exists an N_2 such that $m > N_2$ implies
$$\|x^n(t_m;\omega) - x^n(t_0;\omega)\| < \tfrac{1}{4}\varepsilon.$$

Let $N = \max(N_1, N_2)$. Then for $m > N$ we have

$$\frac{\|x(t_m;\omega) - x(t_0;\omega)\|_{\psi(\Omega,\mathscr{A},\mathscr{P})}}{g(t_0)} = [\|x(t_m;\omega) - x^n(t_m;\omega) + x^n(t_m;\omega) - x^n(t_0;\omega)$$
$$+ x^n(t_0;\omega) - x(t_0;\omega)\|_{\psi(\Omega,\mathscr{A},\mathscr{P})}]/g(t_0)$$
$$\leqslant \frac{\|x(t_m;\omega) - x^n(t_m;\omega)\|_{\psi(\Omega,\mathscr{A},\mathscr{P})}}{g(t_0)}$$
$$+ \frac{\|x^n(t_m;\omega) - x^n(t_0;\omega)\|_{\psi(\Omega,\mathscr{A},\mathscr{P})}}{g(t_0)}$$
$$+ \frac{\|x^n(t_0;\omega) - x(t_0;\omega)\|_{\psi(\Omega,\mathscr{A},\mathscr{P})}}{g(t_0)}$$
$$< \frac{\tfrac{1}{4}\varepsilon g(t_m)}{[g(t_0)]^2} + \frac{\tfrac{1}{4}\varepsilon}{g(t_0)} + \frac{\tfrac{1}{4}\varepsilon}{g(t_0)}.$$

However,
$$g(t_m) = g(t_m) + g(t_0) - g(t_0)$$
$$\leqslant |g(t_m) - g(t_0) + g(t_0)| < g(t_0) + g(t_0)$$
$$= 2g(t_0).$$

Hence
$$\frac{\|x(t_m;\omega) - x(t_0;\omega)\|_{\psi(\Omega,\mathscr{A},\mathscr{P})}}{g(t_0)} < \frac{\tfrac{1}{4}[2g(t_0)]\varepsilon}{[g(t_0)]^2} + \frac{\tfrac{1}{4}\varepsilon}{g(t_0)} + \frac{\tfrac{1}{4}\varepsilon}{g(t_0)}$$
$$= \frac{\tfrac{1}{2}\varepsilon}{g(t_0)} + \frac{\tfrac{1}{4}\varepsilon}{g(t_0)} + \frac{\tfrac{1}{4}\varepsilon}{g(t_0)}.$$

7.1 MATHEMATICAL PRELIMINARIES

This in turn implies that for $m > N$

$$\|x(t_m;\omega) - x(t_0;\omega)\|_{\psi(\Omega,\mathscr{A},\mathscr{P})} < \tfrac{1}{2}\varepsilon + \tfrac{1}{4}\varepsilon + \tfrac{1}{4}\varepsilon = \varepsilon.$$

This is simply the definition of convergence in $\psi(\Omega,\mathscr{A},\mathscr{P})$. Since the choice of t_0 was arbitrary, the same argument will suffice for each t and we have that the function $t \to x(t;\omega)$ is continuous. We can thus conclude that the function $t \to x(t;\omega)$ is an element of $C'_g(R_+,\psi(\Omega,\mathscr{A},\mathscr{P}))$. The conclusion of the first part of the proof that there exists an N_ε such that $n > N_\varepsilon$ implies that

$$\sup_{0 \leq t}\{\|x^n(t;\omega) - x(t;\omega)\|_{\psi(\Omega,\mathscr{A},\mathscr{P})}/g(t)\} < \varepsilon.$$

Then for $n > N_\varepsilon$

$$\|x^n(t;\omega) - x(t;\omega)\|_{C'_g} < \varepsilon,$$

which implies that the Cauchy sequence $x^n(t;\omega) \to x(t;\omega)$ in $C'_g(R_+,\psi(\Omega,\mathscr{A},\mathscr{P}))$. Therefore the space $C'_g(R_+,\psi(\Omega,\mathscr{A},\mathscr{P}))$ is complete, as was to be shown.

Lemma 7.1.13 The space $C'(R_+,\psi(\Omega,\mathscr{A},\mathscr{P}))$ is a normed linear subspace of $C_\psi(R_+,\psi(\Omega,\mathscr{A},\mathscr{P}))$ where the norm in C' is given by

$$\|x(t;\omega)\|_{C'} = \sup_{0 \leq t}\{\|x(t;\omega)\|_{\psi(\Omega,\mathscr{A},\mathscr{P})}\}.$$

The proof is similar to that of Lemma 7.1.10 with $g(t) = 1$.

Lemma 7.1.14 The space $C'(R_+,\psi(\Omega;\mathscr{A},\mathscr{P}))$ is complete.

The proof is analogous to that given for Lemma 7.1.12.

Lemma 7.1.15 The Banach spaces

$$C'_g(R_+,\psi(\Omega,\mathscr{A},\mathscr{P})) \quad \text{and} \quad C'(R_+,\psi(\Omega,\mathscr{A},\mathscr{P}))$$

are stronger than $C_\psi(R_+,\psi(\Omega,\mathscr{A},\mathscr{P}))$.

PROOF Let $x^n(t;\omega) \to x(t;\omega)$ in $C'(R_+,\psi(\Omega,\mathscr{A},\mathscr{P}))$ as $n \to \infty$. Select $Q > 0$ and consider the interval $[0,Q]$. We must show that $\|x^n(t;\omega) - x(t;\omega)\|_{\psi(\Omega,\mathscr{A},\mathscr{P})}$ converges to zero uniformly on $[0,Q]$. Choose $\varepsilon > 0$. Since $x^n(t;\omega) \to x(t;\omega)$ in $C'(R_+,\psi(\Omega,\mathscr{A},\mathscr{P}))$, there exists an N such that $n > N$ implies that $\sup_{0 \leq t}\{\|x^n(t;\omega) - x(t;\omega)\|_{\psi(\Omega,\mathscr{A},\mathscr{P})}\} < \varepsilon$. Hence for every $t \in [0,Q]$ and $n > N_\varepsilon$

$$\|x^n(t;\omega) - x(t;\omega)\|_{\psi(\Omega,\mathscr{A},\mathscr{P})} < \varepsilon,$$

implying that $C'(R_+,\psi(\Omega,\mathscr{A},\mathscr{P}))$ is stronger than $C_\psi(R_+,\psi(\Omega,\mathscr{A},\mathscr{P}))$. Let $x^n(t;\omega) \to x(t;\omega)$ in $C'_g(R_+,\psi(\Omega,\mathscr{A},\mathscr{P}))$. Pick $Q > 0$ and consider the interval $[0,Q]$. Choose $\varepsilon > 0$. Since g is continuous, g assumes a maximum at some point $t_0 \in [0, Q]$. Since $x^n(t; \omega) \to x(t; \omega)$ in $C'_g(R_+, \psi(\Omega, \mathcal{A}, \mathcal{P}))$, there exists an N_ε such that $n > N_\varepsilon$ implies

$$\sup_{0 \leq t} \{\|x^n(t; \omega) - x(t; \omega)\|_{\psi(\Omega, \mathcal{A}, \mathcal{P})}/g(t)\} < \varepsilon/g(t_0).$$

Hence for $t \in [0, Q]$ and $n > N_\varepsilon$ we have

$$\|x^n(t; \omega) - x(t; \omega)\|_{\psi(\Omega, \mathcal{A}, \mathcal{P})}/g(t) < \varepsilon/g(t_0).$$

This in turn implies that for $t \in [0, Q]$ and $n > N_\varepsilon$

$$\|x^n(t; \omega) - x(t; \omega)\|_{\psi(\Omega, \mathcal{A}, \mathcal{P})} < \varepsilon g(t)/g(t_0).$$

Since $g(t) \leq g(t_0)$, $g(t)/g(t_0) \leq 1$ and we conclude that for $t \in [0, Q]$ and $n > N_\varepsilon$

$$\|x^n(t; \omega) - x(t; \omega)\|_{\psi(\Omega, \mathcal{A}, \mathcal{P})} < \varepsilon.$$

Thus the space $C'_g(R_+, \psi(\Omega, \mathcal{A}, \mathcal{P}))$ is stronger than $C_\psi(R_+, \psi(\Omega, \mathcal{A}, \mathcal{P}))$.

7.2 An Existence and Uniqueness Theorem

With respect to the partial aim of this chapter we state and prove the following theorem, which gives sufficient conditions for the existence of a unique random solution of (7.0.1). Also, a special case will be given which is useful when studying applications.

Theorem 7.2.1 Suppose that the random integral equation (7.0.1) satisfies the following conditions:

(i) $B, D \subseteq C_\psi(R_+, \psi(\Omega, \mathcal{A}, \mathcal{P}))$ are Banach spaces stronger than $C_\psi(R_+, \psi(\Omega, \mathcal{A}, \mathcal{P}))$ and the pair (B, D) is admissible with respect to

$$(Tx)(t; \omega) = \int_0^t x(\tau; \omega) \, d\tau.$$

(ii) $k(t, x(t; \omega); \omega)$ is a mapping from the set

$$W = \{x(t; \omega) : x(t; \omega) \in D, \quad \|x(t; \omega)\|_D \leq \rho\}$$

into the space B for some $\rho \geq 0$, such that

$$\|k(t, x(t; \omega); \omega) - k(t, y(t; \omega); \omega)\|_B \leq \lambda \|x(t; \omega) - y(t; \omega)\|_D$$

for $x(t; \omega), y(t; \omega) \in W$ and $\lambda \geq 0$.

(iii) $x(t; \omega) \to h(t, x(t; \omega))$ is a mapping from W into D such that

$$\|h(t, x(t; \omega)) - h(t, y(t; \omega))\|_D \leq \gamma \|x(t; \omega) - y(t; \omega)\|_D$$

for some $\gamma \geq 0$.

7.2 AN EXISTENCE AND UNIQUENESS THEOREM

Then there exists a unique random solution of Eq. (7.0.1) provided that $\gamma + \lambda M < 1$ and

$$\|h(t, x(t;\omega))\|_D + M\|k(t, x(t;\omega);\omega)\|_B \leq \rho,$$

where $M = \|T\|^*$.

PROOF Define the operator U from W into D by

$$(Ux)(t;\omega) = h(t, x(t;\omega)) + \int_0^t k(\tau, x(\tau;\omega);\omega)\, d\tau.$$

We need to show that $U(W) \subseteq W$ and that for some $r \in [0, 1)$

$$\|(Ux)(t;\omega) - (Uy)(t;\omega)\|_D \leq r\|x(t;\omega) - y(t;\omega)\|_D.$$

Let $x(t;\omega), y(t;\omega) \in W$. Since $(Ux)(t;\omega)$ and $(Uy)(t;\omega) \in D$ and D is a Banach space, $(Ux)(t;\omega) - (Uy)(t;\omega) \in D$. Thus we can write

$$\|(Ux)(t;\omega) - (Uy)(t;\omega)\|_D = \left\|h(t, x(t;\omega)) + \int_0^t k(\tau, x(\tau;\omega);\omega)\, d\tau \right.$$

$$\left. - h(t, y(t;\omega)) - \int_0^t k(\tau, y(\tau;\omega);\omega)\, d\tau \right\|_D$$

$$= \left\|h(t, x(t;\omega)) - h(t, y(t;\omega)) \right.$$

$$\left. + \int_0^t [k(\tau, x(\tau;\omega);\omega) - k(\tau, y(\tau;\omega);\omega)]\, d\tau \right\|_D$$

$$\leq \|h(t, x(t;\omega)) - h(t, y(t;\omega))\|_D$$

$$+ \left\|\int_0^t [k(\tau, x(\tau;\omega);\omega) - k(\tau, y(\tau;\omega);\omega)]\, d\tau \right\|_D$$

$$\leq \gamma \|x(t;\omega) - y(t;\omega)\|_D$$

$$+ M\|k(t, x(t;\omega);\omega) - k(t, y(t;\omega);\omega)\|_B,$$

where the last inequality is due to the Lipschitz condition given in (iii) and the fact that T is continuous from B to D by Lemma 7.1.9, and therefore bounded. However,

$$\gamma\|x(t;\omega) - y(t;\omega)\|_D + M\|k(t, x(t;\omega);\omega) - k(t, y(t;\omega);\omega)\|_B$$

$$\leq \gamma\|x(t;\omega) - y(t;\omega)\|_D + M\lambda\|x(t;\omega) - y(t;\omega)\|_D$$

$$= (\gamma + M\lambda)\|x(t;\omega) - y(t;\omega)\|_D$$

by the Lipschitz condition given in (ii). Since $\gamma + M\lambda < 1$, the first condition of the definition of a contraction map is satisfied.

We must now show that the inclusion property holds. Let $x(t;\omega) \in W$. We can thus write

$$\|(Ux)(t;\omega)\|_D = \left\| h(t, x(t;\omega)) + \int_0^t k(\tau, x(\tau;\omega);\omega) \, d\tau \right\|_D$$

$$\leq \|h(t, x(t;\omega))\|_D + \left\| \int_0^t k(\tau, x(\tau;\omega);\omega) \, d\tau \right\|_D$$

$$\leq \|h(t, x(t;\omega))\|_D + M\|k(t, x(t;\omega);\omega)\|_B$$

$$\leq \rho.$$

Hence $(Ux)(t;\omega) \in W$, implying $U(W) \subseteq W$. Applying Banach's fixed-point theorem, we conclude that there exists a unique point $x(t;\omega) \in W$ such that

$$(Ux)(t;\omega) = h(t, x(t;\omega)) + \int_0^t k(\tau, x(\tau;\omega);\omega) \, d\tau = x(t;\omega)$$

and the proof is complete.

The following theorem is a useful special case of Theorem 7.2.1.

Theorem 7.2.2 Assume that Eq. (7.0.1) satisfies the following conditions:

(i) $k(t, x(t;\omega);\omega)$ is a mapping from the set

$$W = \{x(t;\omega) : x(t;\omega) \in C'(R_+, \psi(\Omega, \mathscr{A}, \mathscr{P})), \quad \|x(t;\omega)\|_{C'} \leq \rho\}$$

into the space $C'_g(R_+, \psi(\Omega, \mathscr{A}, \mathscr{P}))$ for some $\rho \geq 0$;

$$\|k(t, x(t;\omega);\omega) - k(t, y(t;\omega);\omega)\|_{C'_g} \leq \lambda \|x(t;\omega) - y(t;\omega)\|_{C'}$$

for $x(t;\omega), y(t;\omega) \in W$, $\lambda \geq 0$ a constant; g is also integrable on R_+.

(ii) $x(t;\omega) \to h(t, x(t;\omega))$ is a mapping from W into C' such that

$$\|h(t, x(t;\omega)) - h(t, y(t;\omega))\|_{C'} \leq \gamma \|x(t;\omega) - y(t;\omega)\|_{C'}$$

for some $\gamma \geq 0$.

Then there exists a unique random solution of Eq. (7.0.1) provided that $\gamma + \lambda M < 1$, where $M = \|T\|^*$, and

$$\|h(t, x(t;\omega))\|_{C'} + M\|k(t, x(t;\omega);\omega)\|_{C'_g} \leq \rho.$$

PROOF The proof consists in showing that under the assumption that g is integrable the pair $(C'_g(R_+, \psi(\Omega, \mathscr{A}, \mathscr{P})), C'(R_+, \psi(\Omega, \mathscr{A}, \mathscr{P})))$ is admissible with respect to the operator T given by

$$(Tx)(t;\omega) = \int_0^t x(\tau;\omega) \, d\tau.$$

Let $x(t;\omega) \in C'_g(R_+, \psi(\Omega, \mathscr{A}, \mathscr{P}))$. We can write

$$|(Tx_i)(t;\omega)| = \left|\int_0^t x_i(\tau;\omega)\,d\tau\right| \leqslant \int_0^t |x_i(\tau;\omega)|\,d\tau \leqslant \int_0^t |||x_i(\tau;\omega)|||\,d\tau$$

$$\leqslant \int_0^t [|||x_i(\tau;\omega)|||/g(\tau)]g(\tau)\,d\tau \leqslant \int_0^t [\|x(\tau;\omega)\|_{\psi(\Omega,\mathscr{A},\mathscr{P})}/g(\tau)]g(\tau)\,d\tau$$

$$\leqslant \int_0^\infty [\|x(\tau;\omega)\|_{\psi(\Omega,\mathscr{A},\mathscr{P})}/g(\tau)]g(\tau)\,d\tau \leqslant \|x(\tau;\omega)\|_{C'_g} \int_0^\infty g(\tau)\,d\tau$$

$$= \beta < \infty, \quad \mathscr{P}\text{-a.e.}$$

By definition of the norm in $L_\infty(\Omega, \mathscr{A}, \mathscr{P})$, we can conclude that

$$|||(Tx_i)(t;\omega)||| \leqslant \beta \quad \text{for each } i.$$

This in turn implies that

$$\|(Tx)(t;\omega)\|_{\psi(\Omega,\mathscr{A},\mathscr{P})} = \max_i\{|||(Tx_i)(t;\omega)|||\} \leqslant \beta,$$

which is the condition needed for $(Tx)(t;\omega)$ to be an element of $C'(R_+, \psi(\Omega, \mathscr{A}, \mathscr{P}))$. Since the remaining conditions are identical to those of Theorem 7.2.1, the proof is complete.

7.3 A Stochastic Chemical Kinetics Model

Gavalas [1] has formulated a deterministic model which characterizes a chemically reacting system. It is the aim of this section to formulate a stochastic version of this model and thus make it describe the physical situation more realistically (Milton and Tsokos [1]). The basic formulations of such a model involve a random or stochastic integral equation of the type discussed theoretically in the previous sections. Thus we shall illustrate the applicability of the theoretical results, mainly Theorems 7.2.1 and 7.2.2, to obtain conditions under which a chemically reacting system will possess a unique random solution.

The stochastic approach to the study of chemical kinetics is relatively new and has been developing rapidly during the last ten years. One of the main reasons for studying the classical chemical kinetics problem from a statistical point of view is that the evolution in time of a chemically reacting system is indeed random rather than deterministic. Blanc-Lapierre and Fortet [1] give a general discussion of how randomness enters into many physical systems. A brief description of how randomness enters into a chemically reacting system is given by McQuarrie [1]. An excellent bibliography of recent work in the area of stochastic chemical kinetics is also given in McQuarrie's paper. Bartholomay [1] gives a strong argument relative to the desirability of viewing chemical kinetics from a statistical point of view.

In Section 7.3.1 we shall give a brief description of the classical chemical kinetics problem and give a basic definition and introduce some notation of the subject area to set the stage for the stochastic formulation. The stochastic interpretation of the rate of reaction of a simple system is given in Section 7.3.2. In Section 7.3.3 we shall give some basic concepts of the rate functions of a general reacting system and describe the manner in which an integral equation arises in chemical kinetics. The stochastic formulation of the chemical kinetics is given in Section 7.3.4.

7.3.1 The Concept of Chemical Kinetics

Chemical kinetics is that branch of chemistry which deals with the rate and mechanism of chemical reactions and attempts to discover and explain those factors which influence the speed and manner by which a reaction proceeds. The reaction of a system under study which takes place in a single phase can be characterized at each point by the following variables: the velocity, the concentrations of all chemical species, and a thermodynamic variable such as the internal energy or temperature. A chemical system is called uniform if there are no space variations within the system. In our brief discussion of the subject area we shall assume that we are dealing with a homogeneous, uniform system at constant volume and constant temperature. We shall be concerned with the evolution of the system in time, and the variables used to study this evolution will be the concentrations of the species involved in the reactions. Such variables are called state variables. Thus chemical kinetics is concerned with the manner by which a reacting system gets from one state to another and with the time required to make the transition.

In what follows we shall give certain notation and ideas which are fairly standard in chemical kinetics and stoichiometry. We shall be concerned with a mixture of N chemical species $M_1, M_2, \ldots, M_N$. For example, the chemical reaction

$$2H_2O \rightarrow 2H_2 + O_2 \tag{7.3.1}$$

consists of three species $M_1 = H_2O$, $M_2 = H_2$, and $M_3 = O_2$. A chemical reaction is usually written as

$$\sum_{i=1}^{N} v_i M_i = 0, \tag{7.3.2}$$

7.3 A STOCHASTIC CHEMICAL KINETICS MODEL

where v_i is called the *stoichiometric coefficient* of species M_i in the balanced equation for the reaction. That is, reaction (7.3.1) can be written as

$$2H_2O - 2H_2 - O_2 = 0.$$

We shall consider a species M_i to be a reactant if $v_i > 0$ and a product if $v_i < 0$, for $i = 1, 2, \ldots, N$. A similar convention is given by Gavalas [1]. Note that from the stoichiometric equation we have

$$\delta n_i / \delta n_k = v_i / v_k, \qquad 1 \leq i \leq N \quad \text{and} \quad 1 \leq k \leq N,$$

where δn_i represents a change in the number of moles of species M_i in the chemical system. For the reaction given by Eq. (7.3.1) we have for each mole of water that is decomposed by electrolysis the generation of one-half mole of oxygen and one mole of hydrogen. This notation, of course, can be generalized to describe a system of R reactions, that is

$$\sum_{i=1}^{N} v_{ij} M_i = 0, \qquad j = 1, 2, \ldots, R, \qquad (7.3.3)$$

where v_{ij} is the stoichiometric coefficient of the species M_i in the jth reaction, $i = 1, 2, \ldots, N$ and $j = 1, \ldots, R$.

Definition 7.3.1 A *rate function* or *rate of reaction* or *reaction rate* $r_i(t)$, $i = 1, 2, \ldots, N$, is the rate of change of the concentration of a fixed species M_i involved in the reaction.

The exact functional expression of the rate function does depend on the species used in defining the term. However, once the rate of reaction at a given time is determined for one species entering into the reaction, the rate of any other species involved may be calculated from stoichiometric considerations. We shall use $r(t)$ to denote any one of the possible N rate expressions. But when we are interested in a particular species being used to define the rate of reaction we shall denote it accordingly. Thus if $c_i(t)$ is the function which represents the concentration of species M_i at time t, then $r_i(t) = dc_i(t)/dt$. In the reaction given by (7.3.1) we have $M_1 = H_2O$, $M_2 = H_2$, $M_3 = O_2$, and

$$r_1(t) = dc_1(t)/dt = -dc_2(t)/dt = -2\,dc_3(t)/dt.$$

That is, $r_1(t) = -r_2(t) = -2r_3(t)$. The rate of reaction at a fixed temperature T is generally a function of the concentrations of the various species of the reaction alone. Thus the reaction rate at a fixed temperature T can almost always be expressed in the form

$$r(t) = K_T f[c_1(t), c_2(t), \ldots, c_N(t)]. \qquad (7.3.4)$$

The subscript T of the constant K indicates that the constant involved is usually dependent on the particular value of T at which the reaction takes place but independent of concentration. In some cases there are also unknown constants involved in expression (7.3.4). For example, some rate functions are of the form

$$r(t) = K_T[c_1(t)]^{\alpha_1}[c_2(t)]^{\alpha_2}\ldots[c_N(t)]^{\alpha_N},$$

where $K_T, \alpha_1, \alpha_2, \ldots, \alpha_N$ are constants that must be determined experimentally. It is a principal task for an experimental chemist to obtain the form of this rate expression and also to estimate the values of the constants involved from laboratory data. It has been shown that in a slightly acidic solution, I^- is oxidized by H_2O yielding I_3^- and H_2O, stoichiometrically represented by

$$H_2O_2 + 3I^- + 2H^+ \rightarrow I_3^- + 2H_2O, \tag{7.3.5}$$

where the rate has the form

$$r(t) = dc_1(t)/dt = K_T[c_1(t)]^m[c_2(t)]^n \tag{7.3.6}$$

with $c_1(t)$ the concentration of H_2O_2 at time t and $c_2(t)$ the concentration of I^- at time t. From experimental data one can determine the values of the constants K_T, m, and n. However, it is a well-known fact that it is difficult if not impossible to outline a procedure for determining both the form and estimates for pertinent constants which will be applicable to all situations.

The following is a basic procedure that could be followed:

(i) If possible, postulate a general form for the rate expression based on any previous information available.

(ii) Set up an experiment which allows either the direct observation of the concentration of some species in time or the observation of some other quantity whose relationship to concentration is known.

(iii) The experiment should be run several times at the temperature of interest with initial concentrations of species varying.

(iv) Using the data obtained from the experiments, the hypothesized form of the rate expression can either be accepted or rejected and estimates for constants involved can be obtained.

(v) The value of a constant finally used in the rate expression is the average of the values of this constant obtained in successive runs of the experiment. However, recently Box [1], Kittrell, Mezaki, and Watson [1], and Lee [1] have given more sophisticated techniques for determining from the data the value of the constant to be used.

Thus it is clear that determining the rate function for a reaction is extremely difficult and of limited accuracy. In what follows we give a stochastic version of the rate function (Milton and Tsokos [1]).

7.3.2 Stochastic Interpretation of the Rate of Reaction

In this section we shall discuss a stochastic interpretation of the rate function $r(t)$ of a single reaction, that is, the function given by Eq. (7.3.4).

As we have mentioned, one of the initial things that the experimenter attempts to decide is the exact form of the function $f[c_1(t), c_2(t), \ldots, c_N(t)]$. This often involves estimating from experimental data such things as the powers to which the concentrations of various species are to be raised and the coefficients of concentration terms. Thus, due to the complexity of the problem, it is reasonable to interpret the function f as random rather than deterministic. In addition, if some form of the function f has been established, one is still faced with the problem of obtaining an estimate of K_T. Such an estimate is obtained from experimental data and it is usually the average value that is used as an estimate of the "true" value. If the experiment is repeated under the identical conditions, the estimates of K_T usually will be different. Hence it is more reasonable to consider K_T not as a constant but rather as a stochastic variable $K_T(\omega)$. Furthermore, Parrott [1] states that it is not possible to know the exact concentration of any given species M_i at a particular time t_0 and it must be estimated from several observations of this concentration. Therefore we shall denote by $c_i(t_0; \omega)$ the stochastic analog of $c_i(t_0)$. Since the argument is independent of the choice of t_0, one can consider $c_i(t; \omega)$ to be a random variable for each $i = 1, 2, \ldots, N$ and each $t \in R_+$. That is, for each i, $c_i(t; \omega)$ is a random function.

In view of these remarks a more realistic form of Eq. (7.3.4) is given by

$$r(t; \omega) = K_T(\omega) f[c_1(t; \omega), c_2(t; \omega), \ldots, c_N(t; \omega)]$$

$$= F[c_1(t; \omega), c_2(t; \omega), \ldots, c_N(t; \omega); \omega]. \quad (7.3.7)$$

7.3.3 Rate Functions of General Reaction Systems

We shall begin by focusing our attention on the mechanism by which the observed chemical change proceeds. This calls for looking at the concept of an *elementary reaction*, a reaction that corresponds in a sense to a single molecular collision (Gavalas [1]). A reaction is generally considered as being made up of a number of elementary reactions each with its own rate function. The overall rate functions are then related to the rates at which the elementary reactions proceed.

Definition 7.3.2 The reaction system given by Eq. (7.3.3) is said to be capable of describing the observed chemical change if with each reaction we can associate a function f_j of the N concentrations $c_i(t)$ such that

$$dc_i(t)/dt = \sum_{j=1}^{R} v_{ij} f_j[c_1(t), c_2(t), \ldots, c_N(t)] = F_i[c_1(t), c_2(t), \ldots, c_N(t)]. \quad (7.3.8)$$

The functions $f_1, f_2, \ldots, f_R$ and $F_1, F_2, \ldots, F_N$ are all referred to as rate functions, reaction rates, or kinetics for the system. One can easily illustrate these concepts by using the classical reaction in which hydrogen and bromine unite to form hydrogen bromide. Gavalas [1] discusses the mechanism of this reaction. That is,

$$H_2 + Br_2 \xrightarrow{f} 2HBr$$

$$Br_2 + M \underset{f_2}{\overset{f_1}{\rightleftharpoons}} 2Br + M$$

$$Br + H_2 \underset{f_4}{\overset{f_3}{\rightleftharpoons}} HBr + H \quad (7.3.9)$$

$$H + Br_2 \xrightarrow{f_5} HBr + Br,$$

with M being either H_2 or Br_2. Experimentally, it has been determined that the rate functions f_j of the elementary reactions are of the forms

$$f_1[c_1(t), c_2(t), \ldots, c_5(t)] = K_1 c_2(t)[c_1(t) + c_2(t)],$$
$$f_2[c_1(t), c_2(t), \ldots, c_5(t)] = K_2[c_5(t)]^2[c_1(t) + c_2(t)],$$
$$f_3[c_1(t), c_2(t), \ldots, c_5(t)] = K_3[c_5(t)][c_1(t)],$$
$$f_4[c_1(t), c_2(t), \ldots, c_5(t)] = K_4[c_3(t)][c_4(t)],$$
$$f_5[c_1(t), c_2(t), \ldots, c_5(t)] = K_5[c_4(t)][c_2(t)],$$

where $M_1 = H_2$, $M_2 = Br_2$, $M_3 = HBr$, $M_4 = H$, and $M_5 = Br$.

Therefore if the reaction system (7.3.9) is capable of describing the observed chemical change, then the reaction rate for the overall reaction expressed in terms of the rate of formation of HBr is given as follows:

$$r_3(t) = dc_3(t)/dt = \sum_{j=1}^{5} v_{3j} f_j[c_1(t), c_2(t), \ldots, c_5(t)]$$

$$= 0 \cdot f_1[c_1(t), c_2(t), \ldots, c_5(t)] + 0 \cdot f_2[c_1(t), c_2(t), \ldots, c_5(t)]$$

$$- f_3[c_1(t), \ldots, c_5(t)] + f_4[c_1(t), c_2(t), \ldots, c_5(t)]$$

$$- f_5[c_1(t), c_2(t), \ldots, c_5(t)]$$

$$= - K_3 c_5(t) c_1(t) + K_4 c_3(t) c_4(t) - K_5 c_4(t) c_2(t)$$

$$= F_3[c_1(t), c_2(t), \ldots, c_5(t)].$$

7.3 A STOCHASTIC CHEMICAL KINETICS MODEL

We shall now look at the concept of extent of reaction in terms of a single reaction $\sum_{i=1}^{N} v_i M_i = 0$ before looking at the extension to a system of reactions.

It was stated that the relationship $(\delta n_i)(t)/(\delta n_k)(t) = v_i/v_k$, where $(\delta n_i)(t)$ represents the change in the number of moles of species M_i from the beginning of the reaction to time t, must hold for any pair (i, k) such that $1 \leq i \leq N$, $1 \leq k \leq N$. This relationship can be written as

$$(\delta n_1)(t)/v_1 = (\delta n_2)(t)/v_2 = \cdots = (\delta n_N)(t)/v_N = x(t).$$

We shall assume for convenience that the reaction always starts at $t = 0$. Thus for each $i = 1, 2, \ldots, N$, $(\delta n_i)(t) = n_i(t) - n_i(0)$ and $x(t) = [n_i(t) - n_i(0)]/\delta_i$. The function $x(t)$ is called the *molar extent* or *degree of advancement* of the reaction. Its usefulness lies in the fact that it is a function linked to the reaction as a whole and not any particular species M_i, the choice of which would be arbitrary. If there are several reactions under study, as in the system $\sum_{i=1}^{N} v_{ij} M_i = 0$, then an *extent* can be defined for each of them. If we let $x_j(t)$ denote the extent of the jth reaction, then it has contributed $v_{ij} x_j(t)$ moles to the total change in the number of moles of species M_i. Thus the total change in the number of moles of M_i in the entire system can be expressed by

$$n_i(t) - n_i(0) = \sum_{j=1}^{R} v_{ij} x_j(t), \qquad i = 1, 2, \ldots, N. \tag{7.3.10}$$

In order to free these equations from any dependence upon the actual size of the sample, it is convenient to divide Eq. (7.3.10) by the volume V to obtain an expression which reflects the extent of reaction in terms of concentration change rather than actual change in number of moles. Hence we arrive at the equation

$$[n_i(t) - n_i(0)]/V = \sum_{j=1}^{R} v_{ij}[x_j(t)/V] \tag{7.3.11}$$

or

$$c_i(t) - c_i(0) = \sum_{j=1}^{R} v_{ij} \xi_j(t), \qquad i = 1, 2, \ldots, N, \tag{7.3.12}$$

where $\xi_j(t) = x_j(t)/V$. Throughout we will be referring to $\xi_j(t)$ when we speak of the *extent* of the jth reaction.

We shall now give a brief discussion of how integral equations arise in chemical kinetics. Recall that in the evolution in time of a reaction system under study we assume that the system is homogeneous, uniform, of constant temperature and volume, and that it can be represented by the following set of equations:

$$dc_i(t)/dt = \sum_{j=1}^{R} v_{ij} f_j[c_1(t), c_2(t), \ldots, c_N(t)], \quad i = 1, 2, \ldots, N. \quad (7.3.13)$$

We shall use the following vector notation:

$$c(t) = [c_1(t), c_2(t), \ldots, c_N(t)], \qquad c(0) = [c_1(0), c_2(0), \ldots, c_N(0)],$$

and refer to $c(0)$ as the initial conditions and $c(t)$ as a trajectory of the system (7.3.13) passing through $c(0)$, which is assumed known. Due to the stoichiometry of the reaction there exist functions c_{im} and c_{iM} such that $c_{im}[c_i(0)]$ is the minimum concentration possible for species M_i and $c_{iM}[c_i(0)]$ is the maximum concentration possible for species M_i during the course of the reaction when the initial concentration is $c_i(0)$. Equations (7.3.12) allow one to express the state of the system at time t in terms of either concentrations $c(t) = [c_1(t), c_2(t), \ldots, c_N(t)]$ or extent of reaction $\xi(t) = [\xi_1(t), \xi_2(t), \ldots, \xi_R(t)]$. They also allow one to define a one-to-one map g from $\Gamma[c(0)]$ onto its image $\tilde{\Gamma}[c(0)]$, where $\Gamma[c(0)]$ is the set of all possible trajectories passing through $c(0)$. This in turn allows us to write

$$d\xi_j(t)/dt = \tilde{f}_j[\xi_1(t), \xi_2(t), \ldots, \xi_R(t)] = \tilde{f}_j[\xi(t)], \quad (7.3.14)$$

where $\tilde{f}_j[\xi(t)] = f_j[g^{-1}(\xi(t))]$ for $j = 1, 2, 3, \ldots, R$. The question of interest is then to show the existence of a solution to the system

$$d\xi_j(t)/dt = \tilde{f}_j[\xi(t)], \quad j = 1, 2, \ldots, R, \quad \xi_j(0) = 0. \quad (7.3.15)$$

This gives rise to a nonlinear integral equation of the form

$$\xi(t) = \int_0^t \tilde{f}[\xi(\tau)] \, d\tau, \quad t \geq 0. \quad (7.3.16)$$

Gavalas [1] discusses the existence and uniqueness of a solution of Eq. (7.3.16) in the deterministic sense. His general method of proof is to call upon a fixed-point method which guarantees the existence of a solution and then he further imposes a Lipschitz condition on the kernel $\tilde{f}[\xi(\tau)]$ in order to guarantee uniqueness.

7.3.4 A Stochastic Integral Equation Arising in Chemical Kinetics

In this section we shall extend the ideas of Section 7.3.2 to a system of reactions and show how Eq. (7.3.16) can be more realistically studied from a stochastic point of view.

It was argued that due to many factors such as estimation of constants involved it would be more realistic to assume that the rate function involved

7.3 A STOCHASTIC CHEMICAL KINETICS MODEL

in a single elementary reaction is stochastic. These arguments can easily be extended to a system of reactions such as Eq. (7.3.3) to imply that we can consider f_j, $j = 1, 2, \ldots, R$ to be random functions. Furthermore, since the variables $\xi_j(t)$ are defined in terms of concentrations, which we have seen can be given a statistical interpretation, we can logically consider these variables to be random $\xi_j(t; \omega)$. Thus, since $\tilde{f}_j$, $j = 1, \ldots, R$, are defined by using random functions, they may also be considered stochastic and the integral equation (7.3.16) is written as

$$\xi(t; \omega) = \int_0^t \tilde{f}[\xi(\tau; \omega); \omega] \, d\tau, \qquad t \geq 0. \tag{7.3.17}$$

We shall assume the following:

(i) $\omega \in \Omega$, where Ω is the supporting set of the probability measure space $(\Omega, \mathscr{A}, \mathscr{P})$.

(ii) $\xi(t; \omega)$ is the unknown R-dimensional, vector-valued random function defined on R_+.

(iii) Under appropriate conditions the stochastic kernel $\tilde{f}[\xi(\tau; \omega); \omega]$ is an R-dimensional, vector-valued random function on R_+.

With respect to the aims of this section, we state the following theorem concerning the stochastic integral equation (7.3.17), which assures the existence of a unique random solution.

Theorem 7.3.1 Suppose that Eq. (7.3.17) satisfies the following conditions:

(i) $B, D \subseteq C_\psi(R_+, \psi(\Omega, \mathscr{A}, \mathscr{P}))$ are Banach spaces stronger than $C_\psi(R_+, \psi(\Omega, \mathscr{A}, \mathscr{P}))$ and the pair (B, D) is admissible with respect to the operator T given by

$$(T\xi)(t; \omega) = \int_0^t \xi(\tau; \omega) \, d\tau.$$

(ii) $\tilde{f}(\xi(\tau; \omega); \omega)$ is a mapping from the set

$$W = \{\xi(t; \omega) : \xi(t; \omega) \in D, \quad \|\xi(t; \omega)\|_D \leq \rho\}$$

into the space B for some $\rho \geq 0$ such that

$$\|\tilde{f}(\xi(t; \omega); \omega) - \tilde{f}(\gamma(t; \omega); \omega)\|_B \leq \lambda \|\xi(t; \omega) - \gamma(t; \omega)\|_D$$

for $\xi(t; \omega), \gamma(t; \omega) \in W$ and $\lambda \geq 0$ a constant.

Then there exists a unique random solution $\xi(t; \omega)$ of Eq. (7.3.17) in W provided that $\lambda M < 1$, where $M = \|T\|^*$, and $M\|\tilde{f}(\xi(t; \omega); \omega)\|_B \leq \rho$.

PROOF   The proof of this theorem is identical to that of Theorem 7.2.1 with the operator U given by

$$(U\xi)(t;\omega) = \int_0^t \tilde{f}(\xi(\tau;\omega);\omega)\,d\tau.$$

Theorem 7.3.2   Suppose that Eq. (7.3.17) satisfies the following conditions:

(i) $\tilde{f}(\xi(t;\omega);\omega)$ is a mapping from the set

$$W = \{\xi(t;\omega) : \xi(t;\omega) \in C'(R_+, \psi(\Omega,\mathscr{A},\mathscr{P})), \quad \|\xi(t;\omega)\|_{C'(R_+,\psi(\Omega,\mathscr{A},\mathscr{P}))} \leqslant \rho\}$$

into the space $C'_g(R_+, \psi(\Omega,\mathscr{A},\mathscr{P}))$ for some $\rho \geqslant 0$;

$$\|\tilde{f}(\xi(t;\omega);\omega) - \tilde{f}(\gamma(t;\omega);\omega)\|_{C'_g} \leqslant \lambda \|\xi(t;\omega) - \gamma(t;\omega)\|_{C'}$$

for $\xi(t;\omega), \gamma(t;\omega) \in W$ and $\lambda \geqslant 0$.

(ii) $\int_0^\infty g(\tau)\,d\tau = \beta < \infty$.

Then there exists a unique random solution of Eq. (7.3.17) in W provided that $\lambda M < 1$, where $M = \|T\|^*$, and $M\|\tilde{f}(\xi(t;\omega);\omega)\|_{C'_g} \leqslant \rho$.

The proof consists of showing that the pair $(C'_g(R_+, \psi(\Omega,\mathscr{A},\mathscr{P}), C'(R_+, \psi(\Omega,\mathscr{A},\mathscr{P}))$ is admissible with respect to the operator T and the remaining arguments are identical to Theorem 7.2.1.

We shall conclude our discussion by looking at the practicality of the assumptions given earlier. The main assumption is that for each $j = 1, 2, 3, \ldots, R$ and each $t \geqslant 0$, $\xi_j(t;\omega) \in L_\infty(\Omega, \mathscr{A}, \mathscr{P})$. That is,

$$|\xi_j(t;\omega)| = |[N_{ij}(t;\omega) - N_{ij}(0;\omega)]/v_{ij}V|$$
$$\leqslant (1/v_{ij}V)[|N_{ij}(t;\omega)| + |N_{ij}(0;\omega)|]. \tag{7.3.18}$$

A kinetic experiment is a carefully controlled laboratory procedure, and hence the number of moles of reactant at time zero is known apart from measurement error. By assumption, V and v_{ij} are known constants. The amount of species i present in reaction j at time t, $N_{ij}(t;\omega)$, is controlled by the initial amount of each reactant present at the start of the experiment. In fact, due to stoichiometric considerations we can say that for each i, j there exists a constant β_{ij} such that for every t

$$|N_{ij}(t;\omega)| \leqslant \beta_{ij}, \qquad \mathscr{P}\text{-a.e.} \tag{7.3.19}$$

From (7.3.19), inequality (7.3.18) becomes

$$|\xi_j(t;\omega)| \leqslant (1/v_{ij}V)[|N_{ij}(t;\omega)| + |N_{ij}(0;\omega)|] \leqslant (1/v_{ij}V) \cdot 2\beta_{ij}, \qquad \mathscr{P}\text{-a.e.}$$

That is, for each $j = 1, 2, \ldots, R$ and $t \geqslant 0$, $\xi_j(t;\omega) \in L_\infty(\Omega, \mathscr{A}, \mathscr{P})$.

CHAPTER VIII

Stochastic Integral Equations of the Ito Type

8.0 Introduction

Another type of stochastic integral equation which has been of considerable importance to applied mathematicians and engineers is that involving the Ito or Ito–Doob form of stochastic integrals. We shall give some historical remarks concerning the development of this type of equation and point out the essential difference between them and random integral equations discussed in the previous chapters.

In 1930 N. Wiener introduced an integral of the form

$$\int_a^b g(\tau)\, d\beta(\tau),$$

where $g(t)$ is a deterministic real-valued function and $\{\beta(\tau), \tau \in [a, b]\}$ is a scalar Brownian motion process. Ito [1] in 1944 generalized Wiener's integral to include those cases where the integrand is random. That is, he obtained an integral of the form

$$\int_0^t g(\tau;\omega)\,d\beta(\tau), \qquad t \in [0,1],$$

which is referred to as the *Ito stochastic integral* or simply the stochastic integral. Since that time many scientists have contributed to the general development of this type of stochastic integral. For example, see Doob [1], Dynkin [1], Jazwinski [1], Ito [2], McKean [1], Saaty [1], Gikhmann and Shorokhod [1], Stratonovich [1], and Wong and Zakai [1].

In 1946 Ito [2] formulated a stochastic integral equation of the form

$$x(t;\omega) = c + \int_0^t f(\tau, x(\tau;\omega))\,d\tau + \int_0^t g(\tau, x(\tau;\omega))\,d\beta(\tau), \qquad (8.0.1)$$

where $t \in [0,1]$, $\{\beta(t): t \in [0,1]\}$ is a scalar Brownian motion process, and c is a constant. Restrictions are usually placed on the functions f and g so that the first integral is interpreted as the usual Lebesgue integral of the sample functions which can then be related to the sample integral of the process $\{f(t, x(t;\omega)): t \in [0,1]\}$ and the second integral is an Ito stochastic integral.

The principal feature which distinguishes the type of equation studied in the previous chapters from an equation of the Ito type is the fact that in the former case each of the integrals involved is interpreted as a Lebesgue integral for almost all $\omega \in \Omega$. That is, almost all sample functions are Lebesgue integrable. Since in the Ito stochastic integral the limit is taken in the mean-square or in the probability sense, the theory of such integrals has been developed as self-contained and self-consistent.

One of the main purposes of subsequent work in connection with the Ito stochastic integral equation has been to construct Markov processes such that their transition probabilities satisfy given Kolmogorov equations and to investigate the continuity of the processes, among other properties of the sample function.

The method of successive approximation was used by Ito and Doob to show the existence and uniqueness of a random solution to Eq. (8.0.1). The objective of this chapter is to attempt to apply the theoretical techniques of probabilistic functional analysis developed in the previous chapters to answer the question of existence of a random solution to (8.0.1).

8.1 Preliminary Remarks

Let $\{\beta(t); t \in [a,b]\}$ be a scalar Brownian motion process. In this section we shall be concerned with the integral

$$\int_a^b \Phi(t;\omega)\,d\beta(t), \qquad a < b, \qquad (8.1.1)$$

8.1 PRELIMINARY REMARKS

for a fairly general class of functions Φ. This integral will be called the Ito stochastic integral, as we mentioned previously. As is well known, almost all the sample functions of the Brownian motion process are of unbounded variation and hence the integral (8.1.1) cannot be defined as an ordinary Stieltjes integral.

First we shall define the integral (8.1.1) for the class of step functions. That is, functions Φ of the form

$$\Phi(t;\omega) = \begin{cases} 0, & t < a, \\ \Phi_i(\omega), & t_i \leq t \leq t_{i+1}, \\ 0, & t > b, \end{cases} \qquad (8.1.2)$$

where $a = t_0 < t_1 < t_2 < \cdots < t_{n-1} < t_n = b$, $\Phi_i(\omega)$ are measurable with respect to the σ-algebra A_{t_i}, and $E\{|\Phi_i(\omega)|^2\} < \infty$. For such functions we define the Ito integral by

$$\int_a^b \Phi(t;\omega)\, d\beta(t) = \sum_{i=0}^{n-1} \Phi_i(\omega)\{\beta(t_{i+1}) - \beta(t_i)\}. \qquad (8.1.3)$$

Now suppose that $\Phi(t;\omega)$ is any function satisfying the following conditions:

(i) $\Phi(t;\omega)$ is a product-measurable function from $[a,b] \times \Omega \to R_+$, assuming the usual Lebesgue measure on R_+.

(ii) For each $t \in [a,b]$, $\Phi(t;\omega)$ is measurable with respect to σ-algebra A_t, where A_t is the smallest σ-algebra on Ω, such that $\beta(s)$, $s \leq t$, is measurable.

(iii) $\int_{-\infty}^{\infty} E|\Phi(t;\omega)|^2\, dt < \infty$.

In view of Eq. (8.1.2), it is evident that the class of step functions satisfy Conditions (i)–(iii).

For the functions $\Phi(t;\omega)$ satisfying Conditions (i)–(iii) we shall define their norm as follows:

$$\|\Phi(t;\omega)\| = \left\{ \int_a^b E[|\Phi(t;\omega)|]^2\, dt \right\}^{\frac{1}{2}}. \qquad (8.1.4)$$

For this case Doob [1] has shown the following:

(i) $\Phi(t;\omega)$ can be approximated in the mean-square sense by a sequence of step functions $\{\Phi_n(t;\omega)\}$. That is,

$$\|\Phi(t;\omega) - \Phi_n(t;\omega)\| \to 0$$

as $n \to \infty$.

(ii) The sequence of integrals

$$\int_a^b \Phi_n(t;\omega)\, d\beta(t)$$

possesses a mean-square limit. That is, there exists a $\theta(\omega)$ such that

$$E\left\{\theta(\omega) - \int_a^b \Phi_n(t;\omega)\,d\beta(t)\right\}^2 \to 0 \qquad (8.1.5)$$

as $n \to \infty$.

Now we shall define the integral (8.1.1) for a class of functions $\{\Phi(t;\omega)\}$ satisfying Conditions (i)–(iii) by

$$\int_a^b \Phi(t;\omega)\,d\beta(t) = \theta(\omega). \qquad (8.1.6)$$

As with the ordinary integrals, we shall define

$$\int_{-\infty}^{\infty} \Phi(t;\omega)\,d\beta(t) = \lim_{a \to -\infty, b \to +\infty} \int_a^b \Phi(t;\omega)\,d\beta(t). \qquad (8.1.7)$$

Definition 8.1.1 Let $G \in L$, where L denotes the collection of Lebesgue-measurable subsets of R_+. Define a function χ_G from $R_+ \times \Omega \to \{0,1\}$ by

$$\chi_G(\tau;\omega) = \begin{cases} 1 & \text{if } (\tau,\omega) \in G \times \Omega, \\ 0 & \text{otherwise.} \end{cases}$$

Lemma 8.1.1 The function $\Phi\chi_G : R_+ \times \Omega \to R_+$ defined by

$$(\Phi\chi_G)(\tau;\omega) = \Phi(\tau;\omega)\chi_G(\tau;\omega),$$

where Φ satisfies Conditions (i)–(iii), and χ_G is as defined earlier, also satisfies Conditions (i)–(iii).

PROOF The proof is a straightforward result of the definition of χ_G and the fact that Φ satisfies Conditions (i)–(iii).

We are now in a position to define exactly what is meant by the expression

$$\int_G \Phi(\tau;\omega)\,d\beta(\tau).$$

Definition 8.1.2 We define $\int_G \Phi(\tau;\omega)\,d\beta(\tau)$ for G a Lebesgue-measurable subset of R_+ by

$$\int_G \Phi(\tau;\omega)\,d\beta(\tau) = \int_{-\infty}^{\infty} (\Phi\chi_G)(\tau;\omega)\,d\beta(\tau).$$

Note that Lemma 8.1.4 guarantees that the expression on the right exists and is well defined.

Definition 8.1.3 We shall denote by $C^*([a,b], L_2(\Omega, \mathscr{A}, \mathscr{P}))$ the space of all continuous functions from $[a,b]$ into $L_2(\Omega, \mathscr{A}, \mathscr{P})$. We shall define the norm of $C^*([a, b], L_2(\Omega, \mathscr{A}, \mathscr{P}))$ by

$$\sup_{a \leq t \leq b} \left\{ \int_\Omega |x(t;\omega)|^2 \, d\mathscr{P}(\omega) \right\}^{\frac{1}{2}}.$$

Lemma 8.1.2

$$E[\theta] = E\left\{ \int_{-\infty}^\infty \Phi(t;\omega) \, d\beta(t) \right\} = 0.$$

Lemma 8.1.3

$$E[|\theta|^2] = \int_{-\infty}^\infty E\{|\Phi(t;\omega)|^2\} \, dt.$$

Lemma 8.1.4 If we define a distance between two functions Φ_1 and Φ_2 each satisfying Conditions (i)–(iii) by

$$\|\Phi_1 - \Phi_2\| = \left\{ \int_{-\infty}^\infty E|\Phi_1(t;\omega) - \Phi_2(t;\omega)|^2 \, dt \right\}^{\frac{1}{2}}$$

and the distance between

$$\theta_1 = \int_{-\infty}^\infty \Phi_1(t;\omega) \, d\beta(t) \quad \text{and} \quad \theta_2 = \int_{-\infty}^\infty \Phi_2(t;\omega) \, d\beta(t)$$

by

$$\|\theta_1 - \theta_2\| = \{E[|\theta_1 - \theta_2|^2]\}^{\frac{1}{2}},$$

then

$$\|\theta_1 - \theta_2\| = \|\Phi_1 - \Phi_2\|.$$

For the proof of these lemmas see Doob [1].

Lemma 8.1.5 Let

$$x(t;\omega) = \int_a^t \Phi(\tau;\omega) \, d\beta(\tau), \quad t \in [a, b].$$

Then $x(t;\omega) \in C^*([a, b], L_2(\Omega, \mathscr{A}, \mathscr{P}))$.

For the proof see Jazwinski [1].

It is easily seen that many of the properties of the stochastic integral are analogous to those of the ordinary integral.

8.2 On an Ito Stochastic Integral Equation

In this section we shall investigate a stochastic integral equation of the type

$$x(t;\omega) = \int_0^t k(t,\tau;\omega) f(\tau, x(\tau;\omega))\, d\tau + \int_0^t \Phi(\tau;\omega)\, d\beta(\tau), \qquad t \geq 0, \quad (8.2.1)$$

where $x(t;\omega)$ is the unknown random process defined for $t \in R_+$ and $\omega \in \Omega$. We shall place the following restrictions on the random functions which constitute the stochastic integral equation (8.2.1).

(i)′ $k(t,\tau;\omega)$ is an element of $L_\infty(\Omega, \mathscr{A}, \mathscr{P})$ and $k(t,\tau;\omega): \Delta \to L_\infty(\Omega, \mathscr{A}, \mathscr{P})$ is continuous, where $\Delta = \{(t,\tau): 0 \leq \tau \leq t < \infty\}$.

(ii)′ $x(t;\omega) \to f(t, x(t;\omega))$ is an operator on the set S with values in the Banach space B satisfying

$$\|f(t, x(t;\omega)) - f(t, y(t;\omega))\|_B \leq \lambda \|x(t;\omega) - y(t;\omega)\|_D$$

for $x(t;\omega), y(t;\omega) \in S$.

(iii)′ Conditions (i)–(iii) of Section 8.1 hold.

Thus with the given assumptions the first integral of (8.2.1) can be interpreted as a Lebesgue integral and the second as an Ito stochastic integral.

We shall now proceed to state and prove a theorem concerning the behavior of the Ito integral. More precisely, if we show that the Ito integral is an element of the space $C_c(R_+, L_2(\Omega, \mathscr{A}, \mathscr{P}))$, we can apply the theory of admissibility to Eq. (8.2.1) to show the existence of a random solution. By a random solution to Eq. (8.2.1) we mean a random function $x(t;\omega)$ from R_+ into $L_2(\Omega, \mathscr{A}, \mathscr{P})$ such that for each $t \in R_+$, $x(t;\omega)$ satisfies the integral equation $\mathscr{P}$-a.e. Showing that the Ito integral is an element of $C_c(R_+, L_2(\Omega, \mathscr{A}, \mathscr{P}))$ will make feasible the assumption that we wish to make that the integral is an element of D, a Banach space contained in the topological space mentioned.

For convenience we shall denote the Ito integral by

$$h(t;\omega) = \int_0^t \Phi(\tau;\omega)\, d\beta(\tau), \qquad t \geq 0.$$

Theorem 8.2.1 For $t \in R_+$, $h(t;\omega) \in C_c(R_+, L_2(\Omega, \mathscr{A}, \mathscr{P}))$.

PROOF Fix $t \in R_+$. Then

$$h(t;\omega) = \int_0^t \Phi(\tau;\omega)\, d\beta(\tau) = \int_{-\infty}^\infty (\Phi \chi_{[0,t]})(\tau;\omega)\, d\beta(\tau).$$

8.2 ON AN ITO STOCHASTIC INTEGRAL EQUATION

Thus

$$E|h(t;\omega)|^2 = E\left|\int_{-\infty}^{\infty} (\Phi\chi_{[0,t]})(\tau;\omega)\,d\beta(\tau)\right|^2 = \int_{-\infty}^{\infty} E|(\Phi\chi_{[0,t]})(\tau;\omega)|\,d\tau,$$

by Lemma 8.1.2. By Condition (vii) and Lemma 8.1.4 we have

$$E|h(t;\omega)|^2 < \infty.$$

Therefore for fixed t, $h(t;\omega) \in L_2(\Omega, \mathcal{A}, \mathcal{P})$. Now let $t_n \to t$ in R_+. To show that $h(t_n;\omega) \to h(t;\omega)$ in $L_2(\Omega, \mathcal{A}, \mathcal{P})$, it is sufficient due to Lemma 8.1.3 to show that

$$\|\Phi\chi_{[0,t_n]} - \Phi\chi_{[0,t]}\|$$

can be made arbitrarily small. That is, we must show that

$$\int_{-\infty}^{\infty} E|(\Phi\chi_{[0,t_n]})(\tau;\omega) - (\Phi\chi_{[0,t]})(\tau;\omega)|^2\,d\tau$$

can be made arbitrarily small. Choose $\varepsilon > 0$. Consider the nonnegative function $q(\tau;\omega) = E|\Phi(\tau;\omega)|^2$. By Condition (iii) $q(\tau;\omega)$ is integrable over R_+. Hence there exists a $\delta > 0$ such that for every set of Lebesgue measure less than δ, $\int_G q(\tau;\omega)\,d\tau < \infty$. Thus

$$\int_{-\infty}^{\infty} E|(\Phi\chi_{[0,t_n]})(\tau;\omega) - (\Phi\chi_{[0,t]})(\tau;\omega)|^2\,d\tau$$

$$= \int_{t}^{t_n} E|(\Phi\chi_{[0,t_n]})(\tau;\omega) - (\Phi\chi_{[0,t]})(\tau;\omega)|^2\,d\tau$$

$$= \int_{t}^{t_n} E|\Phi(\tau;\omega)|^2\,d\tau = \int_{t}^{t_n} q(\tau;\omega)\,d\tau.$$

Since for $n > N_\delta$, $|t_n - t| < \delta$ and since the Lebesgue measure of the interval (t, t_n) is its length, we conclude that the Lebesgue measure of (t, t_n) is less than δ. Hence

$$\int_{t}^{t_n} q(\tau;\omega)\,d\tau < \varepsilon,$$

implying that $t \to h(t;\omega)$ is continuous from R_+ into $L_2(\Omega, \mathcal{A}, \mathcal{P})$ and the proof is complete.

Since we have shown that $h(t;\omega) \in C_c(R_+, L_2(\Omega, \mathcal{A}, \mathcal{P}))$, we can conclude that under the same conditions stated in Theorem 2.1.2 the stochastic integral equation (8.2.1) possesses a unique random solution. Furthermore, one can state the special cases given in Section 2.1.2 concerning Eq. (8.2.1).

8.3 On Ito-Doob-Type Stochastic Integral Equations

In this section we shall study the existence and uniqueness of a random solution to a stochastic integral equation of the form

$$x(t;\omega) = \int_0^t f(\tau, x(\tau;\omega))\, d\tau + \int_0^t \Phi(\tau, x(\tau;\omega))\, d\beta(\tau), \qquad (8.3.1)$$

where $t \in [0, 1]$. As before, the first integral is a Lebesgue integral, while the second is an Ito-type stochastic integral defined with respect to a scalar Brownian motion process $\{\beta(t) : t \in [0, 1]\}$.

Recall that $C^*([0, 1], L_2(\Omega, \mathcal{A}, \mathcal{P})) \subset C_c(R_+, L_2(\Omega, \mathcal{A}, \mathcal{P}))$. We shall define the operators W_1 and W_2 from $C^*([0, 1], L_2(\Omega, \mathcal{A}, \mathcal{P}))$ into $C^*([0, 1], L_2(\Omega, \mathcal{A}, \mathcal{P}))$ by

$$(W_1 x)(t;\omega) = \int_0^t x(\tau;\omega)\, d\tau \qquad (8.3.2)$$

and

$$(W_2 x)(t;\omega) = \int_0^t x(\tau;\omega)\, d\beta(\tau). \qquad (8.3.3)$$

Note that in view of Lemma 8.1.5, $x(t;\omega) \in C^*([0, 1], L_2(\Omega, \mathcal{A}, \mathcal{P}))$. It is clear that W_1 and W_2 are linear operators.

Lemma 8.3.1 The operators W_1 and W_2 defined by (8.3.2) and (8.3.3), respectively, are continuous operators from $C^*([0, 1], L_2(\Omega, \mathcal{A}, \mathcal{P}))$ into $C^*([0, 1], L_2(\Omega, \mathcal{A}, \mathcal{P}))$.

PROOF The fact that W_1 is a continuous operator from $C^*([0, 1], L_2(\Omega, \mathcal{A}, \mathcal{P}))$ into $C^*([0, 1], L_2(\Omega, \mathcal{A}, \mathcal{P}))$ follows from Lemma 2.1.1.

From (8.3.3) we have

$$\|(W_2 x)(t;\omega)\|^2_{L_2(\Omega, \mathcal{A}, \mathcal{P})} = \int_\Omega d\mathcal{P}(\omega) \left\{ \int_0^t x(\tau;\omega)\, d\beta(\tau) \right\}^2$$

$$= \int_0^t d\tau \int_\Omega x^2(\tau;\omega)\, d\mathcal{P}(\omega)$$

by Lemma 8.1.3. Furthermore,

$$\|(W_2 x)(t;\omega)\|^2_{L_2(\Omega, \mathcal{A}, \mathcal{P})} \leq \int_0^t d\tau \sup_{0 \leq \tau \leq t} \|x(\tau;\omega)\|^2_{L_2(\Omega, \mathcal{A}, \mathcal{P})}$$

$$= \int_0^t \{ \sup_{0 \leq \tau \leq t} \|x(\tau;\omega)\|_{L_2(\Omega, \mathcal{A}, \mathcal{P})} \}^2\, d\tau$$

$$= \|x(t;\omega)\|^2 \int_0^t d\tau$$

$$\leq \|x(t;\omega)\|^2.$$

8.3 ON ITO-DOOB-TYPE STOCHASTIC INTEGRAL EQUATIONS

Therefore

$$\|(W_2 x)(t;\omega)\| \leq \|x(t;\omega)\|.$$

Thus W_1 and W_2 are continuous operators from $C^*([0, 1], L_2(\Omega, \mathscr{A}, \mathscr{P}))$ into $C^*([0, 1], L_2(\Omega, \mathscr{A}, \mathscr{P}))$.

8.3.1 An Existence Theorem

We shall assume that Lemma 2.1.1 holds with respect to the operators W_1 and W_2. Therefore there exist positive constants K_1 and K_2 less than one such that

$$\|(W_1 x)(t;\omega)\|_D \leq K_1 \|x(t;\omega)\|_B \quad \text{and} \quad \|(W_2 x)(t;\omega)\|_D \leq K_2 \|x(t;\omega)\|_B.$$

The following theorem gives sufficient conditions for the existence of a unique random solution, a second-order stochastic process, to the Ito–Doob stochastic integral equation (8.3.1).

Theorem 8.3.2 Consider the stochastic integral equation (8.3.1) under the following conditions:

(i) B and D are Banach spaces in $C^*([0, 1], L_2(\Omega, \mathscr{A}, \mathscr{P}))$ which are stronger than $C^*([0, 1], L_2(\Omega, \mathscr{A}, \mathscr{P}))$ such that (B, D) is admissible with respect to the operators W_1 and W_2.

(ii) (a) $x(t;\omega) \to f(t, x(t;\omega))$ is an operator on

$$S = \{x(t;\omega) : x(t;\omega) \in D \quad \text{and} \quad \|x(t;\omega)\|_D \leq \rho\}$$

with values in B satisfying

$$\|f(t, x(t;\omega)) - f(t, y(t;\omega))\|_B \leq \lambda_1 \|x(t;\omega) - y(t;\omega)\|_D.$$

(b) $x(t;\omega) \to \Phi(t, x(t;\omega))$ is an operator on S into B satisfying

$$\|\Phi(t, x(t;\omega)) - \Phi(t, y(t;\omega))\|_B \leq \lambda_2 \|x(t;\omega) - y(t;\omega)\|_D,$$

where λ_1 and λ_2 are constants.

Then there exists a unique random solution to Eq. (8.3.1) provided that

$$K_1 \lambda_1 + K_2 \lambda_2 < 1 \quad \text{and}$$

$$\|f(t, 0)\|_B + \|\Phi(t, 0)\|_B \leq \rho(1 - \lambda_1 K_1 - \lambda_2 K_2).$$

PROOF Define an operator U from the set S into D as follows:

$$(Ux)(t;\omega) = \int_0^t f(\tau, x(\tau;\omega)) \, d\tau + \int_0^t \Phi(\tau, x(\tau;\omega)) \, d\beta(\tau).$$

We need to show that U is a contraction operator on S and that $US \subset S$.

Let $x(t;\omega), y(t;\omega) \in S$. Then $(Ux)(t;\omega) - (Uy)(t;\omega) \in D$, because D is a Banach space. Further, we have

$$\|(Ux)(t;\omega) - (Uy)(t;\omega)\|_D \leq \left\|\int_0^t [f(\tau, x(\tau;\omega)) - f(\tau, y(\tau;\omega))]\, d\tau\right\|_D$$

$$+ \left\|\int_0^t [\Phi(\tau, x(\tau;\omega)) - \Phi(\tau, y(\tau;\omega))]\, d\beta(\tau)\right\|_D$$

$$\leq K_1 \|f(t, x(t;\omega)) - f(t, y(t;\omega))\|_B$$

$$+ K_2 \|\Phi(t, x(t;\omega)) - \Phi(t, y(t;\omega))\|_B$$

$$\leq (\lambda_1 K_1 + \lambda_2 K_2)\|x(t;\omega) - y(t;\omega)\|_D$$

$$< \|x(t;\omega) - y(t;\omega)\|_D.$$

Thus U is a contraction operator.

For any element in S we have

$$\|(Ux)(t;\omega)\|_D \leq \left\|\int_0^t f(\tau, x(\tau;\omega))\, d\tau\right\|_D + \left\|\int_0^t \Phi(\tau, x(\tau;\omega))\, d\beta(\tau)\right\|_D$$

$$\leq K_1 \|f(t, x(t;\omega))\|_B + K_2 \|\Phi(t, x(t;\omega))\|_B$$

$$\leq \lambda_1 K_1 \|x(t;\omega)\|_D + \lambda_2 K_2 \|x(t;\omega)\|_D$$

$$+ K_1 \|f(t,0)\|_B + K_2 \|\Phi(t,0)\|_B.$$

Since $x(t;\omega) \in S$, it follows that

$$\|(Ux)(t;\omega)\|_D \leq \rho(\lambda_1 K_1 + \lambda_2 K_2) + \|f(t,0)\|_B + \|\Phi(t,0)\|_B \leq \rho$$

from the assumptions in the theorem. Thus the existence and uniqueness of a random solution to Eq. (8.3.1) follow from the Banach fixed-point theorem.

One can very easily state the special cases given in Chapter II for the Ito–Doob stochastic integral equation (8.3.1).

CHAPTER IX

Stochastic Nonlinear Differential Systems

9.0 Introduction

In this chapter we shall investigate the existence of a random solution, a second-order stochastic process, and the stability properties of the nonlinear differential systems with random parameters of the form

$$\dot{x}(t;\omega) = A(\omega)x(t;\omega) + b(\omega)\phi(\sigma(t;\omega)) \quad (\cdot = d/dt) \quad (9.0.1)$$

with

$$\sigma(t;\omega) = \langle c(t;\omega), x(t;\omega) \rangle; \quad (9.0.2)$$

$$\dot{x}(t;\omega) = A(\omega)x(t;\omega) + b(\omega)\phi(\sigma(t;\omega)) \quad (9.0.3)$$

with

$$\sigma(t;\omega) = f(t;\omega) + \int_0^t \langle c(t-\tau;\omega), x(\tau;\omega) \rangle \, d\tau; \quad (9.0.4)$$

$$\dot{x}(t;\omega) = A(\omega)x(t;\omega) + \int_0^t b(t-\eta;\omega)\phi(\sigma(\eta;\omega)) \, d\eta \quad (9.0.5)$$

with
$$\sigma(t;\omega) = f(t;\omega) + \int_0^t \langle c(t-\eta;\omega), x(\eta;\omega)\rangle\, d\eta; \qquad (9.0.6)$$
and
$$\dot{x}(t;\omega) = A(\omega)x(t;\omega) + \int_0^t b(t-\tau;\omega)\phi(\sigma(\tau;\omega))\, d\tau$$
$$+ \int_0^t c(t-\tau;\omega)\sigma(\tau;\omega)\, d\tau \qquad (9.0.7)$$
with
$$\sigma(t;\omega) = f(t;\omega) + \int_0^t \langle d(t-\tau;\omega), x(\tau;\omega)\rangle\, d\tau, \qquad (9.0.8)$$

where $A(\omega)$ is an $n \times n$ matrix whose elements are measurable functions; $x(t;\omega)$, $c(t;\omega)$, $b(t;\omega)$, and $d(t;\omega)$ are $n \times 1$ vectors whose elements are random variables; $\sigma(t;\omega)$ and $f(t;\omega)$ are scalar random variables defined for $t \in R_+$ and $\omega \in \Omega$; and $\langle x, y\rangle$ denotes the scalar product in the Euclidean space.

The problem of absolute stability was first formulated in 1944 by two Russian mathematicians, A. I. Lur'e and V. N. Postnikov. Since 1944 many scientists have worked, and are currently working, on the absolute stability of nonlinear control systems. The primary mathematical technique which was used universally to study this type of stability was Lyapunov's direct method. A good summary of the results and methods of such studies can be found in the book by LaSalle and Lefschetz [1]. However, in the late 1950's Lyapunov's method appeared to have been exhausted, and at about that time V. M. Popov introduced the frequency response method, which is currently being used in differential control systems.

The concept of absolute stability is connected with both mathematical and engineering considerations. In engineering problems one is led to this concept because the characteristic function $\phi(\sigma)$ cannot be accurately determined and may even change with time wherever the stability of the system must be preserved. From a mathematical point of view one arrives at the concept of absolute stability from considerations of continuity.

Although Popov's method has been used extensively by many scientists in ordinary control systems, it is only the work of Morozan [1, 2] and Tsokos [1–3, 5] which utilizes this method in differential control systems with random parameters.

The nonlinear stochastic differential systems (9.0.1)–(9.0.2), (9.0.3)–(9.0.4), (9.0.5)–(9.0.6), and (9.0.7)–(9.0.8) will be reduced to a nonlinear stochastic integral equation of the form

$$\sigma(t;\omega) = h(t;\omega) + \int_0^t k(t-\tau;\omega)\phi(\sigma(\tau;\omega))\, d\tau. \qquad (9.0.9)$$

9.1 REDUCTION OF THE STOCHASTIC DIFFERENTIAL SYSTEMS

Utilizing a generalized version of Popov's frequency response method, in Section 9.2 we shall investigate the stochastic absolute stability of the reduced form of each of the systems described. Finally, we shall state the conditions under which the stochastic differential systems are *stochastically absolutely stable*.

In Appendix 9.A we shall give a schematic representation of some of the given nonlinear stochastic differential systems along with their reduced form into a random integral equation.

9.1 Reduction of the Stochastic Differential Systems

9.1.1 Stochastic System (9.0.1)–(9.0.2)

The stochastic differential system (9.0.1)–(9.0.2) may be written as

$$\dot{x}(t;\omega) - A(\omega)x(t;\omega) = b(\omega)\phi(\sigma(t;\omega)). \tag{9.1.1}$$

Multiplying Eq. (9.1.1) by $e^{-A(\omega)t}$ and simplifying, it reduces to

$$(d/dt)\, e^{-A(\omega)t} x(t;\omega) = e^{-A(\omega)t} b(\omega)\phi(\sigma(t;\omega)). \tag{9.1.2}$$

Integrating both sides of (9.1.2) from t_0 to t and simplifying, we have

$$e^{-A(\omega)t}x(t;\omega) - e^{-A(\omega)t_0}x(t_0;\omega) = \int_{t_0}^{t} e^{-A(\omega)\tau} b(\omega)\phi(\sigma(\tau;\omega))\, d\tau. \tag{9.1.3}$$

Multiplying Eq. (9.1.3) by $e^{A(\omega)t}$ and taking $t_0 = 0$, it reduces to

$$x(t;\omega) = e^{A(\omega)t} x_0(\omega) + \int_{0}^{t} e^{A(\omega)(t-\tau)} b(\omega)\phi(\sigma(\tau;\omega))\, d\tau, \tag{9.1.4}$$

where $x_0(\omega) = x(0;\omega)$. Therefore $x(t;\omega)$ as given by Eq. (9.1.4) is a solution of the stochastic differential system (9.0.1)–(9.0.2).

Recall that the nonlinear part of the differential system is a function of

$$\sigma(t;\omega) = \langle c(t;\omega), x(t;\omega) \rangle = c^T(t;\omega) x(t;\omega), \tag{9.1.5}$$

where T denotes the transpose. Now, substituting $x(t;\omega)$ given by Eq. (9.1.4) into Eq. (9.1.5), we have

$$\sigma(t;\omega) = c^T(t;\omega)\, e^{A(\omega)t} x_0(\omega)$$
$$+ c^T(t;\omega) \int_0^t e^{A(\omega)(t-\tau)} b(\omega)\phi(\sigma(\tau;\omega))\, d\tau. \tag{9.1.6}$$

If we let

$$h(t;\omega) = c^T(t;\omega)\, e^{A(\omega)t} x_0(\omega)$$

and
$$k(t - \tau;\omega) = c^T(t;\omega)\, e^{A(\omega)(t-\tau)} b(\omega), \qquad 0 \leqslant \tau \leqslant t < \infty,$$
then Eq. (9.1.6) reduces to a stochastic integral equation of the form (9.0.9). Therefore the stochastic differential system reduces to a special form of the stochastic integral equation of the Volterra type (2.0.1), for which we have given conditions such that a random solution exists in Chapter II.

9.1.2 The Random Differential System (9.0.3)–(9.0.4)

The stochastic system (9.0.3)–(9.0.4) can be written as
$$x(t;\omega) = e^{A(\omega)t} x_0(\omega) + \int_0^t e^{A(\omega)(t-\tau)} b(\omega) \phi(\sigma(\tau;\omega))\, d\tau, \qquad (9.1.7)$$
as was shown, since (9.0.1) and (9.0.3) are identical. Substituting Eq. (9.1.7) into (9.0.4), the nonlinear part of the system, we have
$$\sigma(t;\omega) = f(t;\omega) + \int_0^t c^T(t - \tau;\omega)\, e^{A(\omega)\tau} x_0(\omega)\, d\tau$$
$$+ \int_0^t c^T(t - \tau;\omega) \int_0^\tau e^{A(\omega)(\tau-s)} b(\omega) \phi(\sigma(s;\omega))\, ds\, d\tau. \qquad (9.1.8)$$
Let $h(t;\omega) = f(t;\omega) + \int_0^t c^T(t - \tau;\omega)\, e^{A(\omega)\tau} x_0(\omega)\, d\tau$. Then Eq. (9.1.8) results in the following expression:
$$\sigma(t;\omega) = h(t;\omega) + \int_0^t c^T(t - \tau;\omega) \int_0^\tau e^{A(\omega)(\tau-s)} b(\omega) \phi(\sigma(s;\omega))\, ds\, d\tau. \qquad (9.1.9)$$
By interchanging the order of integration and changing variables, we have
$$\int_0^t c^T(t - \tau;\omega) \int_0^\tau e^{A(\omega)(\tau-s)} b(\omega) \phi(\sigma(s;\omega))\, ds\, d\tau$$
$$= \int_0^t c^T(\tau;\omega) \int_0^{t-\tau} e^{A(\omega)(t-\tau-s)} b(\omega) \phi(\sigma(s;\omega))\, ds\, d\tau$$
$$= \int_0^t \int_0^{t-\tau} c^T(\tau;\omega)\, e^{A(\omega)(t-\tau-s)} b(\omega) \phi(\sigma(s;\omega))\, ds\, d\tau$$
$$= \int_0^t \int_0^{t-s} c^T(\tau;\omega)\, e^{A(\omega)(t-\tau-s)} b(\omega) \phi(\sigma(s;\omega))\, d\tau\, ds. \qquad (9.1.10)$$
Now define
$$k(t;\omega) = \int_0^t c^T(\tau;\omega)\, e^{A(\omega)(t-\tau)} b(\omega)\, d\tau.$$

9.1 REDUCTION OF THE STOCHASTIC DIFFERENTIAL SYSTEMS

Then

$$\int_0^t k(t-s;\omega)\phi(\sigma(s;\omega))ds = \int_0^t \int_0^{t-s} c^T(\tau;\omega)\, e^{A(\omega)(t-s-\tau)}b(\omega)\, d\tau\, \phi(\sigma(s;\omega))\, ds,$$

which is the same as in Eq. (9.1.10). Therefore the equation for $\sigma(t;\omega)$ can be written as

$$\sigma(t;\omega) = h(t;\omega) + \int_0^t k(t-\tau;\omega)\phi(\sigma(\tau;\omega))\, d\tau. \qquad (9.1.11)$$

Equation (9.1.11) is a special case of the stochastic integral equation of the Volterra type stated in Chapter II.

9.1.3 The Stochastic System (9.0.5)–(9.0.6)

Equation (9.0.5) can be written as follows:

$$x(t;\omega) = e^{A(\omega)t}x_0(\omega) + \int_0^t e^{A(\omega)(t-s)} \int_0^s b(t-u;\omega)\phi(\sigma(u;\omega))\, du\, ds, \qquad (9.1.12)$$

where $x(0;\omega) = x_0(\omega)$. From the commutativity property of the convolution product we have

$$\int_0^t e^{A(\omega)(t-s)} \int_0^s b(t-u;\omega)\phi(\sigma(u;\omega))\, du\, ds$$

$$= \int_0^t e^{A(\omega)s} \int_0^{t-s} b(t-s-u;\omega)\phi(\sigma(u;\omega))\, du\, ds$$

$$= \int_0^t \int_0^{t-s} e^{A(\omega)s} b(t-s-u;\omega)\phi(\sigma(u;\omega))\, du\, ds. \qquad (9.1.13)$$

Changing the order of integration and letting

$$k_1(t;\omega) = \int_0^t e^{A(\omega)s} b(t-s;\omega)\, ds,$$

the integral in Eq. (9.1.12) becomes

$$\int_0^t k_1(t-u;\omega)\phi(\sigma(u;\omega))\, du.$$

Hence Eq. (9.1.12) can be written as

$$x(t;\omega) = e^{A(\omega)t}x_0(\omega) + \int_0^t k_1(t-u;\omega)\phi(\sigma(u;\omega))\, du. \qquad (9.1.14)$$

Substituting Eq. (9.1.14) into Eq. (9.1.13), we have

$$\sigma(t;\omega) = f(t;\omega) + \int_0^t c^T(t-s;\omega) e^{A(\omega)s} x_0(\omega) \, ds$$

$$+ \int_0^t c^T(t-s;\omega) \int_0^s k_1(s-u;\omega) \phi(\sigma(u;\omega)) \, du \, ds. \quad (9.1.15)$$

Let $h(t;\omega) = f(t;\omega) + \int_0^t c^T(t-s;\omega) e^{A(\omega)s} x_0(\omega) \, ds$, and applying the commutativity property of the convolution product, we can write Eq. (9.1.15) as

$$\sigma(t;\omega) = h(t;\omega) + \int_0^t \int_0^{t-u} c^T(s;\omega) k_1(t-s-u;\omega) \phi(\sigma(u;\omega)) \, ds \, du. \quad (9.1.16)$$

Define $k(t;\omega) = \int_0^t c^T(s;\omega) k_1(t-s;\omega) \, ds$. Then

$$\int_0^t \int_0^{t-u} c^T(s;\omega) k_1(t-s-u;\omega) \, ds \, \phi(\sigma(u;\omega)) \, du$$

$$= \int_0^t k(t-u;\omega) \phi(\sigma(u;\omega)) \, du. \quad (9.1.17)$$

Therefore the differential system (9.0.5)–(9.0.6) with random parameters reduces to

$$\sigma(t;\omega) = h(t;\omega) + \int_0^t k(t-u;\omega) \phi(\sigma(u;\omega)) \, du, \quad (9.1.18)$$

where

$$k(t;\omega) = \int_0^t c^T(s;\omega) k_1(t-s;\omega) \, ds,$$

$$k_1(t;\omega) = \int_0^t e^{A(\omega)s} b(t-s;\omega) \, ds,$$

and

$$h(t;\omega) = f(t;\omega) + \int_0^t c^T(t-s;\omega) e^{A(\omega)s} x_0(\omega) \, ds.$$

The stochastic integral equation (9.1.18) is a special case of (2.0.1). Thus a random solution exists.

9.1.4 The Random Differential System (9.0.7)–(9.0.8)

Equation (9.0.7) can be written as

$$x(t;\omega) = e^{A(\omega)t} x_0(\omega) + \int_0^t e^{A(\omega)(t-s)} \left\{ \int_0^s b(s-\tau;\omega) \phi(\sigma(\tau;\omega)) \, d\tau \right.$$

$$\left. + \int_0^s c(s-\tau;\omega) \sigma(\tau;\omega) \, d\tau \right\} ds. \quad (9.1.19)$$

9.1 REDUCTION OF THE STOCHASTIC DIFFERENTIAL SYSTEMS

Substituting Eq. (9.1.19) into Eq. (9.0.8), we have

$$\sigma(t;\omega) = h(t;\omega) + \int_0^t d^T(t - \tau;\omega)$$
$$\times \left\{ \int_0^\tau e^{A(\omega)(\tau-s)} \left[\int_0^s b(s - v;\omega)\phi(\sigma(v;\omega))\, dv \right.\right.$$
$$\left.\left. + \int_0^s c(s - v;\omega)\sigma(v;\omega)\, dv \right] \right\} ds, \tag{9.1.20}$$

where $h(t;\omega) = f(t;\omega) + \int_0^t d^T(t - \tau;\omega)\, e^{A(\omega)\tau} x_0(\omega)\, d\tau$. Consider the following portion of Eq. (9.1.20):

$$\int_0^\tau e^{A(\omega)(\tau-s)} \left\{ \int_0^s b(s - v;\omega)\phi(\sigma(v;\omega))\, dv + \int_0^s c(s - v;\omega)\sigma(v;\omega)\, dv \right\} ds$$
$$= \int_0^\tau \int_0^{\tau-s} e^{A(\omega)s} b(s - v - u;\omega)\phi(\sigma(v;\omega))\, dv\, ds$$
$$+ \int_0^\tau \int_0^{\tau-s} e^{A(\omega)s} c(s - \tau - v;\omega)\sigma(v;\omega)\, dv\, ds. \tag{9.1.21}$$

Changing the order of integration and letting

$$k_4(t;\omega) = \int_0^t e^{A(\omega)s} b(t - s;\omega)\, ds$$

and

$$k_3(t;\omega) = \int_0^t e^{A(\omega)s} c(t - s;\omega)\, ds$$

in Eq. (9.1.21), the right-hand side becomes

$$\int_0^\tau k_4(\tau - v;\omega)\phi(\sigma(v;\omega))\, dv + \int_0^\tau k_3(\tau - v;\omega)\sigma(v;\omega)\, dv.$$

Hence Eq. (9.1.20) can be written as

$$\sigma(t;\omega) = h(t;\omega) + \int_0^t d^T(t - \tau;\omega)$$
$$\times \left\{ \int_0^\tau k_4(\tau - v;\omega)\phi(\sigma(v;\omega))\, dv + \int_0^\tau k_3(\tau - v;\omega)\sigma(v;\omega)\, dv \right\} d\tau$$

or

$$\sigma(t;\omega) = h(t;\omega) + \int_0^t \int_0^{t-\tau} d^T(\tau;\omega) k_4(t - \tau - v;\omega)\phi(\sigma(v;\omega))\, dv\, d\tau$$
$$+ \int_0^t \int_0^{t-\tau} d^T(\tau;\omega) k_3(t - \tau - v;\omega)\sigma(v;\omega)\, dv\, d\tau. \tag{9.1.22}$$

Changing the order of integration, letting

$$k_1(t;\omega) = \int_0^t d^T(\tau;\omega)k_4(t-\tau;\omega)\,d\tau$$

and

$$k_2(t;\omega) = \int_0^t d^T(\tau;\omega)k_3(t-\tau;\omega)\,d\tau,$$

and applying the commutativity property of the convolution product, Eq. (9.1.22) becomes

$$\sigma(t;\omega) = h(t;\omega) + \int_0^t k_1(t-v;\omega)\phi(\sigma(v;\omega))\,dv$$
$$+ \int_0^t k_2(t-v;\omega)\sigma(v;\omega)\,dv. \qquad (9.1.23)$$

Utilizing a convolution theorem, we can write Eq. (9.1.23) as follows:

$$\sigma(t;\omega) = \psi(t;\omega) + \int_0^t u(t-s;\omega)\left\{h'(s;\omega) + \int_0^s k'(s-v;\omega)\phi(\sigma(v;\omega))\,dv\right\}ds,$$
$$(9.1.24)$$

where

$$\psi(t;\omega) = f(0;\omega)u(t;\omega) + \int_0^t u(t-s;\omega)h'(s;\omega)\,ds$$

and $u(t;\omega)$ is the random solution of

$$u(t;\omega) = 1 + \int_0^t k_2(t-s;\omega)u(s;\omega)\,ds.$$

Equation (9.1.24) can be reduced to the following form:

$$\sigma(t;\omega) = \psi(t;\omega) + \int_0^t \int_0^{t-s} u(s;\omega)k_1'(t-s-v;\omega)\phi(\sigma(v;\omega))\,dv\,ds. \quad (9.1.25)$$

Changing the order of integration and letting

$$k(t;\omega) = \int_0^t u(s;\omega)k_1'(t-s;\omega)\,ds,$$

Eq. (9.1.24) becomes

$$\sigma(t;\omega) = \psi(t;\omega) + \int_0^t k(t-v;\omega)\phi(\sigma(v;\omega))\,dv, \qquad (9.1.26)$$

which is a special case of Eq. (2.0.1), and we know that a random solution exists. We remark that in the reduction of the random system (9.0.7)–(9.0.8) we assumed that the stochastic kernel $k_2(t;\omega)$ is of an exponential form. However, this condition can be relaxed.

9.2 Stochastic Absolute Stability of the Differential Systems

Recall that a stochastic differential system is said to be *stochastically absolutely stable* if there exists a random solution $x(t;\omega)$ to the system such that

$$\mathscr{P}\{\omega; \lim_{t\to\infty} x(t;\omega) = 0\} = 1.$$

With respect to the aims of this chapter, we state and prove the following theorems.

Theorem 9.2.1 Suppose that the nonlinear stochastic integral equation (9.1.11) satisfies the following conditions:

  (i) $h(t;\omega)$ and $h'(t;\omega) \in L_1(R_+, L_\infty(\Omega, \mathscr{A}, \mathscr{P}))$ for $t \in R_+$ and $\omega \in \Omega$.
  (ii) $k(t;\omega)$ and $k'(t;\omega) \in L_1(R_+, L_\infty(\Omega, \mathscr{A}, \mathscr{P})) \cap L_2(R_+, L_\infty(\Omega, \mathscr{A}, \mathscr{P}))$ for $t \in R_+$ and $\omega \in \Omega$.
  (iii) $\phi(\sigma)$ is a continuous and bounded function for all $\sigma \in R$, R being the real line and $\sigma\phi(\sigma) > 0$ for $\sigma \neq 0$.
  (iv) There exists a $q \geq 0$ such that $\operatorname{Re}\{(1 + i\lambda q)\tilde{k}(i\lambda;\omega)\} \leq 0$ for $\lambda \in R$ and a.e. with respect to ω.

Then every random solution $\sigma(t;\omega), t \geq 0$, of the nonlinear stochastic integral equation (9.1.11) is stochastically absolutely stable.

REMARK $\tilde{k}(i\lambda;\omega) = \int_0^\infty k(t;\omega) e^{-i\lambda t}\, dt$, the Fourier transform of the stochastic kernel with λ the frequency.

PROOF Let $\sigma(t;\omega)$ be a solution of the stochastic integral equation (9.1.11). Following the method of V. M. Popov, let us consider the following function on R_+:

$$\phi_t(\tau;\omega) = \begin{cases} \phi(\sigma(\tau;\omega)), & 0 \leq \tau \leq t, \\ 0, & t < \tau. \end{cases} \quad (9.2.1)$$

Define the following function:

$$\gamma_t(\tau;\omega) = \int_0^\tau (k(\tau - \xi;\omega) + qk'(\tau - \xi;\omega))\phi_t(\xi;\omega)\, d\xi + qk(0;\omega)\phi_t(\tau;\omega).$$

$$(9.2.2)$$

To show the validity of the choice of this function, we proceed as follows: Differentiating Eq. (9.1.11) with respect to t, we have

$$\sigma'(t;\omega) = h'(t;\omega) + k(0;\omega)\phi(\sigma(t;\omega)) + \int_0^t k'(t-\tau;\omega)\phi(\sigma(\tau;\omega))\,d\tau. \quad (9.2.3)$$

Utilizing Eqs. (9.2.1)–(9.2.3), we obtain the following:

$$\gamma_t(\tau;\omega) = \sigma(\tau;\omega) + q\sigma'(\tau;\omega) - [h(\tau;\omega) + qh'(\tau;\omega)], \quad 0 \leq \tau \leq t. \quad (9.2.4)$$

To show that we can write Eq. (9.2.4), substitute Eq. (9.2.3) into Eq. (9.2.4), and the right-hand side of Eq. (9.2.4) becomes

$$\sigma(t;\omega) + qk(0;\omega)\phi(\sigma(\tau;\omega)) + q\int_0^\tau k'(\tau-\xi;\omega)\phi(\sigma(\xi;\omega))\,d\xi - h(\tau;\omega). \quad (9.2.5)$$

Substituting $\sigma(t;\omega)$ from Eq. (9.1.11) into Eq. (9.2.5), we have

$$\int_0^\tau [k(\tau-\xi;\omega) + qk'(\tau-\xi;\omega)]\phi(\sigma(\xi;\omega))\,d\xi + qk(0;\omega)\phi(\sigma(\tau;\omega)). \quad (9.2.6)$$

In expression (9.2.6) replace $\phi(\sigma(\tau;\omega))$ with $\phi_t(\tau;\omega)$ as given in Eq. (9.2.1), and we have

$$\int_0^\tau [k(\tau-\xi;\omega) + qk'(\tau-\xi;\omega)]\phi_t(\xi;\omega)]\,d\xi + qk(0;\omega)\phi_t(\tau;\omega) = \gamma_t(\tau;\omega). \quad (9.2.7)$$

Therefore we have shown that the right-hand side of Eq. (9.2.4) equals the left-hand side as shown in Eq. (9.2.7). Further, for $\tau > t$ we can write Eq. (9.2.7) as

$$\gamma_t(\tau;\omega) = \int_0^t [k(\tau-\xi;\omega) + qk'(\tau-\xi;\omega)]\phi(\sigma(\xi;\omega))\,d\xi, \quad \tau > t. \quad (9.2.8)$$

It now follows from (9.2.7), (9.2.8), the hypothesis of the theorem, and the assumption on σ, that

$$\gamma_t(\tau;\omega) \in L_1(R_+, L_\infty(\Omega, \mathscr{A}, \mathscr{P})) \cap L_2(R_+, L_\infty(\Omega, \mathscr{A}, \mathscr{P}))$$

for $t \in R_+$ and $\omega \in \Omega$.

We shall now consider the Fourier transforms of $\gamma_t(\tau;\omega)$ and $\phi_t(\tau;\omega)$ as follows:

$$\tilde{\gamma}_t(i\lambda;\omega) = \int_0^\infty \gamma_t(\tau;\omega)e^{-i\lambda\tau}\,d\tau \quad \text{and} \quad \tilde{\phi}_t(i\lambda;\omega) = \int_0^\infty \phi_t(\tau;\omega)e^{-i\lambda\tau}\,d\tau.$$

9.2 STOCHASTIC ABSOLUTE STABILITY

Using the fact that if

$$k(t;\omega) \in L_1(R_+, L_\infty(\Omega, \mathscr{A}, \mathscr{P})) \quad \text{and} \quad h(t;\omega) \in L_1(R_+, L_\infty(\Omega, \mathscr{A}, \mathscr{P})),$$

then their convolution product belongs to $L_1(R_+, L_\infty(\Omega, \mathscr{A}, \mathscr{P}))$, and applying the well-known result that the Fourier transform of the convolution product is equal to the product of the Fourier transforms to Eq. (9.2.2), that is,

$$\gamma_t(\tau;\omega) = \int_0^\tau k(\tau - \xi;\omega)\phi_t(\xi;\omega)\,d\xi + q\int_0^\tau k'(\tau - \xi;\omega))\phi_t(\xi;\omega)\,d\xi$$
$$+ qk(0;\omega)\phi_t(\tau;\omega),$$

we have

$$\tilde{\gamma}_t(i\lambda;\omega) = \tilde{k}(i\lambda;\omega)\tilde{\phi}_t(i\lambda;\omega) + q[\tilde{k}'(i\lambda;\omega)\tilde{\phi}_t(i\lambda;\omega)] + qk(0;\omega)\tilde{\phi}_t(i\lambda;\omega). \tag{9.2.9}$$

We know that the Fourier transform of $(d/dt)f(t)$ equals $i\lambda$ [Fourier transform of $f(t)$] $- f(0)$. That is, Eq. (9.2.9) can be written as

$$\tilde{\gamma}_t(i\lambda;\omega) = \tilde{k}(i\lambda;\omega)\tilde{\phi}_t(i\lambda;\omega) + q\tilde{k}'(i\lambda;\omega)\tilde{\phi}_t(i\lambda;\omega) + qk(0;\omega)\tilde{\phi}_t(i\lambda;\omega),$$

but

$$\tilde{k}'(i\lambda;\omega) = i\lambda\tilde{k}(i\lambda;\omega) - \tilde{k}(0;\omega)$$

and

$$\tilde{\gamma}_t(i\lambda;\omega) = \tilde{k}(i\lambda;\omega)\tilde{\phi}_t(i\lambda;\omega)(1 + i\lambda q). \tag{9.2.10}$$

Now we define, for $t \in R_+$ and $\omega \in \Omega$,

$$\rho(t;\omega) = \int_0^t \gamma_t(\tau;\omega)\phi(\sigma(\tau;\omega))\,d\tau,$$

which can be written as

$$\rho(t;\omega) = \int_0^\infty \gamma_t(\tau;\omega)\phi_t(\tau;\omega)\,d\tau. \tag{9.2.11}$$

Applying Parseval's equality or completeness relation, we can write Eq. (9.2.11) as follows:

$$\rho(t;\omega) = (1/2\pi)\int_{-\infty}^\infty \tilde{\gamma}_t(i\lambda;\omega)\overline{\tilde{\phi}_t(i\lambda;\omega)}\,d\lambda, \tag{9.2.12}$$

where $\overline{\tilde{\phi}_t(i\lambda;\omega)}$ is the conjugate of the Fourier transform of $\phi_t(\tau;\omega)$. Substituting $\tilde{\gamma}_t(i\lambda;\omega)$ as given in Eq. (9.2.10) into Eq. (9.2.12), we have

$$\rho(t;\omega) = (1/2\pi)\int_{-\infty}^\infty \tilde{k}(i\lambda;\omega)\tilde{\phi}_t(i\lambda;\omega)[1 + i\lambda q]\overline{\tilde{\phi}(i\lambda;\omega)}\,d\lambda. \tag{9.2.13}$$

Equation (9.2.13) can be written as

$$\rho(t;\omega) = (1/2\pi)\int_{-\infty}^{\infty} \tilde{k}(i\lambda;\omega)(1 + i\lambda q)|\tilde{\phi}(i\lambda;\omega)|^2 \, d\lambda. \qquad (9.2.14)$$

Since we know that $\rho(t;\omega)$ is real, because we have defined it as such, we can take only the real part of (9.2.14), that is,

$$\rho(t;\omega) = (1/2\pi)\int_{-\infty}^{\infty} \operatorname{Re}[\tilde{k}(i\lambda;\omega)(1 + i\lambda q)]|\tilde{\phi}(i\lambda;\omega)|^2 \, d\lambda. \qquad (9.2.15)$$

However, by hypothesis we have

$$\operatorname{Re}[(1 + i\lambda q)\tilde{k}(i\lambda;\omega)] \leq 0,$$

which implies that Eq. (9.2.15) becomes

$$\rho(t;\omega) \leq 0.$$

Recall that

$$\rho(t;\omega) = \int_0^t \gamma_t(\tau;\omega)\phi(\sigma(\tau;\omega)) \, d\tau$$

$$= \int_0^t \{\sigma(\tau;\omega) + q\sigma'(\tau;\omega) - [h(\tau;\omega) + qh'(\tau;\omega)]\phi(\sigma(\tau;\omega)) \, d\tau \leq 0. \qquad (9.2.16)$$

It follows from (9.2.16) that

$$\rho(t;\omega) = \int_0^t \sigma(\tau;\omega)\phi(\sigma(\tau;\omega)) \, d\tau + q\int_0^t \sigma'(\tau;\omega)\phi(\sigma(\tau;\omega)) \, d\tau$$

$$- \int_0^t [h(\tau;\omega) + qh'(\tau;\omega)]\phi(\sigma(\tau;\omega)) \, d\tau \leq 0. \qquad (9.2.17)$$

Let $F(\sigma) = \int_0^\sigma \phi(u) \, du$. Equation (9.2.17) then reduces to

$$\rho(t;\omega) = \int_0^t \sigma(\tau;\omega)\phi(\sigma(\tau;\omega)) \, d\tau + q[F(\sigma(\tau;\omega))$$

$$- F(\sigma(0;\omega))] - \int_0^t [h(\tau;\omega) + qh'(\tau;\omega)]\phi(\sigma(\tau;\omega)) \, d\tau \leq 0$$

or

$$\int_0^t \sigma(\tau;\omega)\phi(\sigma(\tau;\omega)) \, d\tau + qF(\sigma(\tau;\omega))$$

$$- \int_0^t [h(\tau;\omega) + qh'(\tau;\omega)]\phi(\sigma(\tau;\omega)) \, d\tau \leq qF(\sigma(0;\omega)). \qquad (9.2.18)$$

9.2 STOCHASTIC ABSOLUTE STABILITY

However, we know from Eq. (9.1.11) that

$$\sigma(0;\omega) = h(0;\omega);$$

hence Eq. (9.2.18) can be written as

$$\int_0^t \sigma(\tau;\omega)\phi(\sigma(\tau;\omega))\,d\tau + qF(\sigma(t;\omega))$$

$$- \int_0^t [h(\tau;\omega) + qh'(\tau;\omega)]\phi(\sigma(\tau;\omega))\,d\tau \leq qF(h(0;\omega)). \quad (9.2.19)$$

By Condition (iii) of the theorem, $F(\sigma(t;\omega)) > 0$ for $\sigma \neq 0$, which implies that

$$\int_0^t \sigma(\tau;\omega)\phi(\sigma(\tau;\omega))\,d\tau \leq \int_0^t [h(\tau;\omega) + qh'(\tau;\omega)]\phi(\sigma(\tau;\omega))\,d\tau + qF(h(0;\omega)). \quad (9.2.20)$$

From inequality (9.2.20) it follows that

$$\int_0^t \sigma(\tau;\omega)\phi(\sigma(\tau;\omega))\,d\tau \leq \int_0^t |h(\tau;\omega) + qh'(\tau;\omega)|\,|\phi(\sigma(\tau;\omega))|\,d\tau + q|F(h(0;\omega))|. \quad (9.2.21)$$

Let $|\phi(\sigma(\tau;\omega))| \leq \Phi_0(\omega)$, with $\Phi_0(\omega)$ a.e. bounded. Since by hypothesis $\phi(\sigma(\tau;\omega))$ is bounded, inequality (9.2.21) can be written as follows:

$$\int_0^t \sigma(\tau;\omega)\phi(\sigma(\tau;\omega))\,d\tau \leq \Phi_0(\omega) \int_0^t \|h(\tau;\omega) + qh'(\tau;\omega)\|\,d\tau + q|F(h(0;\omega))| \leq M < \infty, \quad \mathscr{P}\text{-a.e.}$$

Let $\Xi(t;\omega) = \int_0^t \phi(\sigma(\tau;\omega))\sigma(\tau;\omega)\,d\tau$. Therefore $\Xi(t;\omega)$ is bounded for $t \in R_+$ and $\omega \in \Omega$, and from Condition (iii) of the theorem it is also a monotonic increasing function for $t \geq 0$, which implies that $\Xi(t;\omega)$ allows a finite limit as $t \to \infty$. Hence, applying the lemma of Barbalat, we conclude that

$$\mathscr{P}\{\omega; \lim_{t \to \infty} \sigma(t;\omega) = 0\} = 1.$$

Theorem 9.2.2 Suppose that the differential system with a random parameter (9.0.1)–(9.0.2) satisfies the following conditions:

(i) (a) the matrix $A(\omega)$ is stochastically stable; (b) the vector function $c(t;\omega)$ is defined for all $t \geq 0$ and $\omega \in \Omega$ such that

$$c(t;\omega), c'(t;\omega) \in L_1(R_+, L_\infty(\Omega, \mathscr{A}, \mathscr{P}));$$

(c) $b(\omega)$ is a scalar random variable.

(ii) $\phi(\sigma)$ is a continuous function for all $\sigma \in R$ and $\sigma\phi(\sigma) > 0$ for $\sigma \neq 0$.

(iii) There exists a $q \geq 0$ such that

$$\text{Re}\{(1 + i\lambda q)\tilde{c}^T(i\lambda;\omega)(i\lambda I - A(\omega))^{-1} b(\omega)\} \leq 0,$$

where I is the identity matrix and $\tilde{c}^T(i\lambda;\omega) = \int_0^\infty c^T(t;\omega) e^{-i\lambda t} dt$.

Then the system (9.0.1)–(9.0.2) is stochastically absolutely stable.

PROOF We shall prove this theorem by demonstrating that the conditions of Theorem 9.2.1 are satisfied.

From system (9.0.1) we have

$$x(t;\omega) = e^{A(\omega)t} x_0(\omega) + \int_0^t e^{A(\omega)(t-\tau)} b(\omega)\phi(\sigma(\tau;\omega)) d\tau, \qquad (9.2.22)$$

where $x_0(\omega) = x(0;\omega)$.

Substituting (9.2.22) into (9.0.2) results in

$$\sigma(t;\omega) = c^T(t;\omega) e^{A(\omega)(t-\tau)} x_0(\omega) + \int_0^t c^T(\tau;\omega) e^{A(\omega)(t-\tau)} b(\omega)\phi(\sigma(\tau;\omega)) d\tau.$$

$$(9.2.23)$$

Since $A(\omega)$ is stochastically stable and $c(t;\omega) \in L_1(R_+, L_\infty(\Omega, \mathscr{A}, \mathscr{P}))$, their convolution product $h(t;\omega) = c^T(t;\omega) e^{A(\omega)(t-\tau)} x_0(\omega) \in L_1(R_+, L_\infty(\Omega, \mathscr{A}, \mathscr{P}))$. Similarly,

$$h'(t;\omega) = (d/dt)\{c^T(t;\omega) e^{A(\omega)(t-\tau)}\}$$

$$= c'^T(t;\omega) e^{A(\omega)(t-\tau)} x_0(\omega) + c^T(t;\omega) A(\omega) e^{A(\omega)(t-\tau)} x_0(\omega)$$

$$\in L_1(R_+, L_\infty(\Omega, \mathscr{A}, \mathscr{P}))$$

for almost all $\omega \in \Omega$. Hence Condition (i) of Theorem 9.2.1 is satisfied. Furthermore,

$k(t - \tau;\omega)$

$$= c^T(\tau;\omega) e^{A(\omega)(t-\tau)} b(\omega) \in L_1(R_+, L_\infty(\Omega, \mathscr{A}, \mathscr{P})) \cap L_2(R_+, L_\infty(\Omega, \mathscr{A}, \mathscr{P}))$$

because $c^T(t;\omega) \in L_1(R_+, L_\infty(\Omega, \mathscr{A}, \mathscr{P}))$,

$$e^{A(\omega)(t-\tau)} \in L_1(R_+, L_\infty(\Omega, \mathscr{A}, \mathscr{P})) \cap L_2(R_+, L_\infty(\Omega, \mathscr{A}, \mathscr{P})),$$

and their convolution product, that is,

$$k(t - \tau;\omega) \in L_1(R_+, L_\infty(\Omega, \mathscr{A}, \mathscr{P})) \cap L_2(R_+, L_\infty(\Omega, \mathscr{A}, \mathscr{P}))$$

for almost all $\omega \in \Omega$. By similar argument

$$k'(t - \tau;\omega) \in L_1(R_+, L_\infty(\Omega, \mathscr{A}, \mathscr{P})) \cap L_2(R_+, L_\infty(\Omega, \mathscr{A}, \mathscr{P})).$$

9.2 STOCHASTIC ABSOLUTE STABILITY

Therefore Condition (ii) of Theorem 9.2.1 is satisfied. Condition (iii) of Theorem 9.2.1 is identical with Condition (ii) of this theorem. Now

$$\tilde{k}(i\lambda;\omega) = \int_0^\infty k(t;\omega)\,e^{-i\lambda t}\,dt = \int_0^\infty c^T(t;\omega)\,e^{A(\omega)(t-\tau)}b(\omega)\,e^{-i\lambda t}\,dt. \quad (9.2.24)$$

Applying the well-known result that the Fourier transform of the convolution product is equal to the product of the Fourier transforms and the fact that the Fourier transform of $e^{A(\omega)t}$ is $(i\lambda I - A(\omega))^{-1}$, we can write Eq. (9.2.24) as follows:

$$\tilde{k}(i\lambda;\omega) = \tilde{c}^T(i\lambda;\omega)(i\lambda I - A(\omega))^{-1}b(\omega).$$

From Condition (iii) of the theorem we have

$$\text{Re}\{(1 + i\lambda q)\tilde{k}(i\lambda;\omega)\} \leq 0.$$

Hence Condition (iii) of Theorem 9.2.1 is satisfied, and the stochastic differential system (9.0.1)–(9.0.2) is stochastically absolutely stable, completing the proof.

Theorem 9.2.3 Suppose that the stochastic differential system (9.0.3)–(9.0.4) satisfies the following conditions:

(i) The matrix $A(\omega)$ is stochastically stable.
(ii) (a) The vector function $c(t;\omega)$ is defined for all $t \geq 0$ and $\omega \in \Omega$ such that

$$c(t;\omega) \in L_1(R_+, L_\infty(\Omega, \mathscr{A}, \mathscr{P})) \cap L_2(R_+, L_\infty(\Omega, \mathscr{A}, \mathscr{P}));$$

(b) $f(t;\omega)$ is defined for $t \geq 0$, $\omega \in \Omega$, with $f(t;\omega)$ and $f'(t;\omega) \in L_1(R_+, L_\infty(\Omega, \mathscr{A}, \mathscr{P}))$;
(c) $b(\omega)$ is a scalar random variable.
(iii) $\phi(\sigma)$ is a continuous function for all $\sigma \in R$ and $\sigma\phi(\sigma) > 0$ for $\sigma \neq 0$.
(iv) There exists a $q \geq 0$ such that

$$\text{Re}\{(1 + i\lambda q)\tilde{c}^T(i\lambda;\omega)(i\lambda I - A(\omega))^{-1}b(\omega)\} \leq 0,$$

where $\tilde{c}^T(i\lambda;\omega) = \int_0^\infty c^T(t;\omega)\,e^{-i\lambda t}\,dt$ and I is the identity matrix.

Then the stochastic differential system (9.0.3)–(9.0.4) is stochastically absolutely stable, that is,

$$\mathscr{P}\{\omega; \lim_{t\to\infty}\sigma(t;\omega) = 0\} = 1.$$

PROOF We shall prove this theorem by demonstrating that the conditions of Theorem 9.2.1 are satisfied.

We have seen from the reduction of the differential system (9.0.3)–(9.0.4) to the random integral equation

$$\sigma(t;\omega) = h(t;\omega) + \int_0^t k(t-\tau;\omega)\phi(\sigma(\tau;\omega))\,d\tau$$

that

$$h(t;\omega) = f(t;\omega) + \int_0^t c^T(\tau;\omega)\,e^{A(\omega)(t-\tau)}x_0(\omega)\,d\tau. \tag{9.2.25}$$

We must show that $h(t;\omega)$ as defined in Eq. (9.2.25) belongs to $L_1(R_+, L_\infty(\Omega, \mathscr{A}, \mathscr{P}))$. It is given in Condition (iib) of the theorem that $f(t;\omega) \in L_1(R_+, L_\infty(\Omega, \mathscr{A}, \mathscr{P}))$. Equation (9.2.25) is a convolution product of $c^T(t;\omega) \in L_1(R_+, L_\infty(\Omega, \mathscr{A}, \mathscr{P}))$ and $e^{A(\omega)(t-\tau)}$ which also belongs to $L_1(R_+, L_\infty(\Omega, \mathscr{A}, \mathscr{P}))$ for almost all $\omega \in \Omega$. We know that if two functions belong to $L_1(R_+, L_\infty(\Omega, \mathscr{A}, \mathscr{P}))$, then their convolution product also belongs to the same space for almost all $\omega \in \Omega$. Hence $h(t;\omega) \in L_1(R_+, L_\infty(\Omega, \mathscr{A}, \mathscr{P}))$. Now we must show that $h'(t;\omega) \in L_1(R_+, L_\infty(\Omega, \mathscr{A}, \mathscr{P}))$. Differentiating Eq. (9.2.25) with respect to t, we have

$$h'(t;\omega) = f'(t;\omega) + \int_0^t c^T(\tau;\omega)A(\omega)\,e^{A(\omega)(t-\tau)}x_0(\omega)\,d\tau + c^T(t;\omega)x_0(\omega). \tag{9.2.26}$$

By hypothesis (ii) we know that $f'(t;\omega) \in L_1(R_+, L_\infty(\Omega, \mathscr{A}, \mathscr{P}))$ and $c^T(t;\omega) \in L_1(R_+, L_\infty(\Omega, \mathscr{A}, \mathscr{P}))$. The convolution product in Eq. (9.2.26), as has previously been shown, belongs to $L_1(R_+, L_\infty(\Omega, \mathscr{A}, \mathscr{P}))$. Therefore $h(t;\omega)$ and $h'(t;\omega) \in L_1(R_+, L_\infty(\Omega, \mathscr{A}, \mathscr{P}))$ and Condition (i) of Theorem 9.2.1 is satisfied.

To show part (ii) of Theorem 9.2.1, we recall that

$$k(t;\omega) = \int_0^t c^T(\tau;\omega)\,e^{A(\omega)(t-\tau)}b(\omega)\,d\tau, \tag{9.2.27}$$

which belongs to $L_1(R_+, L_\infty(\Omega, \mathscr{A}, \mathscr{P}))$ for the same reason as before. Now, differentiating Eq. (9.2.27), we have

$$k'(t;\omega) = \int_0^t c^T(\tau;\omega)\,e^{A(\omega)(t-\tau)}A(\omega)b(\omega)\,d\tau + c^T(t;\omega)b(\omega),$$

which obviously belongs to $L_1(R_+, L_\infty(\Omega, \mathscr{A}, \mathscr{P}))$ for almost all $\omega \in \Omega$. Utilizing the same type of reasoning, it is easy to see that $k(t;\omega)$ and $k'(t;\omega) \in L_2(R_+, L_\infty(\Omega, \mathscr{A}, \mathscr{P}))$. Hence Condition (ii) of Theorem 9.2.1 is satisfied.

9.2 STOCHASTIC ABSOLUTE STABILITY

Condition (iii) of Theorem 9.2.1 is identical with Condition (iii) of this theorem. It remains to be shown that Condition (iv) of Theorem 9.2.1 is satisfied. Let us consider

$$\tilde{k}(i\lambda;\omega) = \int_0^\infty k(t;\omega) e^{-i\lambda t} dt, \qquad (9.2.28)$$

where

$$k(t;\omega) = \int_0^t c^T(\tau;\omega) e^{A(\omega)(t-\tau)} b(\omega) d\tau. \qquad (9.2.29)$$

Substituting Eq. (9.2.29) into Eq. (9.2.28), we have

$$\tilde{k}(i\lambda;\omega) = \int_0^\infty \int_0^t c^T(\tau;\omega) e^{A(\omega)(t-\tau)} b(\omega) e^{-i\lambda t} d\tau \, dt. \qquad (9.2.30)$$

Now, applying the well-known result that the Fourier transform of the convolution product is equal to the product of the Fourier transforms and the fact that the Fourier transform of $e^{A(\omega)t}$ is $(i\lambda I - A(\omega))^{-1}$, we can write Eq. (9.2.30) as follows:

$$\tilde{k}(i\lambda;\omega) = \tilde{c}^T(i\lambda;\omega)(i\lambda I - A(\omega))^{-1} b(\omega). \qquad (9.2.31)$$

From Condition (iv) of the theorem, that is,

$$\text{Re}\{(1 + i\lambda q)\tilde{c}^T(i\lambda;\omega)(i\lambda I - A(\omega))^{-1} b(\omega)\} \leq 0,$$

we can write

$$\text{Re}\{(1 + i\lambda q)\tilde{k}(i\lambda;\omega)\} \leq 0. \qquad (9.2.32)$$

Therefore inequality (9.2.32) shows that Condition (iv) of Theorem 9.2.1 holds. Hence, since Theorem 9.2.3 satisfies Theorem 9.2.1, we conclude that the stochastic differential system (9.0.3)–(9.0.4) admits at least one solution, say $\sigma(t;\omega)$, for $t \geq 0$, such that

$$\mathscr{P}\{\omega; \lim_{t \to \infty} \sigma(t;\omega) = 0\} = 1,$$

which completes the proof.

Theorem 9.2.4 Suppose that the nonlinear stochastic integral equation (9.1.18) satisfies the following conditions:

(i) $h(t;\omega)$ and $h'(t;\omega) \in L_1(R_+, L_\infty(\Omega, \mathscr{A}, \mathscr{P}))$ for $t \in R_+$ and $\omega \in \Omega$.

(ii) $k(t;\omega)$ and $k'(t;\omega) \in L_1(R_+, L_\infty(\Omega, \mathscr{A}, \mathscr{P})) \cap L_2(R_+, L_\infty(\Omega, \mathscr{A}, \mathscr{P}))$ for $t \in R_+$ and $\omega \in \Omega$.

(iii) $\phi(\sigma)$ is a continuous and bounded function for all $\sigma \in R$, R being the real line and $\sigma\phi(\sigma) > 0$ for $\sigma \neq 0$.

(iv) There exists a $q \geq 0$ such that $\operatorname{Re}\{(1 + i\lambda q)\tilde{k}(i\lambda;\omega)\} \leq 0$ for $\lambda \in R$ and a.e. with respect to ω.

Then every random solution $\sigma(t;\omega)$ of the nonlinear stochastic integral equation (9.1.18) is stochastically absolutely stable.

Note that
$$\tilde{k}(i\lambda;\omega) = \int_0^\infty k(t;\omega)e^{-i\lambda t}\,dt,$$

the Fourier transform of the stochastic kernel, with λ the frequency.

PROOF The proof of this theorem is similar to that of Theorem (9.2.1) and is omitted.

By placing the same conditions on $\psi(t;\omega)$ as we have on $h(t;\omega)$, we can conclude that every random solution of the stochastic integral equation (9.1.26) is stochastically absolutely stable.

Theorem 9.2.5 Suppose that the differential system with random parameters (9.0.5)–(9.0.6) satisfies the following conditions:

(i) The matrix $A(\omega)$ is stochastically stable.

(ii) (a) The vector-valued function $c(t;\omega)$ is defined for all $t \geq 0$ and $\omega \in \Omega$ such that
$$c(t;\omega) \in L_1(R_+, L_\infty(\Omega, \mathscr{A}, \mathscr{P})) \cap L_2(R_+, L_\infty(\Omega, \mathscr{A}, \mathscr{P}));$$

(b) $f(t;\omega)$ is defined for $t \geq 0$, $\omega \in \Omega$, with $f(t;\omega)$ and $f'(t;\omega) \in L_1(R_+, L_\infty(\Omega, \mathscr{A}, \mathscr{P}))$;

(c) $b(t;\omega)$ is defined for $t \geq 0$ and $\omega \in \Omega$ such that
$$b(t;\omega) \in L_1(R_+, L_\infty(\Omega, \mathscr{A}, \mathscr{P})) \cap L_2(R_+, L_\infty(\Omega, \mathscr{A}, \mathscr{P})).$$

(iii) $\phi(\sigma)$ is a continuous function for all $\sigma \in R$ and $\sigma\phi(\sigma) > 0$ for $\sigma \neq 0$.

(iv) There exists a $q \geq 0$ such that
$$\operatorname{Re}\{(1 + i\lambda q)\tilde{c}^T(i\lambda;\omega)(i\lambda I - A(\omega))^{-1}\tilde{b}(i\lambda;\omega)\} \leq 0,$$

where $\tilde{c}^T(i\lambda;\omega) = \int_0^\infty c^T(t;\omega)e^{-i\lambda t}\,dt$, $\tilde{b}(i\lambda;\omega) = \int_0^\infty b(t;\omega)e^{-i\lambda t}\,dt$, and I is the identity matrix.

Then the random differential system (9.0.5)–(9.0.6) is stochastically absolutely stable.

PROOF We shall prove this theorem by showing that the conditions of Theorem 9.2.4 are satisfied. We have seen that system (9.0.5)–(9.0.6) reduces to
$$\sigma(t;\omega) = h(t;\omega) + \int_0^t k(t-\tau;\omega)\phi(\sigma(\tau;\omega))\,d\tau,$$

9.2 STOCHASTIC ABSOLUTE STABILITY

where

$$h(t;\omega) = f(t;\omega) + \int_0^t c^T(t-s;\omega) e^{A(\omega)s} x_0(\omega)\, ds$$

$$= f(t;\omega) + \int_0^t c^T(s;\omega) e^{A(\omega)(t-s)} x_0(\omega)\, ds. \quad (9.2.33)$$

We must show that $h(t;\omega)$ as defined in Eq. (9.2.33) belongs to $L_1(R_+, L_\infty(\Omega, \mathscr{A}, \mathscr{P}))$. From Condition (iib) of the theorem we have $f(t;\omega) \in L_1(R_+, L_\infty(\Omega, \mathscr{A}, \mathscr{P}))$. Equation (9.2.33) is a convolution product of $c^T(t;\omega) \in L_1(R_+, L_\infty(\Omega, \mathscr{A}, \mathscr{P}))$ and $e^{A(\omega)(t-s)}$ which also belongs to $L_1(R_+, L_\infty(\Omega, \mathscr{A}, \mathscr{P}))$ for almost all $\omega \in \Omega$, because we know that if two functions belong to $L_1(R_+, L_\infty(\Omega, \mathscr{A}, \mathscr{P}))$, then their convolution product also belongs to the same space for almost all $\omega \in \Omega$. Therefore

$$h(t;\omega) \in L_1(R_+, L_\infty(\Omega, \mathscr{A}, \mathscr{P})).$$

Now differentiating $h(t;\omega)$ with respect to t, we have

$$h'(t;\omega) = f'(t;\omega) + \int_0^t c^T(s;\omega) A(\omega) e^{A(\omega)(t-s)} x_0(\omega)\, ds + c^T(t;\omega) x_0(\omega).$$

$$(9.2.34)$$

By hypothesis (ii) we know that $f'(t;\omega) \in L_1(R_+, L_\infty(\Omega, \mathscr{A}, \mathscr{P}))$ and $c^T(t;\omega) \in L_1(R_+, L_\infty(\Omega, \mathscr{A}, \mathscr{P}))$. The convolution product in Eq. (9.2.34) also belongs to $L_1(R_+, L_\infty(\Omega, \mathscr{A}, \mathscr{P}))$ because of the convolution theorem and the stability of the matrix $A(\omega)$. Hence $h(t;\omega)$ and $h'(t;\omega) \in L_1(R_+, L_\infty(\Omega, \mathscr{A}, \mathscr{P}))$ and Condition (i) of Theorem 9.2.4 is satisfied.

To show Condition (ii) of Theorem 9.2.4, recall that

$$k(t;\omega) = \int_0^t c^T(s;\omega) k_1(t-s;\omega)\, ds, \quad (9.2.35)$$

where

$$k_1(t;\omega) = \int_0^t e^{A(\omega)s} b(t-s;\omega)\, ds = \int_0^t e^{A(\omega)(t-s)} b(s;\omega)\, ds.$$

Since $b(t;\omega)$ and $e^{A(\omega)t}$ both belong to $L_1(R_+, L_\infty(\Omega, \mathscr{A}, \mathscr{P}))$, $k_1(t;\omega)$, the convolution product, also belongs to $L_1(R_+, L_\infty(\Omega, \mathscr{A}, \mathscr{P}))$. Again since both $k_1(t;\omega)$ and $c^T(t;\omega)$ belong to $L_1(R_+, L_\infty(\Omega, \mathscr{A}, \mathscr{P}))$, their convolution product $k(t;\omega)$ also belongs to $L_1(R_+, L_\infty(\Omega, \mathscr{A}, \mathscr{P}))$. Now, differentiating Eq. (9.2.35) with respect to t, we have

$$k'(t;\omega) = \int_0^t c^T(s;\omega) k_1'(t-s;\omega)\, ds + c^T(t;\omega) k_1(0;\omega)$$

with

$$k'_1(t;\omega) = \int_0^t A(\omega) e^{A(\omega)(t-s)} b(s;\omega) \, ds + h(t;\omega),$$

which obviously belongs to $L_1(R_+, L_\infty(\Omega, \mathscr{A}, \mathscr{P}))$ for almost all $\omega \in \Omega$. Utilizing the same type of reasoning, it is easy to see that $k(t;\omega)$ and $k'(t;\omega) \in L_1(R_+, L_\infty(\Omega, \mathscr{A}, \mathscr{P})) \cap L_2(R_+, L_\infty(\Omega, \mathscr{A}, \mathscr{P}))$. Thus Condition (ii) of Theorem 9.2.4 is satisfied.

Condition (iii) of Theorem 9.2.4 is identical with Condition (iii) of this theorem. It remains to be shown that Condition (iv) of Theorem 9.2.4 is satisfied. Let us consider

$$\tilde{k}(i\lambda;\omega) = \int_0^\infty k(t;\omega) e^{-i\lambda t} \, dt, \qquad (9.2.36)$$

where

$$k(t;\omega) = \int_0^t c^T(s;\omega) \left\{ \int_0^t e^{A(\omega)\tau} b(t-\tau;\omega) \, d\tau \right\} ds. \qquad (9.2.37)$$

Substituting Eq. (9.2.37) into Eq. (9.2.36), we have

$$\tilde{k}(i\lambda;\omega) = \int_0^\infty \int_0^t c^T(s;\omega) \left\{ \int_0^t e^{A(\omega)\tau} b(t-\tau;\omega) \, d\tau \right\} ds \, dt. \qquad (9.2.38)$$

Now, applying the well-known result that the Fourier transform of the convolution product is equal to the product of the Fourier transforms and the fact that the Fourier transform of $e^{A(\omega)t}$ is $(i\lambda I - A(\omega))^{-1}$, we can write Eq. (9.2.38) as follows:

$$\tilde{k}(i\lambda;\omega) = \tilde{c}^T(i\lambda;\omega)(i\lambda I - A(\omega))^{-1} b(\omega). \qquad (9.2.39)$$

From Condition (iv) of the theorem, that is,

$$\operatorname{Re}\{(1 + i\lambda q)\tilde{c}^T(i\lambda;\omega)(i\lambda I - A(\omega))^{-1} b(\omega)\} \leqslant 0,$$

we can write

$$\operatorname{Re}\{(1 + i\lambda q)\tilde{k}(i\lambda;\omega)\} \leqslant 0. \qquad (9.2.40)$$

Therefore inequality (9.2.40) shows that Condition (iv) of Theorem 9.2.4 holds. Hence, since Theorem 9.2.5 satisfies Theorem 9.2.4, we conclude that the random solution of system (9.0.5)–(9.0.6) is stochastically stable.

Theorem 9.2.6 Suppose that the random system (9.0.7)–(9.0.8) satisfies the following conditions:

(i) The matrix $A(\omega)$ is stochastically stable.

9.2 STOCHASTIC ABSOLUTE STABILITY

(ii) (a) The vector function $b(t;\omega)$ is defined for $t \geq 0$ and $\omega \in \Omega$ such that, $b(t;\omega) \in L_1(R_+, L_\infty(\Omega, \mathscr{A}, \mathscr{P}))$; (b) $c(t;\omega)$ is defined for $t \geq 0$ and $\omega \in \Omega$ such that $c(t;\omega) \in L_1(R_+, L_\infty(\Omega, \mathscr{A}, \mathscr{P}))$; (c) $d(t;\omega)$ is defined for $t \geq 0$ and $\omega \in \Omega$ such that $d(t;\omega) \in L_1(R_+, L_\infty(\Omega, \mathscr{A}, \mathscr{P}))$; (d) $f(t;\omega)$ is defined for $t \geq 0$ and $\omega \in \Omega$ such that $f(t;\omega) \in L_1(R_+, L_\infty(\Omega, \mathscr{A}, \mathscr{P}))$ and

$$f'(t;\omega) \in L_1(R_+, L_\infty(\Omega, \mathscr{A}, \mathscr{P})) \cap L_2(R_+, L_\infty(\Omega, \mathscr{A}, \mathscr{P})).$$

(iii) $\phi(\sigma)$ is a continuous and bounded function for $\sigma \in R$ and $\sigma\phi(\sigma) > 0$ for $\sigma \neq 0$.

(iv) There exists a $q \geq 0$ such that

$$\mathrm{Re}\{(1 + i\lambda q)[1 - \tilde{d}^T(i\lambda;\omega)(i\lambda I - A(\omega))^{-1}\tilde{c}(i\lambda;\omega)]^{-1}$$
$$\times \tilde{d}^T(i\lambda;\omega)(i\lambda I - A(\omega))^{-1}\tilde{b}(i\lambda;\omega)\} \leq 0,$$

where

$$\tilde{d}^T(i\lambda;\omega) = \int_0^\infty d^T(t;\omega) e^{-i\lambda t}\,dt, \qquad \tilde{c}(i\lambda;\omega) = \int_0^\infty c(t;\omega) e^{-i\lambda t}\,dt,$$

$$\tilde{b}(i\lambda;\omega) = \int_0^\infty b(t;\omega) e^{-i\lambda t}\,dt,$$

and I is the identity matrix.

Then the system is stochastically absolutely stable.

PROOF We shall prove the theorem by demonstrating that the conditions of Theorem 9.2.4 are satisfied. We have defined

$$\psi(t;\omega) = f(0;\omega)u(t;\omega) + \int_0^t u(t-s;\omega)h'(s;\omega)\,ds,$$

where

$$h'(t;\omega) = f'(t;\omega) + \int_0^t d^T(\tau;\omega)A(\omega) e^{A(\omega)(t-\tau)}x_0(\omega) + d^T(t;\omega)x_0(\omega).$$

From Conditions (i), (iic), and (iid), $f'(t;\omega)$, $d^T(t;\omega)$, $e^{A(\omega)t} \in L_1(R_+, L_\infty(\Omega, \mathscr{A}, \mathscr{P}))$ implies that $h'(t;\omega) \in L_1(R_+, L_\infty(\Omega, \mathscr{A}, \mathscr{P}))$. Also, from the manner in which $u(t)$ is defined, it belongs to $L_1(R_+, L_\infty(\Omega, \mathscr{A}, \mathscr{P}))$. Thus

$$\psi(t;\omega) \in L_1(R_+, L_\infty(\Omega, \mathscr{A}, \mathscr{P})).$$

Differentiating $\psi(t;\omega)$ with respect to t, we have

$$\psi'(t;\omega) = f(0;\omega)u'(t;\omega) + \int_0^t u'(t-s;\omega)h'(s;\omega)\,ds.$$

$\psi'(t;\omega)$ belongs to $L_1(R_+, L_\infty(\Omega, \mathscr{A}, \mathscr{P}))$ because each of its terms belongs to $L_1(R_+, L_\infty(\Omega, \mathscr{A}, \mathscr{P}))$. Hence Condition (i) of Theorem 9.2.4 is satisfied. The stochastic kernel is defined by

$$k(t;\omega) = \int_0^t u(s;\omega) k_1'(t-s;\omega)\, ds,$$

where

$$k_1(t;\omega) = \int_0^t d^T(\tau;\omega) k_4(t-\tau;\omega)\, d\tau,$$

$$k_2(t;\omega) = \int_0^t d^T(\tau;\omega) k_3(t-\tau;\omega)\, d\tau,$$

$$k_3(t;\omega) = \int_0^t e^{A(\omega)(t-s)} c(s;\omega)\, ds,$$

and

$$k_4(t;\omega) = \int_0^t e^{A(\omega)(t-s)} b(s;\omega)\, ds.$$

Using Conditions (i) and (iia), we have $e^{A(\omega)t} \in L_2(R_+, L_\infty(\Omega, \mathscr{A}, \mathscr{P}))$ and $b(t;\omega) \in L_1(R_+, L_\infty(\Omega, \mathscr{A}, \mathscr{P})) \cap L_2(R_+, L_\infty(\Omega, \mathscr{A}, \mathscr{P}))$. Again, by hypothesis, $c(t;\omega) \in L_1(R_+, L_\infty(\Omega, \mathscr{A}, \mathscr{P}))$ and $e^{A(\omega)t} \in L_2(R_+, L_\infty(\Omega, \mathscr{A}, \mathscr{P}))$ and their convolution product, $k_3(t;\omega) \in L_1(R_+, L_\infty(\Omega, \mathscr{A}, \mathscr{P})) \cap L_2(R_+, L_\infty(\Omega, \mathscr{A}, \mathscr{P}))$. By similar reasoning it can be seen that $k_2(t;\omega)$ and $k_1(t;\omega) \in L_1(R_+, L_\infty(\Omega, \mathscr{A}, \mathscr{P})) \cap L_2(R_+, L_\infty(\Omega, \mathscr{A}, \mathscr{P}))$. Thus

$$k(t;\omega) \in L_1(R_+, L_\infty(\Omega, \mathscr{A}, \mathscr{P})) \cap L_2(R_+, L_\infty(\Omega, \mathscr{A}, \mathscr{P})).$$

By differentiating $k(t;\omega)$ with respect to t and applying a similar argument as before, it can be shown that

$$k'(t;\omega) \in L_1(R_+, L_\infty(\Omega, \mathscr{A}, \mathscr{P})) \cap L_2(R_+, L_\infty(\Omega, \mathscr{A}, \mathscr{P})),$$

which implies that Condition (ii) of Theorem 9.2.4 is satisfied. Condition (iii) of Theorem 9.2.6 is the same as that of Theorem 9.2.5. To show part (iv) of Theorem 9.2.4, we must find the Fourier transform of $k(t;\omega)$. By lengthy computation it can be seen that the Fourier transform is given by

$$\begin{aligned}\tilde{k}(i\lambda;\omega) &= \int_0^t k(t;\omega) e^{-i\lambda t}\, dt \\ &= [1 - \tilde{d}^T(i\lambda;\omega)(i\lambda I - A(\omega))^{-1} \tilde{c}(i\lambda;\omega)]^{-1} \tilde{d}^T(i\lambda;\omega)(i\lambda I - A(\omega))^{-1} \\ &\quad \times \tilde{b}(i\lambda;\omega).\end{aligned}$$

From Condition (iv) of the theorem we can write

$$\text{Re}\{(1 + i\lambda q)\tilde{k}(i\lambda; \omega)\} \leq 0,$$

which implies that Condition (iv) of Theorem 9.2.4 is satisfied. Hence we can conclude that system (9.0.7)–(9.0.8) admits at least one solution, say $\sigma(t; \omega)$ for $t \geq 0$, such that

$$\mathcal{P}\{\omega; \lim_{t \to \infty} \sigma(t; \omega) = 0\} = 1.$$

Appendix 9.A

9.A.1 Stochastic Differential System (9.0.1)–(9.0.2)

$$\dot{x}(t; \omega) = A(\omega)x(t; \omega) + b(\omega)\phi(\sigma(t; \omega)) \qquad (\cdot = d/dt)$$

with

$$\sigma(t; \omega) = \langle c(t; \omega), x(t; \omega) \rangle$$

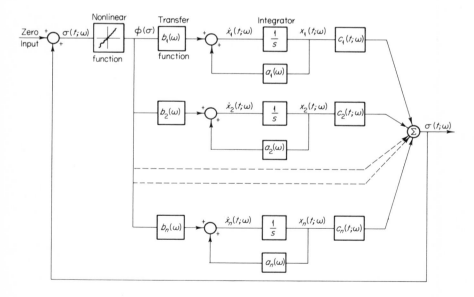

Figure 9.A.1.

9.A.2 Stochastic Differential System (9.0.3)–(9.0.4)

$$\dot{x}(t;\omega) = A(\omega)x(t;\omega) + b(\omega)\phi(\sigma(t;\omega))$$

with

$$\sigma(t;\omega) = f(t;\omega) + \int_0^t \langle c(t-\tau;\omega), x(\tau;\omega)\rangle \, d\tau$$

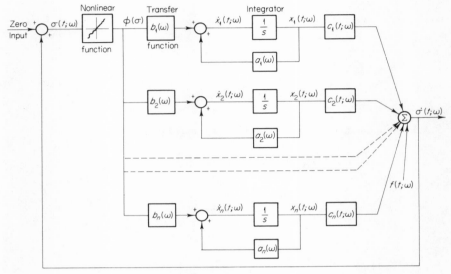

Figure 9.A.2.

9.A.3 The Reduced Stochastic Integral Form of Systems (9.0.1)–(9.0.2) and (9.0.3)

$$\sigma(t;\omega) = h(t;\omega) + \int_0^t k(t-\tau;\omega)\phi(\sigma(\tau;\omega)) \, d\tau$$

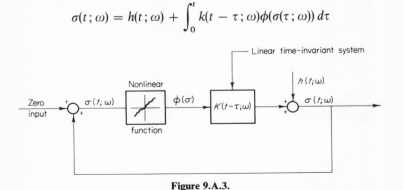

Figure 9.A.3.

CHAPTER X

Stochastic Integrodifferential Systems

10.0 Introduction

The object of this chapter is to study the behavior of a nonlinear stochastic integrodifferential equation of the form

$$\dot{x}(t;\omega) = h[t, x(t;\omega)] + \int_0^t k(t, \tau;\omega) f[x(\tau;\omega)] \, d\tau, \qquad t \geq 0 \quad (10.0.1)$$

and stochastic nonlinear integrodifferential systems with a time lag of the type given by

$$\dot{x}(t;\omega) = A(\omega)x(t;\omega) + B(\omega)x(t - \tau;\omega) + b(\omega)\phi(\sigma(t;\omega)) \quad (10.0.2)$$

with

$$\sigma(t;\omega) = f(t;\omega) + \int_0^t c^T(t - s;\omega)x(s;\omega) \, ds \quad (10.0.3)$$

and

$$\dot{x}(t;\omega) = A(\omega)x(t;\omega) + B(\omega)x(t - \tau;\omega)$$
$$+ \int_0^t \eta(t - u;\omega)\phi(\sigma(u;\omega)) \, du + b(\omega)\phi(\sigma(t;\omega)) \quad (10.0.4)$$

with

$$\sigma(t;\omega) = f(t;\omega) + \int_0^t c^T(u;\omega)x(t-u;\omega)\,du. \qquad (10.0.5)$$

With respect to the random integrodifferential equation (10.0.1), $x(t;\omega)$ is the unknown stochastic process for $t \in R_+$, $h(t, x)$ is a scalar function of $t \in R_+$ and scalar x, $k(t, \tau;\omega)$ is the stochastic kernel defined for t and τ satisfying $0 \leqslant \tau \leqslant t < \infty$, and $f(x)$ is a scalar function of x. For the nonlinear stochastic systems (10.0.2)–(10.0.3) and (10.0.4)–(10.0.5), $x(t;\omega)$, $c(t;\omega)$, and $\eta(t;\omega)$ are n-dimensional vectors whose elements are random variables; $A(\omega)$ and $B(\omega)$ are $n \times n$ matrices whose elements are measurable functions; $\sigma(t;\omega)$ and $f(t;\omega)$ are scalar random variables; $b(\omega)$ is an $n \times 1$ vector whose elements are measurable functions; and $c^T(t;\omega)$ denotes the transpose of $c(t;\omega)$.

In the first part of our presentation we shall give conditions which guarantee the existence and uniqueness of a random solution of the stochastic integrodifferential equation (10.0.1). In addition, we shall study the asymptotic behavior of the random solution in Section 10.1.1. In Section 10.1.2 we shall illustrate the usefulness of the theory with an application to differential systems with random parameters. The second part of this chapter will be concerned with studying the existence and stability of a random solution of the stochastic nonlinear systems (10.0.2)–(10.0.3) and (10.0.4)–(10.0.5). In Section 9.2 we reduced the systems with time lag to a nonlinear stochastic integral equation which was studied in Chapter II. Knowing that a random solution to the system exists, we shall give conditions under which it is stochastically absolutely stable in Section 10.3.

From a deterministic point of view, the concept of stability has been widely used by many scientists under various model formulations. The basic idea, however, is: "If a system has a suitable response for a class of inputs or initial conditions and if small changes in the input or in the initial conditions occur, then the new response should be close to the original one." It is apparent from this formulation that stability is a very basic concept in a great many practical problems. In fact, the conventional design techniques in control theory are all directly or indirectly derived from the stability criteria.

Among the more useful of the concepts of stability is the concept of "absolute stability," which is simply global asymptotic stability for a nonlinearity class. Absolute stability, as we mentioned previously, was originally formulated by Lur'e and Postnikov and is connected both with engineering and mathematical considerations. From a mathematical point of view, one arrives at this concept from considerations of continuity. In engineering problems one is led to this type of stability because system nonlinearities cannot be accurately determined and may even change in time but yet the

10.1 The Stochastic Integrodifferential Equation

The importance of stochastic integrodifferential equations of the form (10.0.1) lies in the fact that they arise in many situations. For example, equations of this kind occur in the stochastic formulation of problems in reactor dynamics which have been investigated from the deterministic point of view by Levin and Nohel [1]. Also, they arise in the study of the growth of biological populations by Miller [1], in the theory of automatic systems resulting in delay-differential equations (Oguztoreli [1]), and in many other problems occurring in the general areas of biology, physics, and engineering.

With respect to the aims of our study we shall assume that $x(t;\omega)$ will be a function of $t \in R_+$ with values in the space $L_2(\Omega, \mathscr{A}, \mathscr{P})$, a second-order stochastic process defined on R_+. The function $h[t, x(t;\omega)]$, the stochastic perturbing term, under certain conditions will also be a function in $L_2(\Omega, \mathscr{A}, \mathscr{P})$, and $f[x(t;\omega)]$ will be considered as a function from R_+ into $L_2(\Omega, \mathscr{A}, \mathscr{P})$.

With respect to the stochastic kernel, we shall assume that for each t and τ such that $0 \leqslant \tau \leqslant t < \infty$, $k(t, \tau; \omega)$ is essentially bounded. As we indicated before, the norm of $k(t, \tau; \omega)$ in $L_\infty(\Omega, \mathscr{A}, \mathscr{P})$ will be denoted by

$$|||k(t, \tau; \omega)||| = \mathscr{P}\text{-ess}\sup_{\omega \in \Omega}|k(t, \tau; \omega)|.$$

It will also be assumed that for each fixed t and τ

$$|||k(s, \tau; \omega)||| \leqslant M_{(t,\tau)} \qquad \text{uniformly for} \quad \tau \leqslant s \leqslant t,$$

where $M_{(t,\tau)} > 0$ is some constant depending on t and τ, $0 \leqslant \tau \leqslant t < \infty$. We shall make use of the integral operators T_1 and T_2 on $C_c(R_+, L_2(\Omega, \mathscr{A}, \mathscr{P}))$, defined as follows:

$$(T_1 x)(t; \omega) = \int_0^t x(\tau; \omega)\, d\tau \qquad (10.1.1)$$

and

$$(T_2 x)(t; \omega) = \int_0^t K(t, \tau; \omega) x(\tau; \omega)\, d\tau, \qquad (10.1.2)$$

where

$$K(t, \tau; \omega) = \int_\tau^t k(s, \tau; \omega)\, ds. \qquad (10.1.3)$$

These integral operators will be needed in obtaining existence and uniqueness of a random solution of Eq. (10.0.1). It is clear from the given conditions and Lemma 4.1.1 that the integral operators T_1 and T_2 are continuous mappings from $C_c(R_+, L_2(\Omega, \mathscr{A}, \mathscr{P}))$ into itself.

If we integrate Eq. (10.0.1) from zero to t, we obtain

$$x(t;\omega) = x_0(\omega) + \int_0^t h(\tau, x(\tau;\omega))\,d\tau + \int_0^t \left[\int_\tau^t k(s,\tau;\omega)\,ds\right]f(x(\tau;\omega))\,d\tau$$

$$= x_0(\omega) + \int_0^t h(\tau, x(\tau;\omega))\,d\tau + \int_0^t K(t,\tau;\omega)f(x(\tau;\omega))\,d\tau, \quad (10.1.4)$$

where $x_0(\omega) = x(0;\omega)$ and $K(t,\tau;\omega)$ is given by Eq. (10.1.3).

We now prove the following existence theorem (Padgett and Tsokos [15]).

Theorem 10.1.1 Suppose the random equation (10.0.1) satisfies the following conditions:

(i) B and D are Banach spaces stronger than $C_c(R_+, L_2(\Omega, \mathscr{A}, \mathscr{P}))$ and the pair (B, D) is admissible with respect to each of the integral operators $(T_1 x)(t;\omega) = \int_0^t x(\tau;\omega)\,d\tau$ and $(T_2 x)(t;\omega) = \int_0^t K(t,\tau;\omega)x(\tau;\omega)\,d\tau$, $t \geq 0$, where $K(t,\tau;\omega)$ is given by (10.1.3).

(ii) $x(t;\omega) \to h(t, x(t;\omega))$ is an operator on

$$S = \{x(t;\omega) \in D : \|x(t;\omega)\|_D \leq \rho\}$$

with values in B satisfying

$$\|h(t, x(t;\omega)) - h(t, y(t;\omega))\|_B \leq \lambda_1 \|x(t;\omega) - y(t;\omega)\|_D$$

for $x(t;\omega), y(t;\omega) \in S$ and λ_1 constant.

(iii) $x(t;\omega) \to f(x(t;\omega))$ is an operator on S with values in B satisfying $f(0) = 0$ and

$$\|f(x(t;\omega)) - f(y(t;\omega))\|_B \leq \lambda_2 \|x(t;\omega) - y(t;\omega)\|_D$$

for $x(t;\omega), y(t;\omega) \in S$ and λ_2 constant.

(iv) $x_0(\omega) \in D$.

Then there exists a unique random solution of (10.1.4), $x(t;\omega) \in S$, provided

$$\lambda_1 K_1 + \lambda_2 K_2 < 1,$$

$$\|x_0(\omega)\|_D + K_1 \|h(t, 0)\|_B \leq \rho(1 - \lambda_1 K_1 - \lambda_2 K_2),$$

where K_1 and K_2 are the norms of T_1 and T_2, respectively.

10.1 THE STOCHASTIC INTEGRODIFFERENTIAL EQUATION

(iii) $x(t;\omega) \to f(x(t;\omega))$ satisfies $f(0) = 0$, $\|f(x(t;\omega))\| \leq \Lambda_3 e^{-\beta t}$, $\Lambda_3 > 0$, and
$$\|f(x(t;\omega)) - f(y(t;\omega))\| \leq \lambda_2 e^{-\beta t}\|x(t;\omega) - y(t;\omega)\|$$
for $\|x(t;\omega)\|$ and $\|y(t;\omega)\| \leq \rho$, $t \geq 0$, and λ_2 constant.

(iv) $x_0(\omega) \in C_1$.

Then there exists a unique random solution of Eq. (10.0.1) satisfying
$$\|x(t;\omega)\| = \{E[|x(t;\omega)|^2]\}^{\frac{1}{2}} \leq \rho, \qquad t \in R_+,$$
(bounded in mean square on R_+), provided that λ_1, λ_2, $\|x_0(\omega)\|_{C_1}$, and $\|h(t,0)\|_{C_g}$ are sufficiently small.

PROOF   It will suffice to show that the pair of spaces $(C_g(R_+, L_2(\Omega, \mathscr{A}, \mathscr{P}))$, $C_1(R_+, L_2(\Omega, \mathscr{A}, \mathscr{P})))$ is admissible with respect to the integral operators defined by equations (10.1.1)–(10.1.3) under Condition (i).

Let $x(t;\omega) \in C_g(R_+, L_2(\Omega, \mathscr{A}, \mathscr{P}))$. Then from Eq. (10.1.1) we have that

$$\|(T_1 x)(t;\omega)\| \leq \int_0^t [\|x(\tau;\omega)\|/e^{-\beta \tau}] e^{-\beta \tau} d\tau$$

$$\leq \sup_{t \geq 0}\{\|x(t;\omega)\|/e^{-\beta t}\} \int_0^t e^{-\beta \tau} d\tau$$

$$= \|x(t;\omega)\|_{C_g}(1/\beta)(1 - e^{-\beta t})$$

$$< \|x(t;\omega)\|_{C_g}(1/\beta) < \infty,$$

by definition of the norm in $C_g(R_+, L_2(\Omega, \mathscr{A}, \mathscr{P}))$. Hence
$$(T_1 x)(t;\omega) \in C_1(R_+, L_2(\Omega, \mathscr{A}, \mathscr{P}))$$
and the pair (C_g, C_1) is admissible with respect to T_1.

Now from Eqs. (10.1.2) and (10.1.3) for $x(t;\omega) \in C_g(R_+, L_2(\Omega, \mathscr{A}, \mathscr{P}))$ we obtain

$$\|(T_2 x)(t;\omega)\| \leq \int_0^t \left\|\int_\tau^t k(s,\tau;\omega)\,ds\right\| \|x(\tau;\omega)\|\,d\tau$$

$$\leq \Lambda_1 \int_0^t [\|x(\tau;\omega)\|/e^{-\beta \tau}] e^{-\beta \tau} d\tau$$

$$\leq \Lambda_1 \|x(t;\omega)\|_{C_g} \int_0^t e^{-\beta \tau} d\tau$$

$$< \Lambda_1 \|x(t;\omega)\|_{C_g}(1/\beta) < \infty$$

from Condition (i). Thus $(T_2 x)(t;\omega) \in C_1(R_+, L_2(\Omega, \mathscr{A}, \mathscr{P}))$ and the pair (C_g, C_1) is admissible with respect to T_2.

Therefore the conditions of Theorem 10.1.1 hold with $B = C_g$, $g(t) = e^{-\beta t}$, $\beta > 0$, and $D = C_1$, and there exists a unique random solution of (10.0.1), $x(t;\omega)$, bounded in mean square by ρ for all $t \in R_+$.

10.1.2 Application to a Stochastic Differential System

Consider the following nonlinear differential system with random parameters:

$$\dot{x}(t;\omega) = A(\omega)x(t;\omega) + b(\omega)\phi(\sigma(t;\omega)), \qquad (10.1.8)$$

$$\dot{\sigma}(t;\omega) = c^T(t;\omega)x(t;\omega), \qquad (10.1.9)$$

where $A(\omega)$ is an $n \times n$ matrix of measurable functions, $x(t;\omega)$ and $c(t;\omega)$ are $n \times 1$ vectors of random variables for each $t \in R_+$, $b(\omega)$ is an $n \times 1$ vector of measurable functions, $\phi(\sigma)$ is a scalar function, $\sigma(t;\omega)$ is a scalar random variable for each $t \in R_+$, and T denotes the transpose of a matrix.

Note that Eq. (10.1.9) can be written as

$$\sigma(t;\omega) = \sigma(0;\omega) + \int_0^t c^T(s;\omega)x(s;\omega)\,ds,$$

which is similar to a system studied in Chapter IX.

The system (10.1.8)–(10.1.9) may be reduced to a stochastic integrodifferential equation of the form (10.0.1). We may write (10.1.8) as

$$x(t;\omega) = e^{A(\omega)t}x_0(\omega) + \int_0^t e^{A(\omega)(t-\tau)}b(\omega)\phi(\sigma(\tau;\omega))\,d\tau.$$

Substituting this expression for $x(t;\omega)$ in (10.1.9), we obtain

$$\dot{\sigma}(t;\omega) = c^T(t;\omega)\,e^{A(\omega)t}x_0(\omega) + \int_0^t c^T(t;\omega)\,e^{A(\omega)(t-\tau)}b(\omega)\phi(\sigma(\tau;\omega))\,d\tau.$$

Assume that $\|\|c^T(t;\omega)\|\| \leq K_1$ for all $t \geq 0$ and $K_1 \geq 0$ a constant. Also let $x_0(\omega) \in C_1$, $\phi(0) = 0$, and $b(\omega) \in L_\infty(\Omega, \mathcal{A}, \mathcal{P})$. If we assume that the matrix $A(\omega)$ is *stochastically stable*, that is, there exists an $\alpha > 0$ such that

$$\mathcal{P}\{\omega : \operatorname{Re} \psi_k(\omega) < -\alpha, \quad k = 1, 2, \ldots, n\} = 1,$$

where $\psi_k(\omega)$, $k = 1, 2, \ldots, n$, are the characteristic roots of the matrix, then it has been shown by Morozan [3] that

$$\|\|e^{A(\omega)t}\|\| \leq K_2\, e^{-\alpha t}$$

for some constant $K_2 > 0$. We also let $\phi(\sigma(t;\omega))$ be in the space $C_g(R_+, L_2(\Omega, \mathcal{A}, \mathcal{P}))$ with $g(t) = e^{-\alpha t}$, $t \geq 0$, and

$$|\phi(\sigma_1(t;\omega)) - \phi(\sigma_2(t;\omega))| \leq \lambda\, e^{-\alpha t}|\sigma_1(t;\omega) - \sigma_2(t;\omega)|.$$

Let
$$h(t, \sigma(t;\omega)) = c^T(t;\omega) e^{A(\omega)t} x_0(\omega).$$
Then,
$$\|h(t, \sigma(t;\omega))\| \leq \|\|c^T(t;\omega)\|\| K_2 e^{-\alpha t} \|x_0(\omega)\| \leq K_1 K_2 e^{-\alpha t} Z,$$
where $Z > 0$ is a constant, since $x_0(\omega) \in C_1$. Thus
$$h(t, \sigma(t;\omega)) \in C_g(R_+, L_2(\Omega, \mathscr{A}, \mathscr{P}))$$
by definition. Also,
$$\|h(t, \sigma_1(t;\omega)) - h(t, \sigma_2(t;\omega))\| = 0$$
so that it satisfies a Lipschitz condition.

Now, by the assumptions we have made on $c^T(t;\omega)$, $b(\omega)$, and $A(\omega)$ we have
$$k(s, \tau;\omega) = c^T(s;\omega) e^{A(\omega)(s-\tau)} b(\omega),$$
satisfying
$$\int_\tau^t \|\|k(s, \tau;\omega)\|\| \, ds \leq \int_\tau^t \|\|c^T(s;\omega)\|\| K_2 e^{-\alpha(s-\tau)} \|\|b(\omega)\|\| \, ds$$
$$\leq K_1 K_2 e^{\alpha \tau} \|\|b(\omega)\|\| \int_\tau^t e^{-\alpha s} \, ds$$
$$= K_1 K_2 \|\|b(\omega)\|\| (1/\alpha)[1 - e^{-\alpha(t-\tau)}]$$
$$< K_1 K_2 \|\|b(\omega)\|\| (1/\alpha).$$

Therefore all conditions of Theorem 10.1.4 are satisfied and there exists a unique random solution of the system (10.1.8)–(10.1.9) which is bounded in mean square on R_+.

10.2 Reduction of the Stochastic Nonlinear Integrodifferential Systems with Time Lag

10.2.1 The Integrodifferential System (10.0.2)–(10.0.3)

The random system with time lag (10.0.2) can be written as a stochastic integral equation of the form
$$x(t;\omega) = X(t;\omega) x_0(\omega) + \int_{-\tau}^0 X(t - \tau - u;\omega) B(\omega) x(u;\omega) \, du$$
$$+ \int_0^t X(t - u;\omega) b(\omega) \phi(\sigma(u;\omega)) \, du, \qquad t > 0, \qquad (10.2.1)$$

where $x_0(\omega) = x(0; \omega)$ and $X(t; \omega)$ is the stochastic fundamental matrix solution of the homogeneous system

$$x'(t; \omega) = A(\omega)x(t; \omega) + B(\omega)x(t - \tau; \omega), \qquad t > 0,$$

with the initial condition $X(0; \omega) = I$, the identity matrix, and $X(t; \omega) = 0$ for $t < 0$. Let

$$\psi(t; \omega) = X(t; \omega)x_0(\omega) + \int_{-\tau}^{0} X(t - \tau - u; \omega)B(\omega)x(u; \omega)\,du.$$

Then the random equation (10.2.1) becomes

$$x(t; \omega) = \psi(t; \omega) + \int_{0}^{t} X(t - u; \omega)b(\omega)\phi(\sigma(u; \omega))\,du \qquad (10.2.2)$$

with

$$\sigma(t; \omega) = f(t; \omega) + \int_{0}^{t} c^{T}(t - u; \omega)x(u; \omega)\,du.$$

Substituting Eq. (10.2.2) into (10.0.3), we get

$$\sigma(t; \omega) = f(t; \omega) + \int_{0}^{t} c^{T}(t - u; \omega)\bigg\{\psi(u; \omega)$$
$$+ \int_{0}^{u} X(u - s; \omega)b(\omega)\phi(\sigma(s; \omega))\,ds\bigg\}\,du. \qquad (10.2.3)$$

Let $h(t; \omega) = f(t; \omega) + \int_{0}^{t} c^{T}(t - u; \omega)\psi(u; \omega)\,du$. Then Eq. (10.2.3) becomes

$$\sigma(t; \omega) = h(t; \omega) + \int_{0}^{t} c^{T}(t - u; \omega) \int_{0}^{u} X(u - s; \omega)b(\omega)\phi(\sigma(s; \omega))\,ds\,du.$$

Using the property of the convolution integral, we have

$$\int_{0}^{t} c^{T}(t - u; \omega) \int_{0}^{u} X(u - s; \omega)b(\omega)\phi(\sigma(s; \omega))\,ds\,du$$
$$= \int_{0}^{t} c^{T}(u; \omega) \int_{0}^{t-u} X(t - u - s; \omega)b(\omega)\phi(\sigma(s; \omega))\,ds\,du$$
$$= \int_{0}^{t}\int_{0}^{t-u} c^{T}(u; \omega)X(t - u - s; \omega)b(\omega)\phi(\sigma(s; \omega))\,ds\,du$$
$$= \int_{0}^{t}\int_{0}^{t-s} c^{T}(u; \omega)X(t - u - s; \omega)b(\omega)\phi(\sigma(s; \omega))\,du\,ds. \qquad (10.2.4)$$

Define

$$k(t; \omega) = \int_{0}^{t} c^{T}(u; \omega)X(t - u; \omega)b(\omega)\,du.$$

10.2 REDUCTION OF SYSTEMS WITH TIME LAG

Then

$$\int_0^t k(t - s; \omega)\phi(\sigma(s; \omega))\, ds$$
$$= \int_0^t \int_0^{t-s} c^T(u; \omega)X(t - s - u; \omega)b(\omega)\, du\; \phi(\sigma(s; \omega))\, ds,$$

which is the same as Eq. (10.2.4). Therefore the equation for the error signal in the presence of a random parameter $\sigma(t; \omega)$ can be written as

$$\sigma(t; \omega) = h(t; \omega) + \int_0^t k(t - u; \omega)\phi(\sigma(u; \omega))\, du. \tag{10.2.5}$$

Thus we know that there exists a unique random solution to Eq. (10.2.5) under the conditions given in Chapter II.

10.2.2 The Random Integrodifferential System (10.0.4)–(10.0.5)

The random equation (10.0.4) can be written as

$$x(t; \omega) = X(t; \omega)x_0(\omega) + \int_{-\tau}^0 X(t - \tau - u; \omega)B(\omega)x(u; \omega)\, du$$
$$+ \int_0^t X(t - u; \omega)b(\omega)\phi(\sigma(u; \omega))\, du$$
$$+ \int_0^t X(t - u; \omega) \int_0^u \eta(t - s; \omega)\phi(\sigma(s; \omega))\, ds\, du, \tag{10.2.6}$$

where $x_0(\omega)$ and $X(t; \omega)$ behave as defined earlier. We shall write

$$\psi(t; \omega) = X(t; \omega)x_0(\omega) + \int_{-\tau}^0 X(t - \tau - u; \omega)B(\omega)x(u; \omega)\, du.$$

Then the stochastic integral system (10.2.6) becomes

$$x(t; \omega) = \psi(t; \omega) + \int_0^t X(t - u; \omega)b(\omega)\phi(\sigma(u; \omega))\, du$$
$$+ \int_0^t X(t - u; \omega) \int_0^u \eta(t - s; \omega)\phi(\sigma(s; \omega))\, ds\, du. \tag{10.2.7}$$

X STOCHASTIC INTEGRODIFFERENTIAL SYSTEMS

Applying the well-known result that the convolution product commutes, we can reduce part of Eq. (10.2.7) as follows:

$$\int_0^t X(t-u;\omega) \int_0^u \eta(t-s;\omega)\phi(\sigma(s;\omega))\,ds\,du$$

$$= \int_0^t X(u;\omega) \int_0^{t-u} \eta(t-u-s;\omega)\phi(\sigma(s;\omega))\,ds\,du$$

$$= \int_0^t \int_0^{t-u} X(u;\omega)\eta(t-u-s;\omega)\phi(\sigma(s;\omega))\,ds\,du$$

$$= \int_0^t \int_0^{t-s} X(u;\omega)h(t-u-s;\omega)\phi(\sigma(s;\omega))\,du\,ds. \qquad (10.2.8)$$

Let

$$k_1(t;\omega) = \int_0^t X(t-s;\omega)\eta(s;\omega)\,ds.$$

Then Eq. (10.2.8) can be written as follows:

$$\int_0^t \int_0^{t-s} X(t-s-u;\omega)\eta(u;\omega)\phi(\sigma(s;\omega))\,du\,ds = \int_0^t k_1(t-s;\omega)\phi(\sigma(s;\omega))\,ds.$$

Therefore Eq. (10.2.7) becomes

$$x(t;\omega) = \psi(t;\omega) + \int_0^t X(t-u;\omega)b(\omega)\phi(\sigma(u;\omega))\,du$$

$$+ \int_0^t k_1(t-u;\omega)\phi(\sigma(u;\omega))\,du$$

or

$$x(t;\omega) = \psi(t;\omega) + \int_0^t K_2(t-u;\omega)\phi(\sigma(u;\omega))\,du, \qquad (10.2.9)$$

where

$$k_2(t;\omega) = X(t;\omega)b(\omega) + k_1(t;\omega).$$

Substituting stochastic Eq. (10.2.9) into Eq. (10.0.5), we have

$$\sigma(t;\omega) = f(t;\omega) + \int_0^t c^T(t-u;\omega)\psi(u;\omega)\,du$$

$$+ \int_0^t c^T(t-u;\omega) \int_0^u k_2(t-s;\omega)\phi(\sigma(s;\omega))\,ds\,du. \qquad (10.2.10)$$

Define

$$h(t;\omega) = f(t;\omega) + \int_0^t c^T(t - u;\omega)\psi(u;\omega)\,du.$$

Then random equation (10.2.10) can be written as

$$\sigma(t;\omega) = h(t;\omega) + \int_0^t \int_0^{t-u} c^T(u;\omega)k_2(t - u - s;\omega)\phi(\sigma(s;\omega))\,ds\,du.$$

(10.2.11)

Let

$$k(t;\omega) = \int_0^t c^T(u;\omega)k_2(t - u;\omega)\,du.$$

Then Eq. (10.2.11) becomes

$$\sigma(t;\omega) = h(t;\omega) + \int_0^t k(t - u;\omega)\phi(\sigma(u;\omega))\,du,$$

which is the same as the nonlinear stochastic integral equation (10.2.5) that we obtain by reducing the stochastic system with time lag (10.0.2)–(10.0.3).

10.3 Stochastic Absolute Stability of the Systems

The following theorems give the conditions under which the stochastic differential systems with lag time (10.0.2)–(10.0.3) and (10.0.4)–(10.0.5) are stochastically absolutely stable.

Theorem 10.3.1 Suppose that the stochastic system with time lag (10.0.2)–(10.0.3) satisfies the following conditions:

(i) The equation $\det\{A(\omega) + e^{-\lambda\tau}B(\omega) - \lambda I\} = 0$ has all its roots in the semiplane $\operatorname{Re}\lambda \leqslant -\alpha < 0$.

(ii) (a) The random vector function $c(t;\omega)$ is defined for all $t \geqslant 0$, $\omega \in \Omega$, and is such that $c(t;\omega) \in L_1(R_+, L_\infty(\Omega, \mathcal{A}, \mathcal{P})) \cap L_2(R_+, L_\infty(\Omega, \mathcal{A}, \mathcal{P}))$;
(b) $f(t;\omega)$ is defined for $t \geqslant 0$, $\omega \in \Omega$, and is such that $f(t;\omega)$, $f'(t;\omega) \in L_1(R_+, L_\infty(\Omega, \mathcal{A}, \mathcal{P}))$.

(iii) There exists a $q \geqslant 0$ such that

$$\operatorname{Re}\{(1 + i\lambda q)\tilde{c}^T(i\lambda;\omega)[i\lambda I - A(\omega) - B(\omega)e^{i\lambda\tau}]^{-1}b(\omega)\} \leqslant 0,$$

where $\tilde{c}^T(i\lambda;\omega) = \int_0^\infty c^T(t;\omega)e^{-i\lambda t}\,dt$.

Then the stochastic system (10.0.2)–(10.0.3) is stochastically absolutely stable.

PROOF We shall prove the theorem by showing that the assumptions of Theorem 9.2.1 are satisfied. By definition,

$$h(t;\omega) = f(t;\omega) + \int_0^t c^T(t-u;\omega)\psi(u;\omega)\,du$$

$$= f(t;\omega) + \int_0^t c^T(u;\omega)\psi(t-u;\omega)\,du.$$

From Condition (ii) we know that the functions

$$f(t;\omega), c(t;\omega) \in L_1(R_+, L_\infty(\Omega, \mathscr{A}, \mathscr{P})).$$

The definition of $\psi(t;\omega)$, together with assumption (i) of the theorem, implies that $\psi(t;\omega) \in L_1(R_+, L_\infty(\Omega, \mathscr{A}, \mathscr{P}))$. Thus the convolution product of $c^T(t;\omega)$ and $\psi(t;\omega)$ also belongs to $L_1(R_+, L_\infty(\Omega, \mathscr{A}, \mathscr{P}))$. Hence

$$h(t;\omega) \in L_1(R_+, L_\infty(\Omega, \mathscr{A}, \mathscr{P})).$$

Differentiating $h(t;\omega)$ with respect to t, we have

$$\dot{h}(t;\omega) = \dot{f}(t;\omega) + \int_0^t c^T(u;\omega)\dot{\psi}(t-u;\omega)\,du. \tag{10.3.1}$$

Each term of Eq. (10.3.1) belongs to $L_1(R_+, L_\infty(\Omega, \mathscr{A}, \mathscr{P}))$. Thus, $\dot{h}(t;\omega) \in L_1(R_+, L_\infty(\Omega, \mathscr{A}, \mathscr{P}))$ and assumption (i) of Theorem 9.2.1 is satisfied. We shall consider the stochastic kernel

$$k(t;\omega) = \int_0^t c^T(s;\omega)X(t-s;\omega)\,ds$$

and

$$\dot{k}(t;\omega) = \int_0^t c^T(s;\omega)\dot{X}(t-s;\omega)\,ds.$$

It is given that

$$c(t;\omega) \in L_1(R_+, L_\infty(\Omega, \mathscr{A}, \mathscr{P})) \cap L_2(R_+, L_\infty(\Omega, \mathscr{A}, \mathscr{P}))$$

and from assumption (i) we conclude that $X(t;\omega)$ belongs to $L_1(R_+, L_\infty(\Omega, \mathscr{A}, \mathscr{P}))$. Using the fact that the convolution product of $c^T(s;\omega)$ and $X(t;\omega)$ belongs to $L_1(R_+, L_\infty(\Omega, \mathscr{A}, \mathscr{P})) \cap L_2(R_+, L_\infty(\Omega, \mathscr{A}, \mathscr{P}))$, we have

$$k(t;\omega) \in L_1(R_+, L_\infty(\Omega, \mathscr{A}, \mathscr{P})) \cap L_2(R_+, L_\infty(\Omega, \mathscr{A}, \mathscr{P})).$$

Applying a similar argument, it can be easily seen that

$$\dot{k}(t;\omega) \in L_1(R_+, L_\infty(\Omega, \mathscr{A}, \mathscr{P})) \cap L_2(R_+, L_\infty(\Omega, \mathscr{A}, \mathscr{P})),$$

and Condition (ii) of Theorem 9.2.1 is satisfied. Assumption (iii) of this theorem is the same as (iii) of Theorem 9.2.1. Knowing the fact that the

10.3 STOCHASTIC ABSOLUTE STABILITY OF THE SYSTEMS

Fourier transform of the convolution product is equal to the product of the Fourier transforms, we have

$$\tilde{k}(i\lambda;\omega) = \tilde{c}^T(i\lambda;\omega)[i\lambda I - A(\omega) - B(\omega)e^{i\lambda\tau}]^{-1}b(\omega).$$

From Condition (iv) of the theorem, we obtain

$$\text{Re}\{(1 + i\lambda q)\tilde{c}^T(i\lambda;\omega)[i\lambda I - A(\omega) - B(\omega)e^{i\lambda\tau}]^{-1}b(\omega)\} \leq 0$$

or

$$\text{Re}\{(1 + i\lambda q)\tilde{k}(i\lambda;\omega)\} \leq 0,$$

which is assumption (iv) of Theorem 9.2.1. Therefore, since Theorem 9.2.1 holds, we conclude that the stochastic system (10.0.2)–(10.0.3) is stochastically absolutely stable.

Theorem 10.3.2 Assume that the random system with time lag (10.0.4)–(10.0.5) satisfies the following conditions:

(i) The equation $\det\{A(\omega) + e^{-\lambda\tau} - \lambda I\} = 0$ has all its roots in the semiplane $\text{Re }\lambda \leq -\alpha < 0$.

(ii) (a) $c(t;\omega)$ is defined for $t \geq 0$, $\omega \in \Omega$, and is such that

$$c(t;\omega) \in L_1(R_+, L_\infty(\Omega, \mathscr{A}, \mathscr{P})) \cap L_2(R_+, L_\infty(\Omega, \mathscr{A}, \mathscr{P}));$$

(b) $f(t;\omega)$ is defined for $t \geq 0$, $\omega \in \Omega$, and is such that

$$f(t;\omega), \dot{f}(t;\omega) \in L_1(R_+, L_\infty(\Omega, \mathscr{A}, \mathscr{P})) \cap L_2(R_+, L_\infty(\Omega, \mathscr{A}, \mathscr{P}));$$

(c) $\eta(t;\omega)$ is defined for $t \geq 0$, $\omega \in \Omega$, and is such that

$$\eta(t;\omega) \in L_1(R_+, L_\infty(\Omega, \mathscr{A}, \mathscr{P})) \cap L_2(R_+, L_\infty(\Omega, \mathscr{A}, \mathscr{P})).$$

(iii) $\phi(\sigma)$ is continuous and bounded for $\sigma \in R$, R being the real line, and $\sigma\phi(\sigma) > 0$ for $\sigma \neq 0$.

(iv) There exists a $q \geq 0$ such that

$$\text{Re}\{(1 + i\lambda q)\tilde{c}^T(i\lambda;\omega)[i\lambda I - A(\omega) - B(\omega)e^{i\lambda\tau}]^{-1}(b(\omega) + \tilde{\eta}(i\lambda;\omega))\} \leq 0,$$

where $\tilde{c}^T(i\lambda;\omega) = \int_0^\infty c^T(t;\omega)e^{-i\lambda t}\,dt$ and $\tilde{\eta}(i\lambda;\omega) = \int_0^\infty \eta(t;\omega)e^{-i\lambda t}\,dt$.

Then the stochastic system of equations (10.0.4)–(10.0.5) is stochastically absolutely stable.

PROOF   This theorem will also be proved by demonstrating that the conditions of Theorem 9.2.1 are satisfied.

By definition

$$h(t;\omega) = f(t;\omega) + \int_0^t c^T(t - u;\omega)\psi(u;\omega)\,du,$$

where
$$\psi(t;\omega) = X(t;\omega)x_0(\omega) + \int_{-\tau}^0 X(t - \tau - u;\omega)B(\omega)x(u;\omega)\,du.$$

From the hypothesis of the theorem, we have
$$f(t;\omega),\quad c(t;\omega) \in L_1(R_+, L_\infty(\Omega, \mathscr{A}, \mathscr{P})) \cap L_2(R_+, L_\infty(\Omega, \mathscr{A}, \mathscr{P})).$$

Also,
$$\psi(t;\omega) \in L_1(R_+, L_\infty(\Omega, \mathscr{A}, \mathscr{P})) \cap L_2(R_+, L_\infty(\Omega, \mathscr{A}, \mathscr{P}))$$
because of the fact that
$$X(t;\omega) \in L_1(R_+, L_\infty(\Omega, \mathscr{A}, \mathscr{P})) \cap L_2(R_+, L_\infty(\Omega, \mathscr{A}, \mathscr{P}))$$
from Condition (i), and $x(t;\omega)$ is a continuous function for $t \in [-\tau, 0]$ and a.e. with respect to ω. Differentiating $h(t;\omega)$ with respect to t, we have
$$\dot{h}(t;\omega) = \dot{f}(t;\omega) + \int_0^t c^T(u;\omega)\psi(t - u;\omega)\,du.$$

Using a similar argument, it is easy to see that
$$\dot{h}(t;\omega) \in L_1(R_+, L_\infty(\Omega, \mathscr{A}, \mathscr{P})) \cap L_2(R_+, L_\infty(\Omega, \mathscr{A}, \mathscr{P})).$$

Thus assumption (i) of Theorem 9.2.1 holds.

We have defined the stochastic kernel $k(t;\omega)$ as follows:
$$k(t;\omega) = \int_0^t c^T(u;\omega)k_2(t - u;\omega)\,du$$
with
$$k_2(t;\omega) = X(t;\omega)b(\omega) + k_1(t;\omega)$$
and
$$k_1(t;\omega) = \int_0^t X(t - s;\omega)\eta(s;\omega)\,ds.$$

It is given that $X(t;\omega)$, $c(t;\omega)$, and $\eta(t;\omega)$ belong to
$$L_1(R_+, L_\infty(\Omega, \mathscr{A}, \mathscr{P})) \cap L_2(R_+, L_\infty(\Omega, \mathscr{A}, \mathscr{P})).$$

Also,
$$k_1(t;\omega) \in L_1(R_+, L_\infty(\Omega, \mathscr{A}, \mathscr{P})) \cap L_2(R_+, L_\infty(\Omega, \mathscr{A}, \mathscr{P}))$$
because each of its terms belongs to $L_1(R_+, L_\infty(\Omega, \mathscr{A}, \mathscr{P})) \cap L_2(R_+, L_\infty(\Omega, \mathscr{A}, \mathscr{P}))$, which implies that
$$k_2(t;\omega) \in L_1(R_+, L_\infty(\Omega, \mathscr{A}, \mathscr{P})) \cap L_2(R_+, L_\infty(\Omega, \mathscr{A}, \mathscr{P})).$$

10.3 STOCHASTIC ABSOLUTE STABILITY OF THE SYSTEMS

Thus
$$k(t;\omega) \in L_1(R_+, L_\infty(\Omega, \mathscr{A}, \mathscr{P})) \cap L_2(R_+, L_\infty(\Omega, \mathscr{A}, \mathscr{P}))$$
because the convolution product of $c^T(t;\omega)$ and $k_2(t;\omega)$ also belongs to
$$L_1(R_+, L_\infty(\Omega, \mathscr{A}, \mathscr{P})) \cap L_2(R_+, L_\infty(\Omega, \mathscr{A}, \mathscr{P})).$$
Further,
$$\dot{k}(t;\omega) = \int_0^t c^T(u;\omega)\dot{k}_2(t-u;\omega)\,du$$
with
$$\dot{k}_2(t;\omega) = \dot{X}(t;\omega) + \dot{k}_1(t;\omega)$$
and
$$\dot{k}_1(t;\omega) = \int_0^t \dot{X}(t-s;\omega)\eta(s;\omega)\,ds.$$
Using a similar argument, it can be seen that
$$\dot{k}(t;\omega) \in L_1(R_+, L_\infty(\Omega, \mathscr{A}, \mathscr{P})) \cap L_2(R_+, L_\infty(\Omega, \mathscr{A}, \mathscr{P})).$$

Thus Condition (ii) of Theorem 9.2.1 is satisfied. Assumption (iii) of Theorem 9.2.1 is the same as Condition (iii) of this theorem. It can be shown that the Fourier transform of the stochastic kernel $k(t;\omega)$ is given by
$$\tilde{k}(i\lambda;\omega) = \tilde{c}^T(i\lambda;\omega)[i\lambda I - A(\omega) - B(\omega)e^{i\lambda\tau}]^{-1}[b(\omega) + \tilde{\eta}(i\lambda;\omega)].$$
From hypothesis (iv) of the theorem, we have
$$\operatorname{Re}\{(1 + i\lambda q)\tilde{k}(i\lambda;\omega)\} \leq 0,$$
which satisfies Condition (iv) of Theorem 9.2.1. Therefore we conclude that the stochastic system with time lag is *stochastically absolutely stable*.

Bibliography

Deterministic

Anselone, P. M.
  [1] Editor, *Nonlinear Integral Equations*. Univ. of Wisconsin Press, Madison, Wisconsin 1964.

Bachman, G., and Narici, L.
  [1] *Functional Analysis*. Academic Press, New York, 1966.

Barbalat, I.
  [1] Systèmes d'équations différentielles d'oscillations non-linéaires, *Rev. Math. Pures Appl.* **4** (1959), 267–270.

Beckenbach, E. F., and Bellman, R.
  [1] *Inequalities* (2nd ed.). Springer-Verlag, Berlin and New York, 1965.

Bellman, R.
  [1] *Introduction to the Mathematical Theory of Control Processes*, Vol. 1. Academic Press, New York, 1967.

Bellman, R., Jacquez, J., and Kalaba, R.
  [1] Mathematical models of chemotherapy, *Proc. Berkeley Symp. Math. Statist. Prob., 4th* Vol. IV, Univ. of California Press, Berkeley, California, 1961, 57–66.
  [2] Some mathematical aspects of chemotherapy. I: One-organ models, *Bull. Math. Biophys.* **22** (1960), 181–198.
  [3] Some mathematical aspects of chemotherapy. II: The distribution of a drug in the body, *Bull. Math. Biophys.* **22** (1960), 309–322.

Bellman, R., Jacquez, J., Kalaba, R., and Kotkin, B.
[1] A Mathematical Model of Drug Distribution in the Body: Implications for Cancer Chemotherapy. The RAND Corp., RM-3463-NIH, 1963.

Bellman, R., and Kotkin, B.
[1] A Numerical Approach to the Convolution Equations of a Mathematical Model of Chemotherapy. The RAND Corp., RM-3716-NIH, 1963.
[2] Chemotherapy, Convolution Equations, and Differential Approximations. The RAND Corp., P-3005, 1964.

Bihari, L.
[1] Notes on a nonlinear integral equation, *Studia Sci. Math. Hungar.* **2** (1967), 1–6.

Bochner, S.
[1] Lectures on Fourier integrals, *Ann. Math. Stud.* No. 42 (1959), 217–218.

Box, G. E. P.
[1] Fitting empirical data, *Ann. N.Y. Acad. Sci.* **86** (1960).

Branson, H.
[1] The kinetics of reactions in biological systems, *Arch. Biochem. Biophys.* **36** (1952).
[2] The use of isotopes in an integral equation description of metabolizing systems, *Cold Spring Harbor Symp. Quant. Biol.* **XIII** (1948), 135–142.

Browder, F. E., de Figueiredo, D. G., and Gupta, C. P.
[1] Maximal monotone operators and nonlinear integral equations of the Hammerstein type, *Bull. Amer. Math. Soc.* **76** (1970), 700–705.

Chandrasekhar, S.
[1] *Radiative Transfer*. Dover, New York, 1960.

Cooper, I., and Jacquez, J.
[1] A Mathematical Model of Chemotherapy Assuming Mixing in the Large Blood Vessels. The RAND Corp., RM-3712-NIH, 1964.

Corduneanu, A.
[1] The stability of the solution of equations of Volterra type, *Bul. Inst. Politehn. Iasi* **15**(19) (1969), fasc. 1–2, 69–73.

Corduneanu, C.
[1] Problèmes globaux dans le théorie des équations intégrales de Volterra, *Ann. Mat. Pura Appl.* **67** (1965), 349–363.
[2] Some perturbation problems in the theory of integral equations, *Math. Systems Theory* **1** (1967), 143–155.
[3] Stability of some linear time-varying systems, *Math. Systems Theory* **3** (1969), 151–155.
[4] Admissibility with Respect to an Integral Operator and Applications, Math. Tech. Rep., Univ. of Rhode Island, 1968.
[5] Nonlinear perturbed integral equations, *Rev. Roumaine Math. Pures Appl.* **13** (1968), 1279–1284.
[6] Sur certaines équations fonctionnelles de Volterra, *Funkcial. Ekvac.* **9** (1966), 119–127.
[7] Sur une équation intégrale non-lineaire, *An. Sti. Univ. "Al. I. Cuza" Iasi Sect. I. (N.S.)* **9** (1963), 369–375.

Davis, H. T.
[1] *Introduction to Nonlinear Differential and Integral Equations*. Dover, New York, 1962.

DePree, J. D.
[1] Reduction of linear integral equations with difference kernels to nonlinear integral equations, *J. Math. Anal. Appl.* **26** (1969), 539–544.

Desoer, C. A., and Tomasian, A. J.
[1] A note on zero-state stability of linear systems, *Proc. Ann. Allerton Conf. Circuit and Systems Theory, 1st* 1963, 50–52.

Dunford, N., and Schwartz, J.
  [1] *Linear Operators*, Part I. Wiley (Interscience), New York, 1958.
Friedman, A.
  [1] Monotonicity of solutions of Volterra integral equations in Banach space, *Trans. Amer. Math. Soc.* **138** (1969), 129–148.
Friedman, A., and Shinbrot, M.
  [1] Volterra integral equations in Banach space, *Trans. Amer. Math. Soc.* **126** (1967), 131–179.
Gavalas, G. R.
  [1] *Nonlinear Differential Equations of Chemically Reacting Systems*. Springer-Verlag, Berlin and New York, 1968.
Goldberg, S.
  [1] *Unbounded Linear Operators: Theory and Applications*. McGraw-Hill, New York, 1966.
Gol'dengeršel', È. I.
  [1] Discrete analog of an integral equation of Volterra type on the half-axis, *Uspehi Mat. Nauk* **21** (1966), no. 2 (128), 223–225.
Green, C. D.
  [1] *Integral Equation Methods*. Nelson, London, 1969.
Grossman, R. I., and Miller, R. K.
  [1] Perturbation theory for Volterra integro-differential systems, *J. Differential Equations* **8** (1970), 457–474.
Grossman, S. I.
  [1] Existence and stability of a class of nonlinear Volterra integral equations, *Trans. Amer. Math. Soc.* **150** (1970), 541–556.
Gupta, C. P.
  [1] On existence of solutions of nonlinear integral equations of Hammerstein type in a Banach space, *J. Math. Anal. Appl.* **32** (1970), 617–620.
Halanay, A.
  [1] *Differential Equations—Stability, Oscillations, Time Lags*. Academic Press, New York, 1966.
Hannsgen, K. B.
  [1] On a nonlinear Volterra equation, *Michigan Math. J.* **16** (1969), 365–376.
Heard, M. L.
  [1] On asymptotic behavior and periodic solutions of a certain Volterra integral equation, *J. Differential Equations* **6** (1969), 172–186.
  [2] On a nonlinear integro-differential equation, *J. Math. Anal. Appl.* **26** (1969), 170–189.
Hearon, J.
  [1] A note on the integral equation description of metabolizing systems. *Bull. Math. Biophys.* **15** (1953).
Hewitt, E., and Stromberg, K.
  [1] *Real and Abstract Analysis*. Springer-Verlag, Berlin and New York, 1965.
Hildebrand, F. B.
  [1] *Finite-Difference Equations and Simulation*. Prentice-Hall, Englewood Cliffs, New Jersey, 1968.
Horváth, John
  [1] *Topological Vector Spaces and Distributions*. Addison-Wesley, Reading, Massachusetts, 1966.
Izé, A. F.
  [1] On an asymptotic property of a Volterra integral equation, *Proc. Amer. Math. Soc.* **28** (1971), 93–99.

John, F.
[1] Integral equations, *Mathematics Applied to Physics*, pp. 316–347. Springer Publ., New York, 1970.

Kartsatos, A. G.
[1] On the relationship between a nonlinear system and its nonlinear perturbation, *J. Differential Equations* **11** (1972), 582–591.
[2] Convergence in perturbed nonlinear systems, *Tôhoku Math. J.* **24** (1972), 539–546.
[3] Positive solutions to linear problems for nonlinear systems, *J. Differential Equations* (in press).
[4] Bounded solutions to perturbed nonlinear systems and asymptotic relationships, *J. Für die Reine und Angewandte Math.* (in press).

Kartsatos, A. G., and Michaelides, G. J.
[1] Existence of convergent solutions of quasi-linear systems and asymptotic equivalence, *J. Differential Equations* **13** (1973), 481–489.

Kermack, W. O., and McKendrick, A. G.
[1] A contribution to the mathematical theory of epidemics, *Proc. Roy. Soc. (A)* **115** (1927), 700–721.

Kirpotina, N. V.
[1] Systems of non-linear integral equations, *Moskov. Oblast. Ped. Inst. Učen. Zap.* **150** (1964), 29–39.

Kittrell, J. R., Mezaki, R., and Watson, C. C.
[1] Estimation of parameters for nonlinear least squares analysis, *Ind. Eng. Chem. Fundamentals* **57** (1965).

Kopal, Z.
[1] *Numerical Analysis.* Chapman & Hall, London, 1961.

Kotkin, B.
[1] A Mathematical Model of Drug Distribution and the Solution of Differential-Difference Equations. The RAND Corp., RM-2907-RC, 1962.

Krasnosel'skii, M. A.
[1] *Topological Methods in the Theory of Nonlinear Integral Equations.* Pergamon, Oxford and Macmillan, New York, 1964.

Krasnosel'skii, M. A., Zabreiko, P. P., Pustyl'nik, E. I., and Sobolevskii, P. E.
[1] *Integral Operators in Spaces of Summable Functions.* Izdat. "Nauka," Moscow, 1966.

Landau, H. G., and Rapoport, A.
[1] Mathematical theory of contagion, *Bull. Math. Biophys.* **15** (1953).

LaSalle, J. P., and Lefschetz, S.
[1] *Stability by Lyapunov's Method with Applications.* Academic Press, New York, 1961.

Lee, E. S.
[1] Quasilinearization and estimation of parameters in differential equations, *Ind. Eng. Chem. Fundamentals* **7** (1968).

Levin, J. J.
[1] The asymptotic behavior of the solution of a Volterra equation, *Proc. Amer. Math. Soc.* **14** (1963), 534–541.
[2] A nonlinear Volterra equation not of the convolution type, *J. Differential Equations* **4** (1968), 176–186.

Levin, J. J., and Nohel, J. A.
[1] On a system of integro-differential equations occurring in reactor dynamics, *J. Math. Mech.* **9** (1960), 347–368.
[2] Perturbations of a nonlinear Volterra equation, *Michigan Math. J.*, **12** (1965), 431–447.
[3] A system of nonlinear integro-differential equations, *Michigan Math. J.* **13** (1966), 257–270.

Levin, J. J., and Shea, D. F.
 [1] On the asymptotic behavior of the bounded solutions of some integral equations, *J. Math. Anal. Appl.* **37** (1972), 42–82.

Londen, Stig-Olof
 [1] On a nonlinear Volterra integro-differential equation, *Commentationes Physico-Mathematicae* **38** (1969), 5–11.
 [2] The qualitative behavior of the solutions of a nonlinear Volterra equation, *Michigan Math. J.* (in press).

Luca, N.
 [1] On the behavior of the solutions of an integro-differential system of equations, *An. Sti. Univ. "Al. I. Cuza" Iasi Sect. I a Mat. (N.S.)* **13** (1967), 299–303.

Lure, A. I., and Postnikov, V. N.
 [1] On the theory of stability of control systems, *Prikl. Mat. i Mehk.* **8** (1944), 3.

Mahmudov, A. P., and Musaev, V. M.
 [1] On the theory of the solutions of nonlinear integral equations of Volterra-Uryson type, *Akad. Nauk Azerbaidžan. SSR Dokl.* **25** (1969), 3–6.

Mikhlin, S. G.
 [1] *Integral Equations*. Macmillan, New York, 1964.

Mikhlin, S. G., and H. L. Smolickii
 [1] *Approximate Methods of Solution of Differential and Integral Equations*. Izdat. "Nauka," Moscow, 1965.

Miller, R. K.
 [1] On Volterra's population equation, *J. SIAM Appl. Math.* **14** (1966), 446–452.
 [2] An unstable nonlinear integro-differential system, *Proc. U.S.-Japan Seminar on Differential and Functional Equations* (*Minneapolis, Minnesota, 1967*), pp. 479–489. Benjamin, New York, 1967.
 [3] Asymptotic stability properties of linear Volterra integro-differential equations, *J. Differential Equations* (in press).
 [4] Admissibility and nonlinear Volterra integral equations, *Proc. Amer. Math. Soc.* **25** (1970), 65–71.
 [5] On the linearization of Volterra integral equations, *J. Math. Anal. Appl.* **23** (1968), 198–208.

Miller, R. K., Nohel, J. A., and Wong, J. S. W.
 [1] A stability theorem for nonlinear mixed integral equations, *J. Math. Anal. Appl.* **25** (1969), 446–449.

Miller, R. K., and Sell, G. R.
 [1] Existence, uniqueness, and continuity of solutions of integral equations, *Ann. Mat. Pura Appl.* **80** (1968), 135–152.
 [2] Volterra integral equations and topological dynamics, *Mem. Amer. Math. Soc.* **102** (1970), 67 p.

Moore, R. H.
 [1] Approximations to nonlinear operator equations and Newton's method, *Numer. Math.* **12** (1968), 23–34.

Muki, R., and Sternberg, E.
 [1] Note on an asymptotic property of solutions to a class of Fredholm integral equations, *Quart. Appl. Math.* **28** (1970), 277–281.

Muldowney, J. S., and Wong, J. S. W.
 [1] Bounds for solutions of nonlinear integro-differential equations, *J. Math. Anal. Appl.* **23** (1968), 487–499.

Nashed, M. Z., and Wong, J. S. W.
 [1] Some variants of a fixed point theorem of Krasnosel'skii and applications to nonlinear integral equations, *J. Math. Mech.* **18** (1969), 767–777.

Natanson, I. P.
 [1] *Theory of Functions of a Real Variable*, Vol. I. Ungar, New York, 1960.
 [2] *Theory of Functions of a Real Variable*, Vol. II. Ungar, New York, 1960.

Nohel, J. A.
 [1] Qualitative behavior of solutions of nonlinear Volterra equations, *Stability Problems of Solutions of Differential Equations*. Proc. NATO Adv. Study Inst., Padua, Italy, 1965.
 [2] Remarks on nonlinear Volterra equations, *Proc. U.S.-Japan Seminar Differential and Functional Equations* (*Minneapolis, Minnesota, 1967*), pp. 249–266. Benjamin, New York, 1967.

Oğuztoreli, M. N.
 [1] *Time-Lag Control Systems*. Academic Press, New York, 1966.

Petrovanu, D.
 [1] Equations Hammerstein intégrales et discrètes, *Ann. Mat. Pura Appl.* **70** (1966), 227–254.

Petry, Walter
 [1] Ein iteratives Verfahren zur Bestimmung einer Lösung gewisser nichtlinearer Operatorgleichungen im Hilbertraum mit Anwendung auf Hammersteinsche Integral-gleichungssysteme, *Math. Ann.* **187** (1970), 127–149.

Poincaré, H.
 [1] Memoire sur les courbes définier par une équation différentiable, *J. Math. Pures Appl.* (3)**7** (1881), 375–422; **8** (1882), 251–296; (4)**1** (1885) 167–244; **2** (1886), 151–217.

Rall, L. B.
 [1] *Computational Solution of Nonlinear Operator Equations*. Wiley, New York, 1969.

Saaty, Thomas L.
 [1] *Modern Nonlinear Equations*. McGraw-Hill, New York, 1967.

Sandberg, I. W.
 [1] On the boundedness of solutions of nonlinear integral equations, *Bell System Tech. J.* **44** (1965), 439–453.

Sauer, R.
 [1] *Ingenieur-Matematik, Band I: Differential-und Integralrechnung*. Springer-Verlag, Berlin and New York, 1969.

Schmeidler, W.
 [1] *Linear Operators in Hilbert Space*. Academic Press, New York, 1965.

Stenger, F.
 [1] The approximate solution of Wiener-Hopf integral equations, *J. Math. Anal. Appl.* (in press).

Stephenson, J.
 [1] Theory of the measurement of blood flow by the dilution of an indicator, *Bull. Math. Biophys.* **10** (1948).

Strauss, A.
 [1] A discussion of the linearization of Volterra integral equations, *Seminar Differential Equations Dynamical Syst. II, Lecture Notes Math.* **144** (1970), 209–217.
 [2] On a perturbed Volterra integral equation, *J. Math. Anal. Appl.* **30** (1970), 564–575.

Titchmarsh, E. C.
 [1] *Introduction to the theory of Fourier Integrals*. Oxford Univ. Press, London and New York, 1959.

Tricomi, F. G.
 [1] *Integral Equations*. Wiley (Interscience), New York, 1957.

Vainberg, M. M.
 [1] Integro-differential equations, *Math. Analysis, Theory of Probability Control* (1962), pp. 5–37. Itogi Nauki, Akad. Nauk SSSR Inst. Naučn. Inform., Moscow, 1964.

Vinokurov, V. R.
[1] Volterra integral equations with infinite interval of integration, *Differencial'nye Uravnenija* **5** (1969), 1894–1898.
[2] Certain questions in the theory of the stability of systems of Volterra integral equations. *Izv. Vysš. Učebn. Zaved. Matematika* **85** (1969), 24–34; **86** (1969), 28–38.

Widom, H.
[1] *Lectures on Integral Equations*, Van Nostrand Math. Stud. No. 17. Van Nostrand Reinhold, Princeton, New Jersey, 1969.

Wijsman, R.
[1] A critical investigation of the integral description of metabolizing systems, *Bull. Math. Biophys.* **15** (1953).

Yosida, K.
[1] *Functional Analysis*. Springer-Verlag, Berlin and New York, 1965.

Zabreiko, P. P., and Povolockii, A. I.
[1] On the theory of the Hammerstein equation, *Ukrain. Math. J.* **22** (1970), 127–138.

Stochastic or Probabilistic

Adomian, G.
[1] Random operator equations in mathematical physics I, *J. Math. Phys.* **11** (1970), 1069–1074.
[2] Linear random operator equations in mathematical physics II, *J. Math. Phys.* **12** (1971), 1944–1948.
[3] Linear random operator equations in mathematical physics III, *J. Math. Phys.* **12** (1971), 1948–1955.
[4] Theory of random systems, *Trans. Prague Conf. on Information Theory, Statistical Decision Functions, Random Processes, 4th, Prague, 1965*, 205–222, Academic, Prague, 1967.

Ahmed, N. U.
[1] A class of stochastic nonlinear integral equations on L^p spaces and its application to optimal control, *Information and Control* **14** (1969), 512–523.

Ahmed, N. U., and Teo, K. L.
[1] On the stability of a class of nonlinear stochastic systems with applications to distributed parameter systems, *Proc. IFAC Symp. Control of Distributed Parameter Syst.*, Banff, Canada, June, 1971.

Anderson, M. W.
[1] *Stochastic Integral Equations*. Ph.D. Dissertation, Univ. of Tennessee (1966).
[2] A stochastic integral equation, *SIAM J. Appl. Math.* **18** (1970), 526–532.

Bailey, N. T. J.
[1] A perturbation approximation to the simple stochastic epidemic in a large population, *Biometrika* **55** (1968), 199–209.
[2] Stochastic birth, death, and migration processes for spatially distributed populations, *Biometrika* **55** (1968), 189–198.
[3] The simple stochastic epidemic: A complete solution in terms of known functions, *Biometrika* **50** (1963), 235–240.

Bartholomay, A.
[1] Stochastic models for chemical reactions I-theory of unimolecular reaction process, *Bull. Math. Biophys.* **20** (1958).

Bartlett, M. S.
 [1] On theoretical models for competitive and predatory biological systems, *Biometrika* **44** (1957), 27–42.
 [2] *Stochastic Processes.* Cambridge Univ. Press, New York and London, 1955.
 [3] *Stochastic Population Models in Ecology and Epidemiology.* Methuen, London, 1960.

Bharucha-Reid, A. T.
 [1] Sur les équations intégrales aléatoires de Fredholm à noyaux séparables, *C.R. Acad. Sci. Paris* **250** (1960), 454–456, 657–658.
 [2] Approximate solutions of random operator equations, *Notices Amer. Math. Soc.* **7** (1960), 361.
 [3] On random solutions of Fredholm integral equations, *Bull. Amer. Math. Soc.* **66** (1960), 104–109.
 [4] On random solutions of integral equations in Banach spaces, *Trans. Prague Conf. Information Theory, Statistical Decision Functions, and Random Processes,* 2nd 27–48. Academic Press, 1960.
 [5] On the theory of random equations, *Proc. Symp. Appl. Math.* **16** (1964), 40–69. Amer. Math. Soc., Providence, Rhode Island.
 [6] Ed., *Probabilistic Methods in Applied Mathematics,* Vol. I. Academic Press, New York, 1968; Vol. II, 1970.
 [7] *Random Integral Equations.* Academic Press, New York, 1972.
 [8] *Elements of the Theory of Markov Processes and Their Applications.* McGraw-Hill, New York, 1960.

Bharucha-Reid, A. T., and Arnold, L.
 [1] On Fredholm integral equations with random degenerate kernels, *Zastos. Mat.* **10** (1969), 85–90.

Bharucha-Reid, A. T., Mukherjea, A., and Tserpes, N. A.
 [1] On the existence of random solutions of some random integral equations (to appear).

Blanc-Lapierre, A., and Fortet, R.
 [1] *Theory of Random Functions.* Gordon and Breach, New York, 1965.

Blum, J. R.
 [1] Approximation methods which converge with probability one, *Ann. Math. Statist.* **25** (1954), 382–386.
 [2] Multidimensional stochastic approximation procedures, *Ann. Math. Statist.* **25** (1954), 737–744.

Buell, J., and Kalaba, R.
 [1] Quasilinearization and the fitting of nonlinear models of drug metabolism to experimental kinetic data, *Math. Biosci.* **5** (1969), 121–132.

Burkholder, D. L.
 [1] On a class of stochastic approximation processes, *Ann. Math. Statist.* **27** (1956), 1044–1059.

Chiang, C. L.
 [1] *Introduction to Stochastic Processes in Biostatistics.* Wiley, New York, 1968.

Chung, K. L.
 [1] On a stochastic approximation method, *Ann. Math. Statist.* **25** (1954), 463–483.

Dawson, D. A.
 [1] Generalized stochastic integrals and equations, *Trans. Amer. Math. Soc.* **147** (1970), 473–506.

Day, W. B., and Boyce, W. E.
 [1] On the relationship between the solution of a stochastic boundary value problem and parameters in the boundary conditions, *SIAM J. Appl. Math.* **19**, No. 1, (1970).

Dence, D.
 [1] Wave propagation in anisotropic media, *Probabilistic Methods in Applied Mathematics* (A. T. Bharucha-Reid, Ed.), Vol. 3. Academic Press, New York, 1972.

Dence, D., and Spence, J. E.
 [1] An effective medium description for a random uniaxial anisotropic medium, *IEEE Trans.* **AP-20** (1972).
 [2] The mean Green's Dyadic for a random uniaxial anisotropic medium with small scale fluctuations, *Proc. Fall Internat. Symp. Antennas and Propagation* (1971).
 [3] Dyson's Equation for a medium possessing uniaxial anisotropic fluctuations, *Electron Lett.* **7** (1971).
 [4] The coherent wave in a random uniaxial anisotropic medium, *IEEE Trans.* **AP-19** (1971).

Derman, C.
 [1] An application of Chung's lemma to the Kiefer-Wolfowitz stochastic approximation procedure, *Ann. Math. Statist* **27** (1956), 532–536.

Distefano, N.
 [1] A Volterra integral equation in the stability of some linear hereditary phenomena, *J. Math. Anal. Appl.* **23** (1968), 365–383.

Doob, J. L.
 [1] *Stochastic Processes*. Wiley, New York, 1953.

Dvoretzsky, A.
 [1] On stochastic approximation, *Proc. Berkeley Symp. Math. Statist. Probability*, 3rd Univ. of California Press, Berkeley, California, 1956.

Dynkin, Y. B.
 [1] *Markov Processes*. Academic Press, New York, 1964.

Feller, W.
 [1] On the integral equation of renewal theory, *Ann. Math. Statist.* **12** (1941), 243–267.
 [2] *An Introduction to Probability Theory and Its Applications*, 3rd ed., Vol. 1. Wiley, New York, 1957.

Fortet, R.
 [1] Random distributions with application to telephone engineering, *Proc. Berkeley Symp. Math. Statist. and Probability*, 3rd Vol. II, pp. 81–88. University of California Press, Berkeley, California, 1956.

Fry, T. C.
 [1] *Probability and Its Engineering Uses*, 2nd ed. Van Nostrand Reinhold, Princeton, New Jersey, 1965.

Gikhmann, I. I., and Skorokhod, A. V.
 [1] *Introduction to the Theory of Random Processes*. Saunders, Philadelphia, Pennsylvania, 1969.

Goodman, L.
 [1] Stochastic models for the population growth of the sexes, *Biometrika* **55** (1968), 469–488.

Hans, O.
 [1] Random operator equations, *Proc. Berkeley Symp. Math. Statist. and Probability*, 4th Vol. II, pp. 185–202. Univ. of California Press, Berkeley, California, 1961.

Hardiman, S. T., and Tsokos, C. P.
 [1] Biological models pertaining to stochastic integral equations, *J. Stat. Physics* (to appear).
 [2] *On a Uryson Type of Stochastic Integral Equations* (to appear).
 [3] Existence theory for nonlinear random integral equations using the Banach–Steinhaus Theorem, *Math. Nachr.* (to appear).

Haskey, H. W.
 [1] A general expression for the mean in a simple stochastic epidemic, *Biometrika* **41** (1954), 272–275.

Hunt, B. R.
[1] Statistical properties of numerical solutions to convolution-type integral equations, preprint, Los Alamos Sci. Lab. of the Univ. of California, Los Alamos, New Mexico.

Ito, K.
[1] Stochastic integral, *Proc. Imp. Acad. Tokyo* **20** (1944), 519–524.
[2] On a stochastic integral equation, *Proc. Japan Acad.* **22** (1946), 32–35.

Jazwinski, A. H.
[1] Stochastic processes and filtering theory, *Mathematics in Science and Engineering* (R. Bellman, ed.), Vol. 64. Academic Press, New York, 1970.

Kallianpur, G.
[1] A note on the Robbins-Monro stochastic approximation method, *Ann. Math. Statist.* **25** (1954), 386–388.

Kallianpur, G., and Striebel, C.
[1] Stochastic differential equations occurring in the estimation of continuous parameter stochastic processes, United States Air Force Office of Aerospace Research, Grant No. AF-AFOSR-885-65; the United States Navy, Grant No. N00014-67-C0307, and the National Science Foundation, Grant No. GP7490.

Kannan, D.
[1] An operator-valued stochastic integral, II, *Ann. Inst. Henri Poincaré* **VIII**, no. 1 (1972), 9–32.
[2] An operator-valued stochastic integral, III, *Ann. Inst. Henri Poincaré* **8** (1972), 217–228.
[3] On enzyme amplifier systems triggered by white noise, *Math. Biosci.* **15** (1972).
[4] Martingales in Banach spaces with Schauder Bases, *J. Math. Phys. Sci.* **7** (1973), 93–100.

Kannan, D., and Bharucha-Reid, A. T.
[1] Probability measures on H_p spaces, *Ann. Inst. Henri Poincaré* **VII**, no. 3 (1971), 205–217.
[2] Note on covariance operators of probability measures on a Hilbert space, *Proc. Japan Acad.* **46**, No. 2 (1970), 124–129.
[3] An operator-valued stochastic integral, *Proc. Japan Acad.* **47**, No. 5 (1971), 472–476.
[4] Random integral equation formulation of a generalized langesian equation, *J. Statist. Phys.* **5**, No. 3 (1972).

Kendall, D. G.
[1] Stochastic processes and population growth, *J. Roy. Statist. Soc.* (1949), 230–267.

Kerr, J. D.
[1] The probability of disease transmission, *Biometrics* **27** (1971), 219–222.

Kesten, H.
[1] Some nonlinear stochastic growth models, *Bull. Amer. Math. Soc.* **77** (1971), 492–511.

Kiefer, J., and Wolfowitz, J.
[1] Stochastic estimation of the maximum of a regression function, *Ann. Math. Statist.* **23** (1953), 462–466.

Kim, Ho Gol
[1] On the parametric dependence of solutions of stochastic integral equations in a Hilbert space, *Cho-sŏn In-min Kong-hwa-kuk Kwa-hak-won T'ong-p'o* (1970), 3–8.

Leslie, P. H.
[1] A stochastic model for studying the properties of certain biological systems by numerical methods, *Biometrika* **45** (1958), 16–31.

Levit, M. V.
[1] Frequency conditions for the absolute stochastic stability of automatic control systems with random external action, *Dokl. Akad. Nauk USSR* **195** (1970), 769–772.

Loève, M.
[1] *Probability Theory*, 3rd ed. Van Nostrand Reinhold, Princeton, New Jersey, 1963.

Lotka, A.
[1] On an integral equation in population analysis, *Ann. Math. Statist.* **10** (1939), 144–161.
Lumley, J. L.
[1] An approach to the Eulerian-Lagrangian problem, *J. Math. Phys.* **3** (1962), 309–312.
McKean, H. P., Jr.
[1] *Stochastic Integrals*. Academic Press, New York, 1969.
McQuarrie, Donald
[1] A stochastic approach to chemical kinetics, *J. Appl. Probability* **4** (1967), 413–478.
Milton, J. S., and C. P. Tsokos
[1] A stochastic model for chemical kinetics (to appear).
[2] Admissibility theory and Ito's stochastic integral equation (to appear).
[3] On a class of nonlinear stochastic integral equations, *Math. Nachr.* (to appear).
[4] Stochastic integral equations in a physiological model (to appear).
[5] On a non-linear perturbed stochastic integral equation, *J. Math. Phys. Sci.* **V**, No. 4 (1971), 361–374.
[6] A random integral equation in a metabolizing system, *J. Statist. Physics* **8** (1973).
[7] A stochastic model for communicable disease, *Int. J. Systems Sci.* (to appear).
[8] On a random solution of a nonlinear perturbed stochastic integral equation of the Volterra type, *Bull. Austral. Math. Soc.* **9** (1973), 227–237.
Milton, J. S., Padgett, W. J., and Tsokos, C. P.
[1] Existence theory of a stochastic integral equation of the Fredholm type with random perturbations, *SIAM J. Appl. Math.*, **22** (1972), 194–208.
Moran, P. A. P.
[1] *The Statistical Processes of Evolutionary Theory*. Oxford Univ. Press, London and New York, 1962.
Morozan, T.
[1] The method of V. M. Popov for control systems with random parameters, *J. Math. Anal. Appl.* **16** (1966), 201–215.
[2] Stability of some linear stochastic systems, *J. Differential Equations* **3** (1967), 153–169.
[3] Stability of linear systems with random parameters, *J. Differential Equations* **3** (1967), 170–178.
[4] *Stabilitatea sistemelor cu parametri aleatori*. Editura Academiei Republicii Socialiste România, Bucarest, 1969.
[5] Stability of stochastic discrete systems, *J. Math. Anal. Appl.* **23** (1968), 1–9.
Mukherjea, A.
[1] Transformations aléatoires séparables: Théorème du point fixe aléatoire, *C.R. Acad. Sci. Paris Ser. A-B* **263** (1966), A393–A395.
[2] Idempotent Probabilities on semigroups, *Z. Wahrscheinlichkeitstheorie verw. Geb.* **11** (1969), 142–146.
[3] Random operators on Lusin probability spaces, *Z. Wahrscheinlichkeitstheorie verw. Geb.* **9** (1968), 232–234.
[4] A Stone-Weierstrass Theorem for random functions, *Bull. Austral. Math. Soc.* **2** (1970), 233–236.
[5] On the convolution equation $P = P * Q$ of Choquet and Deny for probability measures on semigroups, *Proc. Amer. Math. Soc.* **32** (1972), 457–463.
Mukherjea, A., and Bharucha-Reid, A. T.
[1] Separable random operators. I, *Rev. Roumaine Math. Pures Appl.* **14** (1969), 1553–1561.
Mukherjea, A., and Tserpes, N. A.
[1] A problem on r^*-invariant measures on locally compact semigroups, *Indiana Math. J.* **21** (1972), 973–978.

[2] Mesures de probabilite r^*-invariantes sur un semigroupe métrique, *C.R. Acad. Sci. Paris Sér. A.* **268**, (1969), 318–319
[3] A note on countably compact semigroups, *J.A.M.S.* **13** (1972), 180–184.
[4] On the convolution equation $P = P * Q$ of Choquet and Deny for probability measures on semigroups, *Proc. Amer. Math. Soc.* **32** (1972), 457–463.
[5] On certain conjectures on invariant measures on semigroups, *Semigroup Forum* **1** (1970), 260–266.
[6] Some problems on idempotent measures on semigroups, *Bull. Austral. Math. Soc.* **2** (1970), 299–315.
[7] Invariant measures on semigroups with closed translations, *Z. Wahrscheinlichkeitstheorie Verw. Geb.* **17** (1971), 33–38.
[8] Idempotent measures on locally compact semigroups, reprinted from the *Proc. Amer. Math. Soc.*, **29**, No. 1 (1971), 143–150.

Nichols, W. G., and Tsokos, C. P.
[1] Formal solutions for a class of stochastic linear pursuit-evasion games with perfect information, *Internat. J. Systems Sci.* **1** (1970).
[2] An empirical Bayes approach to point estimation in adoptive control, *J. Information and Control* (1972).

Padgett, W. J., and Tsokos, C. P.
[1] On a semi-stochastic model arising in a biological system, *Math. Biosci.* **9** (1970), 105–117.
[2] A stochastic model for chemotherapy: Computer simulation, *Math. Biosci.* **9** (1970), 119–133.
[3] Existence of a solution of a stochastic integral equation in turbulence theory, *J. Math. Phys.* **12** (1971), 210–212.
[4] On a stochastic integral equation of the Volterra type in telephone traffic theory, *J. Appl. Probability* **8** (1971), 269–271.
[5] Random solution of a stochastic integral equation: Almost sure and mean square convergence of successive approximations, *Int. J. Systems Sci.* **4** (1973), 605–612.
[6] On the solution of a random integral equation by a method of stochastic approximation (to appear).
[7] A stochastic discrete Volterra equation with application to stochastic systems, *Proc. Ann. Princeton Conf. Information Sci. and Systems, 5th* March 25–26, 1971, pp. 505–509. Princeton Univ.
[8] On a stochastic integral equation of the Fredholm type, *Z. Wahr. Verw. Geb.* **23** (1972), 22–31.
[9] Existence of a solution of a random discrete Fredholm system (to appear).
[10] A stochastic model for chemotherapy: Two-organ systems, *Int. J. for Biomedical Computing* **3** (1972), 29–41.
[11] A random Fredholm integral equation, *Proc. Amer. Math. Soc.* **33** (1972), 534–542.
[12] The origins and applications of stochastic integral equations, *Int. J. Systems Sci.* **2** (1971), 135–148.
[13] A new stochastic formulation of a population growth problem, *Math. Biosci.* **17** (1973), 105–120.
[14] On the existence of a unique solution of a stochastic integral equation in hereditary mechanics, *J. Math. Phys. Sci.* (to appear).
[15] On a stochastic integro-differential equation of Volterra type. *SIAM J. Applied Math.* **23** (1972), 496–512.
[16] A random discrete equation and non-linear stochastic control system, *Int. J. Systems Sci.* **4** (1973), 77–85.

[17] Stochastic asymptotic stability and approximation of the random solution of a stochastic discrete Fredholm system, *Int. J. Cybernetics* (to appear).
[18] Stochastic integral equations in life sciences and engineering, *Int. Stat. Review* (1973).

Parrott, L. G.
[1] *Probability and Experimental Errors in Science*. Wiley, New York, 1961.

Pincus, M.
[1] Gaussian processes and Hammerstein integral equations, *Trans. Amer. Math. Soc.* **134** (1968), 193–214.

Prabhu, N. U.
[1] *Stochastic Processes*. Macmillan, New York, 1965.

Rabotnikov, Ju. L.
[1] On the theory of linear stochastic differential equations with non-Markov type solutions, *Differencial'nye Uravnenija* **4** (1968), 244–251.
[2] On the theory of nonlinear stochastic systems, *Mat. Zametki* **5** (1969), 607–614.

Rao, A. N. V., and Tsokos, C. P.
[1] On an Ito-Doob type of stochastic integral (to appear).
[2] Sufficient conditions for the existence of a random solution to a mixed Volterra-Fredholm-Ito type of stochastic integral equations (to appear).

Rao, B. L. S. Prakasa
[1] Stochastic integral equations of mixed typed (Abstract), *Bull. Inst. Math. Statist.* **1** (1972), 48.

Robbins, H., and Monro, S.
[1] A stochastic approximation method, *Ann. Math. Statist.* **22** (1951), 400–407.

Roxin, E., and Tsokos, C. P.
[1] On the definition of a stochastic differential game, *Math. Systems Theory* **4** (1970), 60–64.

Sacks, J.
[1] Asymptotic distribution of stochastic approximation procedures, *Ann. Math. Statist.* **29** (1958), 373–405.

Sibul, L. H.
[1] Stochastic Green's functions and their relation to the resolvent kernels of integral equations, *Proc. Ann. Allerton Conf. on Circuit and System Theory, 5th, Monticello, Illinois, 1967*, pp. 356–363. Univ. of Illinois, Urbana, Illinois, 1967.

Stratonovich, R. L.
[1] A new representation for stochastic integrals and equations, *J. SIAM Control* **4** (1966), 362–371.

Tserpes, N. A., and Mukherjea, A.
[1] Invariant measures on semigroups with closed translations, *Z. Wahrscheinlichkeitstheorie Verw. Geb.* **17** (1971), 33–38.
[2] Mesures de probabilité r^*-invarientes sur un semi-group métrique, *C.R. Acad. Sci. Paris* **268** (1969), 318–319.

Tsokos, C. P.
[1] On a nonlinear differential system with a random parameter, Int. Conf. on System Sciences, *IEEE Proc., Honolulu, Hawaii* (1969).
[2] On some nonlinear differential systems with random parameters, *IEEE Proc., Ann. Princeton Conf. on Information Sci. and Systems*, 3rd (1969), 228–234.
[3] On the classical stability theorem of Poincaré-Lyapunov, *Proc. Japan Acad.* **45** (1969), 780–785.
[4] On a stochastic integral equation of the Volterra type, *Math. Systems Theory* **3** (1969), 222–231.
[5] The method of V. M. Popov for differential systems with random parameters, *J. Appl. Probability* **8** (1971), 298–310.

[6] Stochastic absolute stability of a nonlinear differential system, *IEEE Proc. on Circuits and Systems*, Naval Post-graduate School (1968), 586–590.
[7] Stochastic approximation of a random integral equation, *Math. Nachr.* **51** (1971), 101–110.
[8] Stochastic integral equations in engineering and biological science, *An. Stiintifice Tomul.* **XVII** (1971), 407–419.

Tsokos, C. P., and Hamdan, M. A.
[1] Stochastic nonlinear integro-differential systems with time lag, *J. Natur. Sci. and Math.* **10** (1970), 293–303.

Tsokos, C. P., and Nichols, W. G.
[1] On some stochastic differential games, *Probabilistic Methods in Applied Mathematics* (A. T. Bharucha-Reid, ed.), Vol. 3. Academic Press, New York, 1972.

Tsokos, C. P., and Padgett, W. J.
[1] *Random Integral Equations with Applications to Stochastic Systems*, Lecture Notes in Mathematics, Vol. 233. Springer-Verlag, Berlin and New York, 1971.

Tsokos, C. P., and Telionis, D. P.
[1] Stochastic particle trajectories in turbulent flow, *J. Math. Phys.* **14** (1), (1973).

Umegaki, H., and Bharucha-Reid, A. T.
[1] Banach space-valued random variables and tensor products of Banach spaces, reprinted from *J. Math. Anal. Appl.* **31**, No. 1, July (1970).

Wasan, M. T.
[1] *Stochastic Approximation*. Cambridge Univ. Press, London and New York, 1969.

Whittle, P.
[1] A view of stochastic control theory, *J. Roy. Statist. Soc. Ser. A* **132** (1969), 320–334.

Whittle, P., and Gait, P. A.
[1] Reduction of a class of stochastic control problems, *J. Inst. Math. Appl.* **6** (1970), 131–140.

Williams, T.
[1] The simple stochastic epidemic curve for large populations of susceptibles, *Biometrika* **52** (1965), 571–579.

Wolfowitz, J.
[1] On the stochastic approximation method of Robbins and Monro, *Ann. Math. Statist.* **23** (1952), 457–461.

Wong, E., and Zakai, M.
[1] On the relation between ordinary and stochastic differential equations, *Int. J. Engng Sci.* **3** (1965), 213–229.

Yang, Grace Lo
[1] Contagion in stochastic models for epidemics, *Ann. Math. Statist.* **39**, No. 6 (1968), 1863–1889.

Index

Numbers in italics refer to the pages on which the complete references are listed.

A

Admissibility theory, 30, 49, 54, 102
Adomian, G., 3, *266*
Ahmed, N. U., 3, *266*
Almost sure convergence, 68
Anderson, M. W., 3, 4, 79, *266*
Approximate solutions, 65–78, 87–96, 141–148
Arithmetic fixed-point problem, 74
Asymptotic stability, 20–21, 38–39

B

Banach, S., 9
Banach space(s), 8, 14, 193
  admissible, 14, 31
  of sequences, 17–18, 133
Barbalat, I., 10, 27, *260*
Barbalat lemma, 10, 27–28
Bartholomay, A., 198, *266*
Bartlett, M. S., 78, 80, *267*

Beckenbach, E. F., 137, *260*
Bellman, R., 2, 50, 57, 59, 137, *260, 261*
Bharucha-Reid, A. T., 2, 3, 4, 14, 44, 49, 78, *267*
Biological system, 57–64, 165–179
  one-organ, 57–62
  two-organ, 62–64
Bochner, S., 10, *261*
Bounded variation, 18
Branson, H., 165, 166, *261*
Brownian motion, 2
Brownian motion process, 207–209, 214
Burkholder, D. L., 66, 87, 88, *267*

C

Cauchy sequence, 7
Chemical kinetics, 180, 198–200
  rate functions, 201–204
  rate of reaction, 201
  stochastic integral equation in, 204–206

Chemotherapy models, 57–64
Circulatory system, *see* Physiological
  models *and* Chemotherapy models
Closed-graph theorem, 9, 31, 134
Communicable disease model, 176–179
  random integral equation in, 178
Contraction operator, 9, 22, 32
Contraction mapping, 66
Control systems
  discrete, 148–155
  feedback, 115
  stochastic, 115–119, 148–155
Convolution, 9, 227, 256
Corduneanu, C., 5, 98, 115, *261*

D

Desoer, C. A., 115, *261*
Discretized equation, 132
Distefano, N., 46, *268*
Doob, J. L., 208, 209, 211, *268*
Dunford, N., 11, *262*
Dynkin, Y. B., 208, *268*

E

Error signal, 253
Essentially bounded function, 14, 18.
Eulerian velocity field, 56
Existence and uniqueness theorems,
  30–39, 49–55, 84–87, 98–113,
  120–131, 157–164, 194–197, 215, 244
  for random discrete equations, 133–141

F

Feller, W., 79, 80, *268*
Fixed-point theorem(s), 9, 22–27
  Banach's, 9, 22–23, 54, 109
  Brouwer's, 24, 26
  Krasnosel'skii's, 10, 102
  Schauder's, 10, 25–26, 106
Fortet integral equation, 44
Fortet, R., 2, 43, 197, *267, 268*
Fourier transforms, 10, 225
Fréchet space, 8, 99, 101
Fredholm random integral equation, 18,
  97–131
  discrete version, 132–155
Free random variable, 18

G

Gavalas, G. R., 197, 201, 204, *262*
Gikhmann, I. I., 2, 208, *268*

H

Halanay, A., 10, *262*
Hamdan, M. A., 4, *273*
Hans, O., 3, 4, *268*
Hardiman, S., 79, *268*
Hearon, J., 165, 166, *262*
Hereditary mechanics, 46–48
Hilbert space, 11, 16, 101, 102
Hildebrand, F. B., 149, 153, *262*

I

Inner product, 11, 16
Ito, K., 2, 208, *269*
Ito stochastic integral, 208–211

J

Jacquez, J. A., 2, 57, 59, 260, *261*
Jazwinski, A. H., 208, 211, *269*

K

Kalaba, R., 2, 57, 59, 260, *261*
Kendall, D. G., 79, 80, *269*
Kernel, 18, 33, 180, 243
Kotkin, B., 59, *261*

L

L_2, 7
Landau, H. G., 165, 176, 177, *263*
Levin, J. J., 243, *263*
Linear hereditary phenomenon, 46
Linear space, 8
  normed, 8
  topological, 8
Lipschitz condition, 32, 52, 103
Loève, M., 68, *269*
Lumley, J. L., 2, 56, *270*
Lur'e, A. I., 242, *264*
Lyapunov, 4, 40

M

Markov inequality, 68
McKean, H. P., 208, *270*
McQuarrie, D., 198, *270*
Mean-square convergence, 71
Mean-square integral, 3, 18
Metabolizing systems, 165–170
  random integral equation in, 167
Method of successive approximations, 60, 66–78, 208
Metric space, 7
Mikhlin, S. G., 59, *264*
Miller, R. K., 98, 243, *264*
Milton, J. S., 2, 3, 79, 120, 129, 156, 157, 165, 170, 176, 181, 197, 200, *270*
Minkowski's inequality, 9, 13, 50, 137
Morozan, T., 2, 4, 88, 218, *270*
Mukherjea, A., 14, 49, *270, 271*

N

Nohel, J. A., 98, 243, *263, 265*
Norm, 7, 181
  of operator, 157, 185

O

Operator, 9
  bounded, 31, 102, 134
  closed, 14, 31
  completely continuous, 10, 102, 105, 106, 121, 125
  continuous, 14, 30–31, 99, 133, 185, 186, 214
  contraction, 9, 22, 32
  linear, 31, 99, 115, 134, 185, 186
  nonlinear, 115

P

Padgett, W. J., 2, 3, 4, 57, 79, 120, 244, *271, 272*
Parseval equality, 10, 227
Petrovanu, D., 18, 98, 115, *265*
Physiological models, 170–175
  random integral equation in, 172
Poincaré–Lyapunov stability theorem, 40–42

Popov, V. M., 4, 218
Popov frequency response method, 4, 149, 152, 225
Population growth problem, 78–87
  numerical solution, 94–96
Postnikov, V. N., 242, *264*
Probability measure space, 12
  complete, 12

R

Rall, L. B., 72, 74, *265*
Random Arzela–Ascoli theorem, 15, 51
Random differential system, 40, 116
Random equations, 3
  algebraic, 3
  difference, 3
  differential, 3
  discrete, 20–21, 132–155
Random integral equation(s),
  approximate solutions, 65–78, 87–96
  Fredholm, 18, 97–131
  Ito, 207–216
  mixed Volterra–Fredholm, 19, 98, 101, 116–117, 120
  perturbed, 19, 120–131, 156–164
  vector, 180–197
  Volterra, 18, 29–64, 156–164
Random integrodifferential equation, 241, 243–251
Random Lipschitz condition, 52
Random solution, 20, 186
  asymptotically stable in mean-square, 20, 249
  existence and uniqueness theorems, 30–39, 49–55, 84–87, 98–113, 120–131, 133–141, 157–164, 194–197, 244–247
  stochastically asymptotically exponentially stable, 20, 38–39, 113–115, 117, 247
  stochastically geometrically stable, 21, 139, 141, 149, 152
Random vector, 181
Rao, A. N. V., 79, *272*
Rao, B. L. S. Prakasa, 3, *272*
Rapoport, A., 165, 176, 177, *263*
Renewal equation, 79
Retraction, 24

S

Saaty, T. L., 208, *265*
Schwartz inequality, 162
Schwartz, J., 11, *262*
Semi-norm, 11–13, 182–184
Semirandom solution, 57, 59
Semistochastic integral equation, 59
Skorokhod, A. V., 2, 208, *268*
Spacek, A., 3
Square-summable function, 7
Stability, 20–21, 113–115, 139–141, 149, 225, 247, 255–259
Stephenson, J., 170, 175, *265*
Stochastic absolute stability, 4, 225–240, 255–259
Stochastic approximation, 87–94
Stochastic chemical kinetics, 180
  model, 197–206
Stochastic control, *see* Control systems
Stochastic differential systems, 21–22, 116, 250
  absolute stability of, 225–239
  nonlinear, 148, 152, 217–240
  reduction of, 219–225
  schematic representations of, 239–240
  stochastically absolutely stable, 22
Stochastic discrete equations, 20–21
Stochastic free term, 18
Stochastic fundamental matrix solution, 252
Stochastic integral, *see* Ito, K.
Stochastic integral equations, *see also* Random integral equations
  Ito, 212–213
  Ito–Doob, 214–216
Stochastic integrodifferential systems, 21, 241–259
  with time lag, 251–259
Stochastic kernel, 18, 33, 180, 243
Stochastic process, 14
  continuous in mean-square, 14
  second-order, 14, 133, 157, 215, 243
Stochastically stable matrix, 22, 41, 229, 231, 234, 236
Stoichiometry, 198

Stratonovich, R. L., 208, *272*
Successive approximations, 22–23, 60, 145
  almost sure convergence of, 68–71
  error of approximations, 71–74, 147
  rate of convergence of, 71–74, 145
  sequence of, 22, 66, 75, 77, 94

T

Telephone exchange, 44
Telephone traffic, 42–46
Telionis, D. P., *273*
Tomasian, A. J., 115, *261*
Topological space, 8
  linear, 8, 181
  locally convex, 9, 12, 183
Tserpes, N., 14, 49, *270, 272*
Tsokos, C. P., 2, 3, 4, 40, 57, 79, 120, 129, 149, 152, 156, 157, 165, 170, 176, 181, 197, 200, 218, 244, *272, 273*
Turbulence theory, 55–56

V

Vector space, 8
Volterra random integral equation, 18, 29–64
  discrete version, 133

W

Wasan, M. T., 88, *273*
Wijsman, R., 165, 166, 167, *266*
Wong, E., 208, *273*
Wong, J. S. W., 98, *264*

Y

Yosida, K., 17, *266*

Z

Zakai, M., 208, *273*